Experiments Manual

to accompany

Grob's Basic Electronics: Fundamentals of DC & AC Circuits

Frank Pugh
Santa Rosa Junior College

Wes Ponick
Agilent Technologies

Boston   Burr Ridge, IL   Dubuque, IA   Madison, WI   New York   San Francisco   St. Louis
Bangkok   Bogotá   Caracas   Kuala Lumpur   Lisbon   London   Madrid   Mexico City
Milan   Montreal   New Delhi   Santiago   Seoul   Singapore   Sydney   Taipei   Toronto

The **McGraw·Hill** Companies

Experiments Manual to accompany
GROB'S BASIC ELECTRONICS: FUNDAMENTALS OF DC & AC CIRCUITS
FRANK PUGH AND WES PONICK

Published by McGraw-Hill Higher Education, an imprint of The McGraw-Hill Companies, Inc., 1221 Avenue of the Americas, New York, NY 10020. Copyright © 2007 by The McGraw-Hill Companies, Inc. All rights reserved.

No part of this publication may be reproduced or distributed in any form or by any means, or stored in a database or retrieval system, without the prior written consent of The McGraw-Hill Companies, Inc., including, but not limited to, network or other electronic storage or transmission, or broadcast for distance learning.

 Recycled/acid free paper
This book is printed on recycled, acid-free paper containing 10% postconsumer waste.

1 2 3 4 5 6 7 8 9 0 QPD/QPD 0 9 8 7 6

ISBN-13: 978-0-07-320693-6
ISBN-10: 0-07-320693-8

www.mhhe.com

Contents

Preface V

INTRODUCTION TO MATHEMATICS FOR ELECTRONICS

EXPERIMENT I-1	Powers of Ten I-1
EXPERIMENT I-2	Scientific Notation I-9
EXPERIMENT I-3	Engineering Notation I-19
EXPERIMENT I-4	Significant Figures (Accuracy and Precision) I-29
EXPERIMENT 1-1	Lab Safety, Equipment, and Components 1
EXPERIMENT 2-1	Resistance Measurements 9
EXPERIMENT 2-2	Resistor *V* and *I* Measurements 19
EXPERIMENT 3-1	Ohm's Law 29
EXPERIMENT 3-2	Applying Ohm's Law 35
EXPERIMENT 4-1	Series Circuits 41
EXPERIMENT 4-2	Series Circuits—Resistance 47
EXPERIMENT 4-3	Series Circuits—Analysis 53
EXPERIMENT 4-4	Series Circuits—with Opens 59
EXPERIMENT 4-5	Series-Aiding and Series-Opposing Voltages 65
EXPERIMENT 5-1	Parallel Circuits 71
EXPERIMENT 5-2	Parallel Circuits—Resistance Branches 77
EXPERIMENT 5-3	Parallel Circuits—Analysis 83
EXPERIMENT 5-4	Parallel Circuits—Opens and Shorts 89
EXPERIMENT 6-1	Series-Parallel Circuits 97
EXPERIMENT 6-2	Series-Parallel Circuits—Resistance 103
EXPERIMENT 6-3	Series-Parallel Circuits—Analysis 111
EXPERIMENT 6-4	Series-Parallel Circuits—Opens and Shorts 117
EXPERIMENT 6-5	The Wheatstone Bridge 127
EXPERIMENT 6-6	Positive and Negative Voltages to Ground 133
EXPERIMENT 6-7	Additional Series-Parallel Circuits 139
EXPERIMENT 6-8	Additional Series-Parallel Opens and Shorts 143
EXPERIMENT 7-1	Voltage Dividers with Loads 151
EXPERIMENT 7-2	Current Dividers 157
EXPERIMENT 7-3	Potentiometers and Rheostats as Dividers 163
EXPERIMENT 7-4	Voltage Divider Design 169
EXPERIMENT 8-1	Analog Ammeter Design 175
EXPERIMENT 8-2	Analog Voltmeter Design 179
EXPERIMENT 8-3	Analog Ohmmeter Design 185
EXPERIMENT 9-1	Kirchhoff's Laws 193
EXPERIMENT 10-1	Network Theorems 201
EXPERIMENT 11-1	Conductors and Insulators 205

EXPERIMENT 12-1	Battery Internal Resistance 215	**EXPERIMENT 19-1**	Inductors 315
EXPERIMENT 12-2	Load Match and Maximum Power 223	**EXPERIMENT 20-1**	Inductive Reactance 323
EXPERIMENT 13-1	Magnetism 229	**EXPERIMENT 21-1**	Inductive Circuits 331
EXPERIMENT 13-2	Electromagnetism and Coils 233	**EXPERIMENT 22-1**	*RC* Time Constant 341
EXPERIMENT 14-1	Relays 239	**EXPERIMENT 23-1**	*AC* Circuits: *RLC* Series 349
EXPERIMENT 15-1	AC Voltage and Ohm's Law 247	**EXPERIMENT 24-1**	Complex Numbers for AC Circuits 355
EXPERIMENT 15-2	Basic Oscilloscope Measurements for AC Circuits 253	**EXPERIMENT 25-1**	Series Resonance 363
EXPERIMENT 15-3	Additional AC Oscilloscope Measurements 261	**EXPERIMENT 25-2**	Parallel Resonance 369
EXPERIMENT 15-4	Oscilloscope Lissajous Patterns 269	**EXPERIMENT 26-1**	Filters 375
EXPERIMENT 15-5	Oscilloscope Measurements: Superposing AC on DC 277	**EXPERIMENT 26-2**	Filter Applications 381
EXPERIMENT 16-1	Capacitors 285	**APPENDIX A**	Applicable Color Codes 389
EXPERIMENT 17-1	Capacitive Reactance 293	**APPENDIX B**	Lab Report Preparation 392
EXPERIMENT 18-1	Capacitive Coupling 301	**APPENDIX C**	Blank Graph Paper 396
EXPERIMENT 18-2	Capacitive Phase Measurements: Using an Oscilloscope 307	**APPENDIX D**	How to Make Graphs 406
		APPENDIX E	Oscilloscope Graticules 407
		APPENDIX F	The Oscilloscope 412
		APPENDIX G	Component List 415

Preface

Experiments in Basic Electronics is a lab manual for the beginning electronics student who does not have any previous experience in electricity or electronics. It has been developed especially for use with *Basic Electronics,* Tenth Edition, by Bernard Grob and Mitchel Schultz. The experiments are coordinated with the text chapter-by-chapter. In total, there are over 75 experiments, starting with a review of mathematical concepts important for the understanding of the fundamental underlining principles of electronics, then progressing through basic safety, lab equipment, identification of electronic components and concluding with reactive circuits. All basic aspects of circuit theory are covered, ranging from Ohm's law through series-parallel circuits, and reactance circuits. The emphasis throughout is on basic concepts and validation of theory through collecting data and reporting the results. Although the experiments build on one another, with simple concepts being developed before more complex subjects are introduced, minor modifications in the sequence can be made as necessary to suit the requirements of an individual electronic technology program. All experiments have been student-tested to ensure they can be followed with little or no assistance in the laboratory classroom.

Four experiments have been added for this edition to make the coverage of this manual more comprehensive. These additions are the result of instructor responses to a questionnaire distributed before work on the revision was begun. We thank all the instructors who provided us with opinions and suggestions for this edition.

Also in this edition, the order in which the experiments appear has been changed so that they directly correlate with the Grob text, chapter-by-chapter, and all of the chapters in the text are represented by at least one experiment. For the convenience of students and instructors alike, report sheets are supplied for each experiment.

Each experiment, which takes approximately two to three hours to perform, is organized as follows. First, the basic principles and laboratory objectives are explained in detail; then, the student is encouraged to apply electronics theory to troubleshooting; and finally, the student prepares a comprehensive report. Techniques for good technical report preparation are discussed in the Appendixes along with additional blank graph paper to be used for preparing reports.

A CD is included as a bonus with this edition. This disk contains the MultiSIM textbook edition program and 40 simulation activities. These simulation activities provide students with extra experience using the prelabs, and with additional exercises including critical thinking and troubleshooting practice related to select hands-on experiments. The prelabs are intended to help students to better prepare for their hands-on experience. The Critical Thinking and Troubleshooting labs are designed to challenge students by applying their knowledge to a variety of situations.

The authors would like to thank the instructors and students at Santa Rosa Junior College, Electronics Department, for their support and assistance throughout the years. We also wish to thank all the instructors at technical schools throughout the U.S.A. and the world who continue to offer suggestions and improvements. And finally, we thank our wives Ann and Jeanne for their continued patience and support.

Frank Pugh
Wes Ponick

INTRODUCTION

EXPERIMENT I-1

INTRODUCTION TO MATHEMATICS FOR ELECTRONICS—POWERS OF TEN

LEARNING OBJECTIVES

At the completion of this experiment, you will be able to:
- Place numbers in power of ten notation.
- Multiply using powers of ten notation.
- Divide using powers of ten notation.

SUGGESTED READING

Introduction, *Basic Electronics,* Grob/Schultz, Tenth Edition

INTRODUCTION

Powers of Ten: The study of electronics is often confronted by large or small numerical values. An example of a particularly large value would be the number of electrons in a coulomb: 6,280,000,000,000,000,000. Another example would be the number of amperes in 3.12 microamperes: 0.00000312. Since the presentation of these numbers can be confusing, electronic technicians attempt to simplify data by using a notation referred to as a "power of 10."

By using this concept, the following numbers can be expressed in multiples of 10 by attaching the correct exponent. For example:

10	=	10	or	10^1
100	=	10×10	or	10^2
1,000	=	$10 \times 10 \times 10$	or	10^3
10,000	=	$10 \times 10 \times 10 \times 10$	or	10^4
100,000	=	$10 \times 10 \times 10 \times 10 \times 10$	or	10^5
1,000,000	=	$10 \times 10 \times 10 \times 10 \times 10 \times 10$	or	10^6

In a similar way other numbers can also be expressed in multiples of 10. For example:

46.8	=	4.68×10	or	4.68×10^1
468	=	$4.68 \times 10 \times 10$	or	4.68×10^2
4,680	=	$4.68 \times 10 \times 10 \times 10$	or	4.68×10^3
46,800	=	$4.68 \times 10 \times 10 \times 10 \times 10$	or	4.68×10^4
468,000	=	$4.68 \times 10 \times 10 \times 10 \times 10 \times 10$	or	4.68×10^5
4,680,000	=	$4.68 \times 10 \times 10 \times 10 \times 10 \times 10 \times 10$	or	4.68×10^6

Adding Powers of Ten: To add quantities in power of ten notation, first by moving the decimal point such that you place both quantities in the same power of ten notation. Then add the quantities and place the final answer in the common power of ten notation. This process is shown below:

Example 1: Add the quantities 5.69×10^8 and 3.67×10^7.

$5.69 \times 10^8 + 3.67 \times 10^7$ Convert to a common power of ten notation (10^8).
$5.69 \times 10^8 + 0.367 \times 10^8$ Add the quantities.
$\boxed{6.057 \times 10^8}$ The answer.

Note that there are alternative answers for the same problem, See Example 2 and Example 3:

Example 2: Add the quantities 5.69×10^8 and 3.67×10^7.

$5.69 \times 10^8 + 3.67 \times 10^7$ Convert to a common power of ten notation (10^7).
$56.9 \times 10^7 + 3.67 \times 10^7$ Add the quantities.
$\boxed{60.57 \times 10^7}$ The answer.

Or:

Example 3: Add the quantities 5.69×10^8 and 3.67×10^7.

$5.69 \times 10^8 + 3.67 \times 10^7$ Convert to a common power of ten notation (10^9).
$0.569 \times 10^9 + 0.0367 \times 10^9$ Add the quantities.
$\boxed{0.6057 \times 10^9}$ The answer.

Note that the three previous answers 6.057×10^8, 60.57×10^7 and 0.6057×10^9 are equivalent numbers that are written with different power of ten notations.

Subtracting Powers of Ten: To subtract quantities in power of ten notation, first adjust the decimal point place of both quantities so that each quantity will have the same power of ten notation. Then appropriately subtract the quantities and place each number in the common power of ten notation. This process is shown below:

Example 4: Subtract 5.69×10^8 and 3.67×10^7.

$5.69 \times 10^8 - 3.67 \times 10^7$ Convert to a common power of ten notation (10^8).
$5.69 \times 10^8 - 0.367 \times 10^8$ Subtract the quantities.
$\boxed{5.323 \times 10^8}$ The answer.

Note that there are alternative answers for the same problem, See Example 5 and Example 6:

Example 5: Subtract 5.69×10^8 and 3.67×10^7.

$5.69 \times 10^8 - 3.67 \times 10^7$ Convert to a common power of ten notation (10^7).
$56.9 \times 10^7 - 3.67 \times 10^7$ Subtract the quantities.
$\boxed{53.23 \times 10^7}$ The answer.

Or:

Example 6: Subtract 5.69×10^8 and 3.67×10^7.

$5.69 \times 10^8 - 3.67 \times 10^7$ Convert to a common power of ten notation (10^9).
$0.569 \times 10^9 - 0.0367 \times 10^9$ Subtract the quantities.
$\boxed{0.5323 \times 10^9}$ The answer.

Again, note that each of the three answers are equivalent quantities that are written with different power of ten notation.

Multiplying Powers of Ten: To multiply quantities in powers of ten notation, first multiply the quantities and then algebraically add the exponents. Note that the answers shown below are rounded to the fourth decimal place.

Example 7: Multiply (5.69×10^8) and (3.67×10^7).

$(5.69 \times 10^8)(3.67 \times 10^7)$ Add the exponents *and* multiply the quantities.
$\boxed{20.8823 \times 10^{15}}$ The answer.

Example 8: Multiply (2.19×10^{-7}) and (1.97×10^9).

$(2.19 \times 10^{-7})(1.97 \times 10^9)$ Add the exponents *and* multiply the quantities.
$\boxed{4.3143 \times 10^2}$ The answer.

Example 9: Multiply (3.54×10^6) and (2.45×10^{-3}).

$(3.54 \times 10^6)(2.45 \times 10^{-3})$ Add the exponents *and* multiply the quantities.
$\boxed{8.673 \times 10^3}$ The answer.

Example 10: Multiply (8.98×10^{-5}) and (7.78×10^{-8}).

$(8.98 \times 10^{-5})(7.78 \times 10^{-8})$ Add the exponents *and* multiply the quantities.

$\boxed{69.8644 \times 10^{-13}}$ The answer.

Example 11: Multiply (1.77×10^{-4}) and (2.32×10^{4}).

$(1.77 \times 10^{-4})(2.32 \times 10^{4})$ Add the exponents *and* multiply the quantities.
4.1064×10^{0} Note that 10^{0} equals the number 1.

$\boxed{4.1064}$ The answer.

Dividing Powers of Ten: To divide quantities in power of ten notation, first subtract the exponents and then divide the quantities as shown below. It is always the practice to subtract the power of ten notation value located in the denominator from the power of ten notation found in the numerator. The denominator is the bottom part of a fraction. In the fraction of $10^{6}/10^{4}$, the denominator is 10^{4}. The numerator is the top part of a fraction. In another example, $10^{7}/10^{5}$, the numerator is 10^{7} and the denominator is 10^{5}. Additional examples of dividing power of ten quantities are shown below. (Note that the answers presented below have been rounded to three digits).

Example 12: Divide the following quantities.

$\dfrac{3.00 \times 10^{6}}{1.50 \times 10^{3}}$ Subtract the exponents (10^{6-3}) and divide the quantities.

$\boxed{2.00 \times 10^{3}}$ The answer.

Example 13: Divide the following quantities.

$\dfrac{6.34 \times 10^{-4}}{2.17 \times 10^{3}}$ Subtract the exponents (10^{-4-3}) and divide the quantities.

$\boxed{2.92 \times 10^{-7}}$ The answer.

Example 14: Divide the following quantities.

$\dfrac{37.0 \times 10^{-8}}{6.51 \times 10^{-2}}$ Subtract the exponents (10^{-8--2}) and divide the quantities. Note the exponent's *double* negative signs.

$\boxed{5.68 \times 10^{-6}}$ The answer.

Example 15: Divide the following quantities.

$\dfrac{8.00 \times 10^{2}}{2.00 \times 10^{-7}}$ Subtract the exponents (10^{2--7}) and divide the quantities. Note the exponent's *double* negative signs.

$\boxed{4.00 \times 10^{9}}$ The answer.

Example 16: Divide the following quantities.

$\dfrac{9.99 \times 10^{1}}{3.00 \times 10^{8}}$ Subtract the exponents (10^{1-8}) and divide the quantities.

$\boxed{3.33 \times 10^{-7}}$ The answer.

EXERCISE PROBLEMS

Perform the indicated operation for the following power of ten quantities. Record your answers on the Results page. Round your answers to the tenth place, and note that the Results page has already assigned a power of ten notation to conform your answer to.

1. $4.24 \times 10^{2} + 6.87 \times 10^{3}$
2. $3.21 \times 10^{7} + 8.66 \times 10^{9}$
3. $5.35 \times 10^{4} + 4.33 \times 10^{3}$
4. $72.8 \times 10^{1} + 8.62 \times 10^{2}$
5. $6.45 \times 10^{5} + 0.12 \times 10^{7}$
6. $18.4 \times 10^{6} + 7.34 \times 10^{6}$
7. $34.5 \times 10^{9} - 15.1 \times 10^{8}$
8. $7.24 \times 10^{2} - 6.37 \times 10^{1}$
9. $35.8 \times 10^{5} - 23.7 \times 10^{5}$
10. $3.84 \times 10^{7} - 1.21 \times 10^{7}$

11. $42.2 \times 10^1 - 0.02 \times 10^3$
12. $7.65 \times 10^8 - 0.27 \times 10^9$
13. $1.00 \times 10^2 \times 2.00 \times 10^5$
14. $3.46 \times 10^{-2} \times 7.25 \times 10^4$
15. $18.3 \times 10^4 \times 20.1 \times 10^{-5}$
16. $2.67 \times 10^{-7} \times 16.8 \times 10^{-2}$
17. $5.55 \times 10^1 \times 3.46 \times 10^2$
18. $3.14 \times 10^{-3} \times 1.22 \times 10^{-3}$
19. $\dfrac{4.84 \times 10^2}{2.22 \times 10^4}$
20. $\dfrac{8.00 \times 10^3}{2.00 \times 10^3}$
21. $\dfrac{7.50 \times 10^{-4}}{0.50 \times 10^3}$
22. $\dfrac{6.40 \times 10^{-5}}{2.00 \times 10^{-4}}$
23. $\dfrac{12.6 \times 10^2}{3.00 \times 10^{-3}}$
24. $\dfrac{8.88 \times 10^3}{1.00 \times 10^{-4}}$

| NAME | DATE |

RESULTS FOR EXERCISE PROBLEMS FROM INTRODUCTION I-1

1. _____ $\times 10^3$
2. _____ $\times 10^9$
3. _____ $\times 10^3$
4. _____ $\times 10^3$
5. _____ $\times 10^6$
6. _____ $\times 10^6$
7. _____ $\times 10^9$
8. _____ $\times 10^0$
9. _____ $\times 10^6$
10. _____ $\times 10^6$
11. _____ $\times 10^0$
12. _____ $\times 10^6$
13. _____ $\times 10^7$
14. _____ $\times 10^2$
15. _____ $\times 10^{-1}$
16. _____ $\times 10^{-9}$
17. _____ $\times 10^3$
18. _____ $\times 10^{-6}$
19. _____ $\times 10^{-2}$
20. _____ $\times 10^0$
21. _____ $\times 10^{-7}$
22. _____ $\times 10^{-1}$
23. _____ $\times 10^5$
24. _____ $\times 10^7$

QUESTIONS

1. Why is power of ten notation commonly used in the study of electronics?

2. What is the numerator of a fraction?

3. Give an example of a denominator.

4. In your own words describe the process used in multiplying power of ten notation quantities?

5. In your own words describe the process used in dividing power of ten notation quantities?

REPORT

Attach the above results to the following results form. Include your conclusion with this report.

RESULTS REPORT FORM

Experiment No: _____   Name: _____
                                          Date: _____
Experiment Title: _____   Class: _____
_____   Instr: _____

Explain the purpose of the experiment:

List the first Learning Objective:
OBJECTIVE 1:

After reviewing your results, describe how this objective was validated by this experiment?

List the second Learning Objective:
OBJECTIVE 2:

After reviewing your results, describe how this objective was validated by this experiment?

List the third Learning Objective:
OBJECTIVE 3:

After reviewing your results, describe how this objective was validated by this experiment?

Conclusion:

Attach to this Experiment Report Form: ☐ Answers to Questions

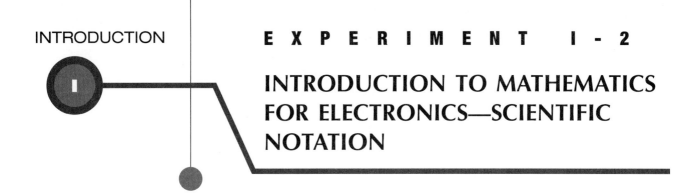

EXPERIMENT I-2

INTRODUCTION TO MATHEMATICS FOR ELECTRONICS—SCIENTIFIC NOTATION

LEARNING OBJECTIVES

At the completion of this experiment, you will be able to:
- Identify numbers that are placed in scientific notation.
- Convert numbers from regular notation to scientific notation.
- Add, subtract, multiply and divide numbers in scientific notation.

SUGGESTED READING

Introduction, *Basic Electronics*, Grob/Schultz, Tenth Edition

INTRODUCTION

Scientific Notation: Numbers can be written in a variety of ways and still be in a correct power of ten notation. One type of notation is referred to as "regular notation." An example of regular notation would be the number 17,645. However, the number 17,645 may also be written in various powers of ten notation. For example the regular notation of the number 17,645 could be rewritten as:

$$17,645 \times 10^0$$
$$1,764.5 \times 10^1$$
$$176.45 \times 10^2$$
$$17.645 \times 10^3$$
$$1.7645 \times 10^4$$
$$0.17645 \times 10^5$$
$$0.017645 \times 10^6$$

Note that all of the numbers listed above are equivalent to each other. To avoid confusion in recording data it has become standard practice in some technical fields to record collected data in a form called "scientific notation." Scientific notation is based on powers of the base number 10. Scientific notation is identified as a number between 1 and 9 inclusive, acquired through the possible adjustment of the decimal point and by multiplying by the appropriate power of ten notation.

For example, the power of ten notation numbers listed above can be rewritten in scientific notation as shown below:

$$\left. \begin{array}{r} 17,645 \times 10^0 \\ 1,764.5 \times 10^1 \\ 176.45 \times 10^2 \\ 17.645 \times 10^3 \\ 1.7645 \times 10^4 \\ 0.17645 \times 10^5 \\ 0.017645 \times 10^6 \end{array} \right\} = 1.7645 \times 10^4$$

For the number 1.7645×10^4, the 1.7645 is referred to as the coefficient and is always greater than or equal to one, and less that ten. The value 10^4 is called the base and the base must always be 10 in scientific notation. The base number 10 is always written in exponent form. For the number 1.7645×10^4 the number 4 is referred to as the exponent or the power of ten.

Converting a Number to Scientific Notation: The process to convert a number from regular notation to scientific notation is accomplished as shown below:

> First: Move the decimal point of the number being converted to the *right* of the first non-zero digit. Count the number of places that you moved the decimal point.
>
> Second: Multiply this number by the correct power of ten notation. This is accomplished by multiplying this number by 10 whose exponent is equal to the number of places that the decimal point was moved. The exponent will be positive (+) if the decimal point was moved to the *left*, and the exponent will be negative (−) if the decimal point was moved to the *right*.

Example 1: Convert the number 257 to scientific notation.

> $257 = \boxed{2.57 \times 10^2}$ This is accomplished by moving the decimal point two places to the *left* and multiplying the number by the correct power of ten notation, in this case 10^2.

Example 2: Convert the number 0.00135 to scientific notation.

> $0.00135 = \boxed{1.35 \times 10^{-3}}$ Move the decimal point three places to the *right* and multiplying the number by the correct power of ten notation, in this case 10^{-3}.

Example 3: Convert the number 159×10^4 to scientific notation.

> $159 \times 10^4 = 1{,}590{,}000$ Convert the quantity 159×10^4 to regular notation of 1,590,000.
>
> $1{,}590{,}000 = \boxed{1.59 \times 10^6}$ Convert 1,590,000 to scientific notation. This is accomplished by moving the decimal point six places to the *left* and multiplying the number by the correct power of ten notation, in this case 10^6.

Example 4: Convert the number 0.778×10^{-2} to scientific notation.

> $0.778 \times 10^{-2} = 0.00778$ First convert the number 0.778×10^{-2} to the regular notation of 0.00778. Then convert 0.00778 to scientific notation. This is accomplished by moving the decimal point three places to the *right* and multiplying the number by the correct power of ten notation, in this case 10^{-3}.
>
> $0.00778 = \boxed{7.78 \times 10^{-3}}$ The answer.

Using Exponents: Exponents are used as a form mathematical shorthand for the operation of multiplication. Remember for power of ten notation, $(10)(10)(10) = 10^3$, and $(10)(10)(10)(10)(10)(10) = 10^6$. The exponent indicates how many times the base number (10) is being multiplied as indicated by the value of the exponent. For example, 10^6 has a base of 10 and an exponent of 6, and will sometimes be referred to as "10 raised to the 6th power", where word "power" is another term for exponent.

Converting Scientific Notation to Regular Notation: The process to convert a number in scientific notation to regular notation is accomplished as shown below:

> First: If the number in scientific notation has a positive (+) exponent on 10's value, for example: 10^{+6} or as it is typically written 10^6, move the decimal point to the *right* the same number of places as the exponent. Add zeros as necessary.
>
> Second: If the number in scientific notation has a negative (−) exponent on 10's value, for example: 10^{-2}, move the decimal point to the *left* the same number of places as the exponent. Add zeros as necessary.

Example 5: Convert the number 6.757×10^3 to regular notation.

> $6.757 \times 10^3 = \boxed{6{,}757}$ Move the decimal point three places to the *right* and eliminate the power of ten notation of 10^3.

Example 6: Convert the number 2.498×10^{-8} to regular notation.

> $2.498 \times 10^{-8} = \boxed{0.00000002498}$ Move the decimal point eight places to the *left* and eliminate the power of ten notation of 10^{-8}.

Adding Numbers in Scientific Notation: The process for adding scientific notation numbers is to make certain that the power of ten exponents are the same. After adjusting the decimal point so that the exponents are equivalent, *add* the quantities and then assign the *common* power of ten notation exponential value.

Example 7: Add (2.98×10^3) and (4.11×10^4) and place the answer in scientific notation.

$(2.98 \times 10^3) + (4.11 \times 10^4)$	Convert to a common power of ten notation, in this case 10^4.
$(0.298 \times 10^4) + (4.11 \times 10^4)$	Add $0.298 + 4.11$ and keep the power of ten notation of 10^4 intact.
$(0.298 \times 10^4) + (4.11 \times 10^4) = \boxed{4.408 \times 10^4}$	The answer.

Alternatively, the problem could have been solved by converting to different power of ten notation of 10^3. If 10^3 were used instead of 10^4, then the example would resemble the following:

$(2.98 \times 10^3) + (4.11 \times 10^4)$	Convert to a common power of ten notation, in this case 10^3.
$(2.98 \times 10^3) + (41.1 \times 10^3)$	Add $2.98 + 41.1$ and keep the power of ten notation of 10^3 intact.
$(2.98 \times 10^3) + (41.1 \times 10^3) = 44.08 \times 10^3$	Note that 44.08×10^3 is not in scientific notation, and will need to be converted as shown below by moving the decimal point one place to the *left* and increasing the exponent by one.
$44.08 \times 10^3 = \boxed{4.408 \times 10^4}$	The answer.

Example 8: Add (2.22×10^{-2}) and (2.48×10^{-1}) and place the answer in scientific notation.

$2.22 \times 10^{-2} + 2.48 \times 10^{-1}$	Convert to a common power of ten notation, in this case 10^{-2}.
$2.22 \times 10^{-2} + 24.8 \times 10^{-2}$	Add the quantities $2.22 + 24.8$ and keep the power of ten notation intact.
$2.22 \times 10^{-2} + 24.8 \times 10^{-2} = 27.02 \times 10^{-2}$	Note that 27.02×10^{-2} is not in scientific notation, and will need to be converted as shown below by moving the decimal point one place to the *left* and decreasing the exponent by one.
$27.02 \times 10^{-2} = \boxed{2.702 \times 10^{-1}}$	The answer.

Example 9: Add (4.61×10^5) and (7.31×10^7) and place the answer in scientific notation.

$(4.61 \times 10^5) + (7.31 \times 10^7)$	Convert to a common power of ten notation, in this case 10^7.
$(0.0461 \times 10^7) + (7.31 \times 10^7)$	Add $0.0461 + 7.31$ and keep the power of ten notation of 10^7 intact.
$(0.0461 \times 10^7) + (7.31 \times 10^7) = \boxed{7.3561 \times 10^7}$	The answer.

Example 10: Add (1.5×10^2) and (1.6×10^{-1}) and place the answer in scientific notation.

$(1.5 \times 10^2) + (1.6 \times 10^{-1})$	Convert to a common power of ten notation, in this case 10^2.
$(1.5 \times 10^2) + (0.0016 \times 10^2)$	Add $1.5 + 0.0016$ and keep the power of ten notation of 10^2 intact.
$(1.5 \times 10^2) + (0.0016 \times 10^2) = \boxed{1.5016 \times 10^2}$	The answer.

Subtracting Numbers in Scientific Notation: The process for subtracting scientific notation numbers is similar to the process used for addition. To subtract make sure that the power of ten exponents are the same. After adjusting the decimal point so that the exponents are equivalent, subtract the quantities and then assign the *common* power of ten notation exponent.

Example 11: Subtract (2.77×10^3) and (2.68×10^4) and place the answer in scientific notation.

$(2.77 \times 10^3) - (2.68 \times 10^4)$	Convert to a common power of ten notation, in this case 10^3.
$(2.77 \times 10^3) - (26.8 \times 10^3)$	Subtract $2.77 - 26.8$ and keep the power of ten notation of 10^3 intact.

$(2.77 \times 10^3) - (26.8 \times 10^3) = -24.03 \times 10^3$ Note that -24.03×10^3 is not in scientific notation, and will need to be converted as shown below by moving the decimal point one place to the *left* and increasing the exponent by one.

$24.03 \times 10^3 = \boxed{-2.403 \times 10^4}$ The answer.

Or, as an alternative method, the problem could have been solved by converting to a different power of ten notation of 10^4. If 10^4 were used instead of 10^3, then the example would resemble the following:

$(2.77 \times 10^3) - (2.68 \times 10^4)$ Convert to a common power of ten notation, in this case 10^4.

$(0.277 \times 10^4) - (2.68 \times 10^4)$ Subtract 2.77 − 26.8 and keep the power of ten notation of 10^4 intact.

$(0.277 \times 10^4) - (2.68 \times 10^4) = \boxed{-2.403 \times 10^4}$ The answer.

Example 12: Subtract (1.5×10^1) and (7.6×10^{-1}) and place the answer in scientific notation.

$(1.5 \times 10^1) - (7.6 \times 10^{-1})$ Convert to a common power of ten notation, in this case 10^1.

$(1.5 \times 10^1) - (0.076 \times 10^1)$ Subtract 1.5 − 0.076 and keep the power of ten notation of 10^1 intact.

$(1.5 \times 10^1) - (0.076 \times 10^1) = \boxed{1.424 \times 10^1}$ The answer. Note that 1.424×10^1 is in scientific notation form and no further adjustment is needed.

Example 13: Subtract (5.6×10^{-2}) and (7.8×10^1) and place the answer in scientific notation.

$(5.6 \times 10^{-2}) - (7.8 \times 10^1)$ Convert to a common power of ten notation, in this case 10^{-2}.

$(5.6 \times 10^{-2}) - (7800 \times 10^{-2})$ Subtract 5.6 − 7800 and keep the power of ten notation of 10^{-2} intact. Therefore:

$(5.6 \times 10^{-2}) - (7800 \times 10^{-2}) = -7794.4 \times 10^{-2}$ Note that -7794.4×10^{-2} is not in scientific notation, and will need to be converted as shown below by moving the decimal point three place to the *left* and increasing the exponent by three.

$7794.4 \times 10^{-2} = \boxed{-7.7944 \times 10^1}$ The answer.

Example 14: Subtract (8.70×10^{-2}) and (6.42×10^{-3}) and place the answer in scientific notation.

$(8.70 \times 10^{-2}) - (6.42 \times 10^{-3})$ Convert to a common power of ten notation, in this case 10^{-2}.

$(8.70 \times 10^{-2}) - (0.642 \times 10^{-2})$ Subtract 8.70 − 0.642 and keep the power of ten notation of 10^{-2} intact.

$(8.70 \times 10^{-2}) - (0.642 \times 10^{-2}) = \boxed{8.058 \times 10^{-2}}$ The answer. Note that 8.058×10^{-2} is in scientific notation form and no further adjustment is needed.

Multiplying Numbers in Scientific Notation: There are two steps used when multiplying numbers in scientific notation.

First: Multiply the quantities.

Second: Algebraically add the exponents of the base 10 values. If the resultant answer is not in scientific notation it will then be necessary to adjust the decimal point and change the base 10 exponent to the correct value.

Example 15: Multiply $(4.367 \times 10^1)(6.367 \times 10^3)$ and place the answer in scientific notation. Round to the fourth decimal place.

$(4.367 \times 10^1)(6.367 \times 10^3) = 27.8047 \times 10^4$ Multiply the quantities and add the power of ten exponents. Note that while the answer is correct it is not correctly written in scientific notation, since the quantity is not between 1 and 10. The decimal point must be moved and the base 10 exponent adjusted.

$27.8047 \times 10^4 = \boxed{2.7805 \times 10^5}$ The answer.

Example 16: Multiply $(8.632 \times 10^1)(6.641 \times 10^{-2})$ and place the answer in scientific notation. Round to the fourth decimal place.

$(8.632 \times 10^1)(6.641 \times 10^{-2}) = 57.3251 \times 10^{-1}$ Multiply the quantities and add the power of ten exponents. The value is not in scientific notation and further adjustment is necessary.

$57.3251 \times 10^{-1} = 5.73251 \times 10^0 = \boxed{5.7325}$ The answer.

Example 17: Multiply $(1.13 \times 10^{-6})(9.66 \times 10^5)$ and place the answer in scientific notation. Round to the fourth decimal place.

$(1.13 \times 10^{-6})(9.66 \times 10^5) = 10.9158 \times 10^{-1}$ Multiply the quantities and add the power of ten exponents. The value is not in scientific notation and further adjustment is necessary.

$10.9158 \times 10^{-1} = \boxed{1.0916}$ The answer.

Example 18: Multiply $(1.61 \times 10^{-3})(1.11 \times 10^{-7})$ and place the answer in scientific notation.

$(1.61 \times 10^{-3})(1.11 \times 10^{-7}) = \boxed{1.7871 \times 10^{-10}}$ The value above is in scientific notation and no adjustment is necessary.

Dividing Numbers in Scientific Notation: There are also two steps used when dividing numbers in scientific notation.

First: Divide the numbers or coefficients.

Second: Algebraically subtract the exponents of the base 10 values. If the resultant answer is not in scientific notation it will then be necessary to adjust the decimal point and change the base 10 exponent to the correct value.

Example 19: Divide (3.67×10^3) by (5.31×10^5) and place the answer in scientific notation.

$(3.67 \times 10^3)/(5.31 \times 10^5) = 0.6911 \times 10^{(3)-(5)} = 0.6911 \times 10^{-2}$

The value above is not in scientific notation and further adjustment is necessary.

$0.6911 \times 10^{-2} = \boxed{6.911 \times 10^{-3}}$

Example 20: Divide (1.42×10^{-7}) by (8.10×10^6) and place the answer in scientific notation.

$(1.42 \times 10^{-7})/(8.10 \times 10^6) = 0.1753 \times 10^{(-7)-(6)} = 0.1753 \times 10^{-13}$

The value above is not in scientific notation and further adjustment is necessary.

$0.1753 \times 10^{-13} = \boxed{1.753 \times 10^{-14}}$

Example 21: Divide (7.77×10^2) by (1.29×10^{-1}) and place the answer in scientific notation.

$(7.77 \times 10^2)/(1.29 \times 10^{-1}) = 6.0233 \times 10^{(2)-(-1)} = \boxed{6.0233 \times 10^3}$

The value above is in scientific notation and no adjustment is necessary.

Example 22: Divide (5.61×10^{-5}) by (4.36×10^{-8}) and place the answer in scientific notation.

$(5.61 \times 10^{-5})/(4.36 \times 10^{-8}) = 1.2867 \times 10^{(-5)-(-8)} = \boxed{1.2867 \times 10^3}$

The value above is in scientific notation and no adjustment is necessary.

EXERCISE PROBLEMS

Convert the following numbers to scientific notation. Record your answers on the Results page.

1. 34
2. 139
3. 1,296
4. 12.65
5. 0.067

Convert the following numbers to regular notation. Record your answers on the Results page.

6. 2.04×10^6
7. 3.45×10^{-8}
8. 7.24×10^2
9. 3.779×10^{-7}
10. 3.66×10^6

Perform the indicated operation for the following power of ten quantities. All results should be in scientific notation. Record your answers on the Results page.

11. $4.28 \times 10^1 + 1.02 \times 10^1$
12. $7.45 \times 10^8 + 1.17 \times 10^9$
13. $3.00 \times 10^2 + 2.00 \times 10^3$
14. $5.46 \times 10^{-2} + 7.25 \times 10^{-4}$
15. $1.93 \times 10^1 + 2.91 \times 10^{-1}$
16. $3.65 \times 10^7 - 1.68 \times 10^6$
17. $5.33 \times 10^3 - 3.46 \times 10^2$
18. $4.24 \times 10^{-3} - 1.22 \times 10^{-3}$
19. $2.90 \times 10^{-3} - 1.90 \times 10^{-4}$
20. $5.73 \times 10^2 - 3.67 \times 10^{-1}$

21. $(3.91 \times 10^2)(4.67 \times 10^5)$
22. $(2.19 \times 10^3)(7.63 \times 10^{-4})$
23. $(4.45 \times 10^{-5})(3.83 \times 10^{-6})$
24. $(6.34 \times 10^7)(2.56 \times 10^{-3})$
25. $(5.55 \times 10^4)(2.78 \times 10^4)$
26. $(9.00 \times 10^2)/(3.00 \times 10^1)$
27. $(2.56 \times 10^{-2})/(5.44 \times 10^5)$
28. $(7.71 \times 10^4)/(1.00 \times 10^{-4})$
29. $(9.31 \times 10^{-3})/(2.99 \times 10^{-3})$
30. $(1.10 \times 10^2)/(3.35 \times 10^6)$

NAME _____ **DATE** _____

RESULTS FOR EXERCISE PROBLEMS FROM INTRODUCTION I-2

1. _____   11. _____   21. _____
2. _____   12. _____   22. _____
3. _____   13. _____   23. _____
4. _____   14. _____   24. _____
5. _____   15. _____   25. _____
6. _____   16. _____   26. _____
7. _____   17. _____   27. _____
8. _____   18. _____   28. _____
9. _____   19. _____   29. _____
10. _____   20. _____   30. _____

QUESTIONS

1. Why is scientific notation used in taking data?

2. What is regular notation?

3. What is a coefficient?

4. In your own words describe the process used in multiplying numbers is scientific notation.

5. In your own words describe the process used in dividing numbers in scientific notation.

REPORT

Attach the above results to the following results form. Include your conclusion with this report.

RESULTS REPORT FORM

Experiment No: _____ Name: _____
                                         Date: _____
Experiment Title: _____ Class: _____
_____ Instr: _____

Explain the purpose of the experiment:

List the first Learning Objective:
OBJECTIVE 1:

After reviewing your results, describe how this objective was validated by this experiment?

List the second Learning Objective:
OBJECTIVE 2:

After reviewing your results, describe how this objective was validated by this experiment?

List the third Learning Objective:
OBJECTIVE 3:

After reviewing your results, describe how this objective was validated by this experiment?

Conclusion:

Attach to this Experiment Report Form: ☐ Results and Answers to Questions

INTRODUCTION

EXPERIMENT I-3

INTRODUCTION TO MATHEMATICS FOR ELECTRONICS—ENGINEERING NOTATION

LEARNING OBJECTIVES

At the completion of this experiment, you will be able to:
- Identify metric prefixes.
- Place numbers in correct engineering notation form.
- Perform basic mathematical operations in engineering notation.

SUGGESTED READING

Introduction, *Basic Electronics*, Grob/Schultz, Tenth Edition

INTRODUCTION

Engineering Notation: In the previous experiment, it was learned that scientific notation is a convenient method to write either very large or very small numbers. A variation on scientific notation is engineering notation. In engineering notation the exponent is generally grouped in multiples of three, for example, 10^3, 10^6, 10^9, 10^{12}.

In electronics, as an alternative to writing powers of 10 notation for calculations or measurements, metric prefixes are utilized. These metric prefixes are also grouped in powers of 1000; however the metric prefix is substituted for the equivalent power of 10 values.

Table I-3.1 details metric prefix values that are used in electronics. Note the relationship between the metric name, metric symbol and the corresponding power of 10 values.

TABLE I-3.1 Metric Prefixes and Their Equivalent Power of 10 Values.

Metric Name	Metric Symbol	Power of 10 Value
yotta	Y	10^{24}
zetta	Z	10^{21}
exa	E	10^{18}
peta	P	10^{15}
tera	T	10^{12}
giga	G	10^{9}
mega	M	10^{6}
kilo	k	10^{3}
None (Base Unit)	None	10^{0}
milli	m	10^{-3}
micro	μ	10^{-6}
nano	n	10^{-9}
pico	p	10^{-12}
femto	f	10^{-15}
atto	A	10^{-18}
zepto	z	10^{-21}
yocto	y	10^{-24}

In electronics among the most common measurements taken are Voltage, Current, and Resistance. Each measurement has its own measurement term and symbol. Voltage is measured in Volts and uses the symbol V in describing that measurement. Current is measured in Amperes and uses the symbol A in identifying its measurement. Resistance is measured in Ohms and uses the symbol Ω in identifying that measurement. For example:

24.3 Volts would be written as 24.3 V
1.5 Amperes would be written as 1.5 A
220 Ohms would be written as 220 Ω

When measuring or calculating either large or small electrical values, metric prefixes are used to conveniently simplify the process. For example:

$0.0000165 \text{ A} = 16.5 \times 10^{-6} \text{ A} = 16.5 \text{ μA}$ Where: $10^{-6} = \mu$
$320,000 \text{ Ω} = 320 \times 10^{3} \text{ Ω} = 320 \text{ kΩ}$ Where: $10^{3} = k$
$60,000,000 \text{ V} = 60 \times 10^{6} \text{ V} = 60 \text{ MV}$ Where: $10^{6} \text{ V} = M$

Converting to Engineering Notation: When converting a number to engineering notation. First, adjust the decimal point so that there is a convenient power of 10 notations where the exponent is a multiple of 3. Note that there are typically more than one correct answers when converting to engineering notation.

Example 1: Convert 0.00005 A to engineering notation.

$0.00005 \text{ A} = \boxed{0.05 \times 10^{-3} \text{ A}}$

Or alternatively:

$0.00005 \text{ A} = \boxed{50 \times 10^{-6} \text{ A}}$

Example 2: Convert 12,000 V to engineering notation.

$12,000 \text{ V} = \boxed{12 \times 10^{3} \text{ V}}$

Or alternatively:

$12,000 \text{ V} = \boxed{0.012 \times 10^{6} \text{ V}}$

Placing Numbers in Metric Prefixes and Engineering Notation: When placing numbers in metric prefixes for numbers in engineering notation, first place the number in a convenient power of 10 notation where the exponents is a multiple of 3. Then exchange the power of 10 notation for its equivalent metric symbol.

Example 3: Convert 0.00005 A to engineering notation and use the appropriate metric prefix notation.

$0.00005 \text{ A} = 0.05 \times 10^{-3} \text{ A} = \boxed{0.05 \text{ mA}}$ Where: $10^{-3} = m$

Or alternatively:

$0.00005 \text{ A} = 50 \times 10^{-6} \text{ A} = \boxed{50 \text{ μA}}$ Where: $10^{-6} = \mu$

Example 4: Convert 12,000 V to engineering notation and use the appropriate metric prefix notation.

$12,000 \text{ V} = 12 \times 10^{3} \text{ V} = \boxed{12 \text{ kV}}$ Where: $10^{3} = k$

Or alternatively:

$12,000 \text{ V} = 0.012 \times 10^{6} \text{ V} = \boxed{0.012 \text{ MV}}$ Where: $10^{6} = M$

Adding in Engineering Notation with a Prefix Notation: Adding numbers in engineering notation with a prefix notation can be accomplished as follows:

First: Place both numbers to be added in engineering notation.
Second: Place the numbers in the same (common) power of 10 engineering notation by moving the decimal point and adjusting the exponent.
Third: *Add* the numerical coefficients and assign the common power of ten engineering notation.
Fourth: Replace the coefficient's power of 10 exponent with the correct metric prefix notation. See Table I-3.1

Example 5: Two electrical components, resistors, are connected in series. The total circuit resistance is the simple addition of these two resistances. If the first resistor has a value of 1.2 kΩ, and the second resistor has a value of 12 kΩ, find the total resistance.

1.2 kΩ + 12 kΩ	Place the numbers in engineering notation.
$(1.2 \times 10^3) + (12 \times 10^3)$	Add the quantities.
13.2×10^3	Place the number in the correct prefix notation.
$13.2 \times 10^3 = \boxed{13.2 \text{ kΩ.}}$	The answer.

Example 6: Two resistors are connected in series. Again, the total circuit resistance is the simple addition of these two resistances. If the first resistor has a value of 150 kΩ, and the second resistor has a value of 1.2 MΩ. Find the total resistance.

150 kΩ + 1.2 MΩ	Place the numbers in engineering notation.
$(150 \times 10^3) + (1.2 \times 10^6)$	Place the numbers to be added in the same power of 10 engineering notation by moving the decimal point and adjusting the exponent.
$(0.150 \times 10^6) + (1.2 \times 10^6)$	Add the coefficients.
1.350×10^6	Place the number in the correct prefix notation.
$1.350 \times 10^6 = \boxed{1.350 \text{ MΩ}}$	The answer.

Subtracting in Engineering Notation: Subtracting numbers in engineering notation with a prefix notation can be accomplished as follows:

- First: Place the numbers to be subtracted in engineering notation.
- Second: Place the numbers in the same (common) power of 10 engineering notation by moving the decimal point and adjusting the exponent.
- Third: *Subtract* the numerical coefficients and assign the common power of ten engineering notation.
- Fourth: Replace the coefficient's power of 10 exponent with the correct metric prefix notation.

Example 7: There is 1.0 MΩ measured across an electrical component called a potentiometer. A potentiometer is a variable resistor. If the potentiometer is adjusted to remove 220 kΩ from its total, how much resistance is remaining?

1.0 MΩ − 220 kΩ	Place the numbers in engineering notation.
$(1.0 \times 10^6) - (220 \times 10^3)$	Place the numbers to be subtracted in the same power of 10 engineering notation by moving the decimal point and adjusting the exponent.
$(1{,}000 \times 10^3) - (220 \times 10^3)$	Subtract the coefficients and assign the common power of ten engineering notation.
780×10^3	Place the number in the correct prefix notation.
$780 \times 10^3 = \boxed{780 \text{ kΩ}}$	The answer.

Or alternatively, note that the metric prefix could be changed to the unit of MΩ and the answer would then be:
$\boxed{0.780 \text{ MΩ}}$

Example 8: Two voltages oppose each other in an electrical circuit. If one voltage is 1.1 kV and the other voltage is 200 V, find the net voltage (difference) and present the answer in kV.

1.1 kV − 200 V	Place the numbers in engineering notation.
$(1.1 \times 10^3) - (200 \times 10^0)$	Place the numbers to be subtracted in the same power of 10 engineering notation by moving the decimal point and adjusting the exponent.
$(1.1 \times 10^3) - (0.200 \times 10^3)$	Subtract the coefficients and assign the common power of ten engineering notation.
0.9×10^3	Place the number in the correct prefix notation.
$0.9 \times 10^3 = \boxed{0.9 \text{ V}}$	The answer.

Or alternatively, note that the metric prefix could be changed to the base unit of volts and the answer would be:
$\boxed{900 \text{ V}}$

Multiplying in Engineering Notation: Multiplying numbers in engineering notation with a prefix notation can be accomplished as follows:

> First: Place the numbers to be multiplied in engineering notation.
>
> Second: *Multiply* the numerical coefficients and then algebraically add the exponents of the power of 10 notation.
>
> Third: Replace the coefficient's power of 10 exponent with the correct metric prefix notation.

Example 9: Ohm's Law states that voltage is equal to current multiplied by the resistance. Find the voltage if the current is 15 mA and the resistance is 10 kΩ.

15 mA $\times$ 10 kΩ	Place the numbers in engineering notation.
$(15 \times 10^{-3}) \times (10 \times 10^{3})$	Multiply the numerical coefficients and then algebraically add the exponents of the power of 10 notation.
150×10^{0}	Place the number in the correct prefix notation.
150×10^{0} = $\boxed{150 \text{ V}}$	The answer.

Example 10: Again by using Ohm's Law, find the voltage if the current is 85 μA and the resistance is 15 kΩ.

85 μA $\times$ 15 kΩ	Place the numbers in engineering notation.
$(85 \times 10^{-6}) \times (15 \times 10^{3})$	Multiply the numerical coefficients and then algebraically add the exponents of the power of 10 notation.
$1{,}275 \times 10^{-3}$	Place the number in the correct prefix notation.
$1{,}275 \times 10^{-3}$ = $\boxed{1{,}275 \text{ mV}}$	The answer.

Or alternatively, note that the metric prefix could be changed to the base unit of volts and the answer would then be:

$\boxed{1.275 \text{ V}}$

Dividing in Engineering Notation: Dividing numbers in engineering notation with a prefix notation can be accomplished as follows:

> First: Place the numbers to be divided in engineering notation.
>
> Second: *Divide* the numerical coefficients and then algebraically subtract the exponents of the power of 10 notation. Algebraically *subtract* the exponent of the denominator from the exponent of the numerator.
>
> Third: Replace the coefficient's power of 10 exponent with the correct metric prefix notation.

Example 11: Ohm's Law states that in an electrical circuit, current (amperes) is equal to voltage divided by the resistance. Find the current in a circuit if the voltage is 120 kV and the resistance is 15 MΩ.

(120 kV)/(15 MΩ)	Place the numbers is engineering notation.
$(120 \times 10^{3})/(15 \times 10^{6})$	Divide the numerical coefficients and then algebraically subtract the exponents of the power of 10 notation.
8×10^{-3}	Place the number in the correct prefix notation.
8×10^{-3} = $\boxed{8 \text{ mA}}$	The answer.

Example 12: Again by using Ohm's Law, find the current in a circuit if the voltage is 2.2 kV and the resistance is 220 Ω.

(2.2 kV)/(220 Ω)	Place the numbers is engineering notation.
$(2.2 \times 10^{3})/(220 \times 10^{0})$	Divide the numerical coefficients and then algebraically subtract the exponents of the power of 10 notation.
0.01×10^{3}	Place the number in the correct prefix notation.
0.01×10^{3} = $\boxed{0.01 \text{ kA}}$	The answer.

Or alternatively, note that the metric prefix could be changed to the base unit of amperes and the answer would then be:

$\boxed{10 \text{ A}}$

EXERCISE PROBLEMS

For each of the following problems, perform the indicated operation and record your answers on the results page.

Place the following numbers in engineering notation.

1. 7.94×10^2
2. 3.21×10^7
3. 5.35×10^4
4. 72.8×10^1

Replace the power of ten notation with the appropriate metric prefix.

5. 6.55×10^6
6. 98.1×10^{-3}
7. 32.7×10^3
8. 9.24×10^{-9}

Add the following numbers and place the answer in engineering notation.

9. $56.8 \times 10^{-3} + 33.7 \times 10^{-6}$
10. $384 \times 10^3 + 121 \times 10^3$
11. $42.2 \times 10^{-6} + 1.02 \times 10^{-3}$
12. $7.65 \times 10^{-12} + 247 \times 10^{-9}$

Subtract the following numbers and place the answer in engineering notation.

13. $1.00 \times 10^3 - 0.10 \times 10^3$
14. $3.46 \times 10^{-6} - 2.25 \times 10^{-3}$
15. $18.3 \times 10^9 - 16.1 \times 10^{12}$
16. $2.67 \times 10^{-3} - 1.68 \times 10^0$

Multiply the following numbers and place the answer in engineering notation.

17. $(5.55 \times 10^6) \times (3.46 \times 10^3)$
18. $(3.44 \times 10^{-3}) \times (1.22 \times 10^{-3})$
19. $(6.84 \times 10^9) \times (2.12 \times 10^9)$
20. $(8.00 \times 10^{-3}) \times (2.00 \times 10^{-3})$

Divide the following numbers and place the answer in engineering notation.

21. $(7.50 \times 10^{-3})/(0.50 \times 10^3)$
22. $(6.00 \times 10^{-9})/(2.00 \times 10^{-6})$
23. $(12.6 \times 10^3)/(3.00 \times 10^{-3})$
24. $(8.88 \times 10^3)/(1.00 \times 10^{-6})$

RESULTS FOR EXERCISE PROBLEMS FROM INTRODUCTION I-3

1. _____   9. _____   17. _____
2. _____   10. _____   18. _____
3. _____   11. _____   19. _____
4. _____   12. _____   20. _____
5. _____   13. _____   21. _____
6. _____   14. _____   22. _____
7. _____   15. _____   23. _____
8. _____   16. _____   24. _____

QUESTIONS

1. What is the relation between engineering notation and metric prefixes?

2. Describe the process of subtracting numbers in engineering notation?

3. 60,000,000 V is how many mega volts?

4. What is the numerical difference between 10^{-3} and 10^{-9}?

5. In your own words describe the process used in multiplying numbers in engineering notation. Give an example.

REPORT

Attach the above results to the following results form. Include your conclusion with this report.

RESULTS REPORT FORM

Experiment No: _____ Name: _____
                                        Date: _____
Experiment Title: _____ Class: _____
_____   Instr: _____

Explain the purpose of the experiment:

List the first Learning Objective:
OBJECTIVE 1:

After reviewing your results, describe how this objective was validated by this experiment?

List the second Learning Objective:
OBJECTIVE 2:

After reviewing your results, describe how this objective was validated by this experiment?

List the third Learning Objective:
OBJECTIVE 3:

After reviewing your results, describe how this objective was validated by this experiment?

Conclusion:

Attach to this Experiment Report Form: ☐ Answers to Questions and the Results Page

EXPERIMENT I-4

INTRODUCTION TO MATHEMATICS FOR ELECTRONICS—SIGNIFICANT FIGURES (ACCURACY AND PRECISION)

LEARNING OBJECTIVES

At the completion of this experiment, you will be able to:
- Describe the difference between exact and approximate numbers.
- Describe the differences between accuracy and precision.
- Perform basic calculations with measured quantities.

SUGGESTED READING

Introduction, *Basic Electronics,* Grob/Schultz, Tenth Edition

INTRODUCTION

Measurement: For the purpose of the study electronics, measurement is defined as the comparison of some quantity to a commonly accepted standard unit. When studying quantities, it is important to understand that numerical values are primarily divided into two categories, exact numbers and approximate numbers.

Exact Numbers: Exact numbers are those which are determined to be finite in value. Examples of exact numbers are the number of pages in this laboratory manual, the temperature at which water will boil, the cost to purchase a stamp for a letter, the number of centimeters in a meter, etc.

Approximate Numbers: Approximate numbers are the result of the use of a measurement device or instrument. In electronics most values are measured and because of that fact approximate numbers are very common. Measuring instruments include voltmeters, ammeters, ohmmeters, oscilloscopes, etc. It is understood by the electronics technician that the better the measuring device the better the measurement, but that *all* measurements provide only *approximate* numbers.

In electronics, when studying measurement it is also important to understand the relationship between the measurement's accuracy and the measurement's precision.

Measurement and Accuracy: Meter readings and other measurements are usually accurate to a certain level and are generally determined by the application and the type of equipment used. The greater the number of significant figures a measurement contains, the greater the accuracy of the measuring instrument. Only the figures (numbers) that can be determined from the measuring instrument are considered significant.

There are rules for determining the number of significant figures that a measurement contains. The five rules are:

Rule 1: All nonzero figures (numbers) in a measurement are considered significant.

    Example 1:   215 contains three significant figures.
                            18.2 contains three significant figures.
                            5.6×10^3 contains two significant figures.

Rule 2: All zeros between significant (nonzero) figures in a measurement are considered significant.

    Example 2:   1405 contains four significant figures.
                            20005 contains five significant figures.
                            10.06×10^4 contains four significant figures.

Rule 3: All zeros to the right of a decimal point *and* a significant figure in a measurement are considered significant.

Example 3: 94.0 contains three significant figures.
34.00 contains four significant figures.
10.0×10^{-3} contains three significant figures.

Note: Depending upon the application, zeros to the right of a nonzero figure in a measurement may or may not be considered significant. If zeros are seen on the instrument, then they are considered significant. If the zeros are used only as placeholders, they are not considered significant. Your instructor will provide further information about the application of significant figures to the types of instrumentation that will be used in your laboratory.

Rule 4: Zero to the right in a whole number measurement are *not* considered significant.

Example 4: 600 contains one significant figure.
12000 contains two significant figures.
150 contains two significant figures.

Rule 5: Zeros to the left of a measurement that is less than one are *not* considered significant.

Example 5: 0.2 contains one significant figure.
0.056 contains two significant figures.
0.44×10^{-3} contains two significant figures.

Measurement and Precision: When a measurement is taken, the precision of that measurement is identified as the smallest unit in which a measurement is made. The smallest unit is determined by the location of the last significant figure.

Example 6: A voltmeter measures potential difference also know as voltage. Determine the precision of a voltmeter measurement of 12000 Volts.

12000 Volts   This measurement has an accuracy of two significant figures. The least significant figure is the number 2. This figure occupies the 1000's place. The precision of this measurement is to 1000 volts.
$\boxed{1000 \text{ Volts}}$ The answer.

Example 7: An Ohmmeter measures resistance. Determine the precision of an Ohmmeter measurement of 625 Ω.

625 Ω   This measurement has an accuracy of three significant figures. The least significant figure is the number 5. This figure occupies the 1's place. The precision of this measurement is to 1 Ω.
$\boxed{1 \text{ Ω}}$ The answer.

Example 8: An ammeter measures current. The ammeter you are using provides a measurement of 0.007896 Amperes. Determine the precision of the measurement 0.007896 Amperes.

0.007896 Amperes   This measurement has an accuracy of four significant figures. The least significant figure is the number 6. This figure occupies the one-millionth place. The precision of this measurement is to one-millionth of an Ampere, or 1×10^{-6} Ampere.
$\boxed{1 \times 10^{-6} \text{ Ampere}}$ The answer.

Confusion: It is not uncommon to hear the term accuracy and precision used interchangeably. However it needs to be understood that when comparing measurements, one measurement could have good accuracy but poor precision in comparison to another measurement with poor accuracy yet good precision.

Example 9: A student using a digital voltmeter takes a measurement and determines its value to be 0.002 Volts. The student then takes another measurement in a different part of the circuit and determines the value to be 126 Volts. Determine the accuracy and precision of each of the two measurements.

First Voltage Measurement:

0.002 Volts — This measurement has an accuracy of one significant figure. The least significant figure is the number 2. This figure occupies the one-thousandth place. The precision of this measurement is to the one-thousandth of Volt, or 1×10^{-3} Volt.

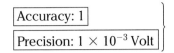

The answers.

Second Voltage Measurement:

126 Volts — This measurement has an accuracy of three significant figures. The least significant figure is the number 6. This figure occupies the one's place. The precision of this measurement is to the ones of a Volt.

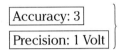

The answers.

Note that the first measurement has *less* accuracy (1 significant figure) and *greater* precision (one-thousandth of a volt). The second measurement displayed *greater* accuracy (3 significant figures) but *less* precision (1-volt).

Adding or Subtracting with Measured Quantities: When adding or subtracting measured quantities the answer can have no more digits to the right of the decimal point than which are contained in the measurement with the *least number* of digits to the right of the decimal point. Another way of analyzing this process is to say that the answer can be no more precise than the *least precise* of the measurements taken when adding or subtracting measurements.

Example 10: An electronics technician wants to add three voltages that are in series. The voltages are 10.1 volt, 123.41 volt and 56.01 volt.

$10.1 + 123.41 + 56.01$ Add the three voltages.

189.52 The least precise of the three voltages is 10.1, and the final answer will need to be limited to the tenth place.

$\boxed{189.5 \text{ Volts}}$ The answer. Note that the technique of rounding a number will be presented in a later section in this experiment.

Multiplying or Dividing with Measured Quantities: When multiplying or dividing measured quantities the answer can contain no more significant figures than the measurement with the least number of significant figures. Another way of analyzing this process is to say that the answer can be no more accurate than the *least accurate* of the measurements taken when multiplying or dividing measurements.

Example 11: An electronics technology student has measured the current and resistance of a particular circuit. She would like to determine the voltage from these measurements and understands that Ohm's law states that the voltage is determined by multiplying the measured current by the measured circuit resistance. The circuit's current is 1.4 amperes and the circuit's resistance is 15.68 ohms. Find the voltage.

1.4×15.68 Multiply the current and resistance measurements.

21.952 The measurement of 1.4 amperes contains two significant figures. The measurement of 15.68 ohms contains 4 significant figures. The less accurate of the two measurements is 1.4, and will limit the final answer to two significant figures.

$\boxed{22 \text{ Volts}}$ The answer. Note that the technique of rounding a number will be presented in the next section in this experiment.

Rounding: There are several commonly used techniques to round off a calculation or measurement that contains too many significant figures. One of the options to round off calculations or measurements is shown below.

If the significant figure to be eliminated is less than 5, then eliminate this figure and leave the remaining figures.

If the significant figure to be eliminated is equal to or greater than 5, then add the number 1 to the preceding figure.

Note: Your Electronics Technology program or instructor may have already established an accepted protocol for rounding significant figures. Talk with your instructor about any mathematical preferences that you may need to follow when recording measurements and performing calculations.

Example 12: A student makes a resistance measurement of 146.7 ohms. Round this measurement to three significant figures.

> 146.7 ohms Since the least significant figure is 7, then add the number 1 to the preceding figure.
>
> 147 ohms The answer.

Example 13: A student makes a voltage measurement of 12.654 volts. Round this measurement to four significant figures.

> 12.654 volts Since the least significant figure is 4 and is less than 5, then eliminate the 4 and leave the remaining figures.
>
> 12.65 volts The answer.

Summary: The following table summarizes the main concepts of this experiment.

Sample Problem	Measurement Value	Significant Figures	Significant Figure Rule	Measurement's Precision
1	23 V	2	1	1 V
2	2002 Ω	4	2	1 Ω
3	568 V	3	1	1 V
4	91.668 V	5	1	1 thousandth of a Volt
5	581.80 V	5	3	1 hundredth of a Volt
6	0.002580 A	4	5	1 millionth of an Ampere
7	0.003 V	1	5	1 thousandth of a Volt
8	178.6 Ω	4	1	1 tenth of an Ω
9	89.4 V	3	1	1 tenth of a Volt
10	3060.7 V	5	2	1 tenth of a Volt
11	60.560 Ω	5	3	1 thousandth of a Ω
12	4.007 A	4	2	1 thousandth of an Ampere
13	0.0087 Ω	2	5	1 ten-thousandth of an Ω
14	69.8×10^5 Ω	3	1	1 Tenth of an Ω
15	75×10^{-3} V	2	1	1 V
16	0.05 A	1	5	1 tenth of an Ampere
17	0.0010 A	2	5	1 ten-thousandth of an Ampere
18	0.1500 A	4	3	1 ten-thousandth of an Ampere
19	1000 Ω	1	4	1 thousand Volts
20	1500 V	2	4	1 hundred Volts

EXERCISE PROBLEMS

For each of the following problems, perform the indicated operation and record your answers on the Results page.

How many significant figures are contained in the following measured quantities?

1. 8.001
2. 721
3. 0.632

What is the precision of each of the following measured quantities?

4. 14.11
5. 738.6
6. 0.0012

What is the accuracy of each of the following measured quantities?

7. 12.63
8. 7684
9. 0.6117

Add the following measured quantities and reduce the answer accordingly.

10. 16.3 + 17.14
11. 10031 + 6763
12. 101 + 12

Subtract the following measured quantities and reduce the answer accordingly.

13. 15.6 − 3.33
14. 1002 − 863
15. 12000 − 11110

Multiply the following measured quantities and reduce the answer accordingly.

16. 6.23 × 1.23
17. 6.1 × 21.21
18. 34 × 12.1

Divide the following measured quantities and reduce the answer accordingly.

19. 73/6.1
20. 108/12.4
21. 17.3/10.03

Round each of the following measured quantities to three significant figures.

22. 763.62
23. 64.762
24. 1.27721

| NAME | DATE |

RESULTS FOR EXERCISE PROBLEMS FROM INTRODUCTION I-4

1. _____   9. _____   17. _____
2. _____   10. _____   18. _____
3. _____   11. _____   19. _____
4. _____   12. _____   20. _____
5. _____   13. _____   21. _____
6. _____   14. _____   22. _____
7. _____   15. _____   23. _____
8. _____   16. _____   24. _____

QUESTIONS

1. What is the difference between exact and approximate numbers? Give an example of each.

2. Describe the differences between accuracy and precision.

3. Describe the main difference between adding or subtracting measured quantities and multiplying or dividing measured quantities.

4. For the two measured values of 15.698 and 17.89 volts, which is considered to be the more precise? Explain.

5. For the two measurements in problem 4, which is considered the more accurate? Explain.

REPORT

Attach the above results to the following results form. Include your conclusion with this report.

RESULTS REPORT FORM

Experiment No: _____   Name: _____
                                          Date: _____
Experiment Title: _____   Class: _____
_____   Instr: _____

Explain the purpose of the experiment:

List the first Learning Objective:
OBJECTIVE 1:

After reviewing your results, describe how this objective was validated by this experiment?

List the second Learning Objective:
OBJECTIVE 2:

After reviewing your results, describe how this objective was validated by this experiment?

List the third Learning Objective:
OBJECTIVE 3:

After reviewing your results, describe how this objective was validated by this experiment?

Conclusion:

Attach to this Experiment Report Form: ☐ Answers to Questions

CHAPTER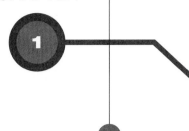

EXPERIMENT 1-1

LAB SAFETY, EQUIPMENT, AND COMPONENTS

OBJECTIVES

At the completion of this experiment, you will:
- Understand the basic safety concepts used in a lab.
- Be familiar with lab equipment and components.
- Understand the resistor color code.

SUGGESTED READING

Chapters 1 and 2, *Basic Electronics,* Grob/Schultz, Tenth Edition

INTRODUCTION

An electronics lab combines a classroom with a technician's workbench. Most classroom labs have similar equipment. You should become familiar with the equipment in your classroom lab, including safety equipment, safety policies, and practices.

SAFETY

Safety always comes first—in a lab class and at work. In fact, most large companies train their technicians in safety measures, including good ergonomic practices. You should not only know how to use the equipment safely but also how to sit at a lab station without stressing your arms, wrists, or eyes when working. Most of the time, this is a matter of common sense—avoid any stressful position for a prolonged time.

But most safety issues in an electronics lab come from the possibility of getting an electric shock. It is your responsibility to avoid this hazard by following these basic rules:

1. Do not plug in or turn on the power to any equipment without knowing how to use the equipment. This means asking for help when you do not know how to use something safely. Do not be afraid to ask for help.
2. Always check power cords. They should not be frayed, loose, or damaged in any way. In general, always inspect your equipment before using it.
3. Never wear loose jewelry or rings when working on equipment. Most jewelry is made of metal and may conduct hazardous current accidentally.
4. Do not keep beverages, open liquids, or food near electronic equipment.
5. Do not touch any circuit or components unless you know that it is safe to do so. For example, touching some components (especially some integrated circuits) can damage them. Other components (large capacitors) can discharge a dangerous current through your body if you touch them—even if they are not connected to a circuit.
6. Never use a soldering iron without proper training.
7. Never disturb another person who is using equipment.
8. Always wear safety glasses or any other required safety equipment.

Your lab class may have a safety test, signs, or some other information that you should read or respond to as required. Although most electronics courses are not known to cause injury, it is still a good idea to be cautious and prepared. Be sure you know where first aid is available and what to do in an emergency.

EQUIPMENT

Every electronics lab, and every school, has different types of equipment. However, there are some basic meters that are common to all electronics labs. In one form or another, your school will have these meters. They may appear to be different only because they are made by different manufacturers. Typical equipment is shown in Grob's *Basic Electronics.*

Simple Meters (Single Purpose)

Ohmmeter: Used to measure resistance (in ohms). Remember that resistance is the opposition to current flow in a circuit. Resistors usually have color stripes on them to identify their values. One way to be sure your resistor is the correct value is to measure it with an ohmmeter.

Ammeter: Used to measure current (in amperes). Remember that current is an electric charge in motion.

Ammeters usually measure the current in milliamperes (0.001 A), because most electronics lab courses do not require large amounts of current.

Voltmeter: Used to measure voltage (in volts). Remember that voltage (force) refers to units of potential difference. Most voltmeters use more than one range, or scale, because voltage, unlike current, is often measured in larger values.

Multimeters (Multipurpose)

VOM (VOLT-OHM-MILLIAMMETER): A multipurpose meter used to measure dc and ac voltage, dc current, and resistance (see Fig. 1-1.1a).

VTVM (VACUUM TUBE VOLTMETER): Another, older type of multipurpose meter used to measure voltage and resistance only.

DMM (DIGITAL MULTIMETER): A digital multipurpose meter, also known as a DVM (digital voltmeter) (see Fig. 1-1.1b).

Other Equipment

DC Power Supply: A source of potential difference, like a battery. It supplies voltage and current, and it can be adjusted to provide the required voltage for any experiment.

Soldering Iron: A pointed electric appliance that heats electric connections so that solder (tin and lead) will melt on those connections. Used to join components and circuits.

Breadboard/Springboard/Protoboard: Used to assemble basic circuits, by either soldering, inserting into springs, or joining together in sockets. These boards are the tools of designers, students, and hobbyists.

Component Familiarization (Symbols)

—⋀⋀⋀— The symbol for resistance or a resistor.

—|⊦|⊦|+ The symbol for a battery (cells) or a power supply.

+ Positive potential symbol (associated with the color red).

− Negative potential symbol (associated with the color black).

—)|— Capacitor (or capacitance) symbol. Used in later study.

—⌒⌒⌒— Inductor (or inductance) symbol. Used in later study.

Leads are simply insulated wires used to join the meters to the circuits, the power supply to the circuit, etc. They are conductors and have no polarity of their own. Leads have different types of connectors on the ends, such as alligator clips, banana plugs, and BNC connectors.

Ω = ohms (unit of resistance), Greek letter omega

∞ = infinity (used to indicate infinite resistance)

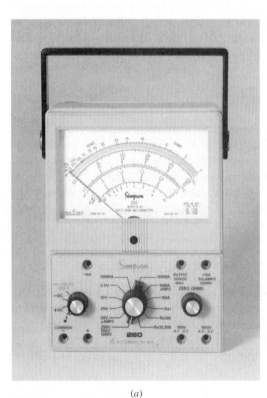

(a)

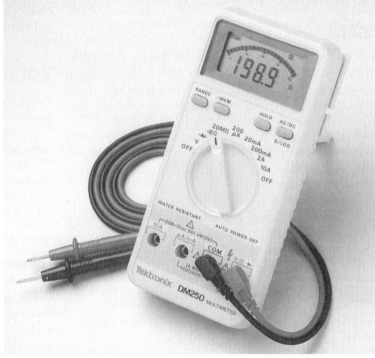

(b)

Fig. 1-1.1 Basic meters. (*a*) VOM. (*b*) DMM.

Common Symbols for Multipliers

Lowercase k = kilo = 1000 or 1×10^3
Uppercase M = mega = 1,000,000 or 1×10^6
Lowercase m = milli = 0.001 or 1×10^{-3}
Greek letter mu (μ) = micro = 0.000 001 or 1×10^{-6}

COMPONENTS

Certain basic components are used in electronics all the time. They are the resistor, the capacitor, the inductor (see Fig. 1-1.2), the diode, and the transistor. Along with these are also hundreds of variations and customized devices and chips.

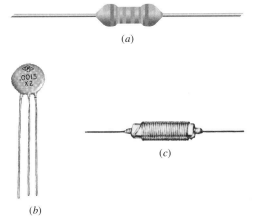

Fig. 1-1.2 Typical resistor (*a*), capacitor (*b*), and inductor (*c*).

Capacitors have the ability (capacity) to store electrical energy, and inductors have the ability to induce a voltage; these will be studied later. But the practical use of resistors is studied early in most lab classes. Learning about resistors and learning (memorizing) the resistor color code are critical for any technician.

In general, resistors limit or resist the flow of current in a circuit. That is why they are called *resistors*. Made of a carbon compound, resistors come in various sizes and power ratings. The most common resistors on circuit boards have the value of resistance color-coded (painted) on them. The value of resistance is the *ohm*, named for Georg Simon Ohm, the scientist who determined a law for resistance in 1828. The resistance value in ohms refers to the resistor's ability to resist the flow of electricity. The lower the value, the less resistance. In practical applications, one ohm (1 Ω) of resistance is very small and has little effect on most circuits. However, 1 million Ω has so much resistance that it resists all but the smallest amount of current flow.

PROCEDURE

Four-Band Resistor Color Code Investigation

1. Answer questions 1 to 15.
2. After studying the resistor color code in Fig. 1-1.3, fill in Tables 1-1.1 and 1-1.2.
3. List the types of meters that you already know how to use and compare them to any meters in your lab.

After you finish, turn in your answers to procedure steps 1, 2, and 3.

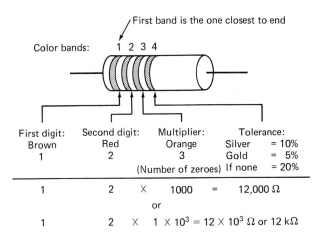

Fig. 1-1.3 How to read color bands (stripes) on carbon resistors. Resistors come in various shapes and sizes. This is only one common form. The larger the physical size, the greater the wattage rating. Refer to Appendix C for other component codes, including the five-band carbon film resistors.

EXPERIMENT 1-1

NAME	DATE

QUESTIONS FOR EXPERIMENT 1-1

Fill in the blanks (1–15) with the letter of the correct answer.

    ____ 1. An instrument used to measure potential difference.
    ____ 2. An instrument used to measure current.
    ____ 3. An instrument used to measure resistance.
    ____ 4. A passive component that opposes the flow of current.
    ____ 5. An instrument used to heat solder and join components.
    ____ 6. A source of dc voltage other than a battery.
    ____ 7. A symbol for dc voltage source.
    ____ 8. A symbol for resistance.
    ____ 9. A color used to represent negative polarity.
    ____ 10. An item used to temporarily build circuits on.
    ____ 11. A Greek letter used to represent a unit of resistance.
    ____ 12. An English letter used to represent 1000.
    ____ 13. An English letter used to represent 0.001.
    ____ 14. An instrument capable of measuring both voltage and current.
    ____ 15. A symbol for infinity.

a. ammeter
b. —⟋⟍⟋⟍—
c. power supply
d. voltmeter
e. —|||—
f. black
g. ohmmeter
h. breadboard
i. soldering iron
j. resistor
k. VOM/VTVM
l. m
m. k
n. Ω
o. ∞

TABLES FOR EXPERIMENT 1-1

TABLE 1-1.1 Resistor Color Codes

First Digit Band 1	Second Digit Band 2	Multiplier Band 3	Tolerance Band 4	Resistor Value
Red	Brown	Brown	Gold	_____
Brown	Brown	Black	Gold	_____
Green	Blue	Red	Silver	_____
Blue	Green	Yellow	Silver	_____
Red	Red	Orange	Silver	_____
Orange	White	Brown	Gold	_____
Blue	Green	Black	Silver	_____
Brown	Black	Red	Gold	_____
Yellow	Violet	Green	Gold	_____
Brown	Black	Orange	Silver	_____
Orange	Orange	Orange	Silver	_____
Brown	Black	Gold	Gold	_____
White	Blue	Red	Silver	_____
Brown	Black	Yellow	Silver	_____
Brown	Green	Green	Gold	_____

TABLE 1-1.2 Resistor Color Codes

Band 1 Color	Band 2 Color	Band 3 Color	Band 4 Color	Resistor Value
_____	_____	_____	_____	680 kΩ, 5%
_____	_____	_____	_____	10 kΩ, 10%
_____	_____	_____	_____	100 kΩ, 5%
_____	_____	_____	_____	3.3 MΩ, 5%
_____	_____	_____	_____	1.2 kΩ, 10%
_____	_____	_____	_____	820 Ω, 10%
_____	_____	_____	_____	47 kΩ, 5%
_____	_____	_____	_____	330 Ω, 10%
_____	_____	_____	_____	470 kΩ, 5%
_____	_____	_____	_____	560 Ω, 10%
_____	_____	_____	_____	1.5 MΩ, 10%
_____	_____	_____	_____	220 Ω, 5%
_____	_____	_____	_____	56 Ω, 10%
_____	_____	_____	_____	12 kΩ, 5%
_____	_____	_____	_____	560 kΩ, 5%

EXPERIMENT RESULTS REPORT FORM

Experiment No: _____   Name: _____
                                          Date: _____
Experiment Title: _____   Class: _____
_____   Instr: _____

Explain the purpose of the experiment:

List the first Learning Objective:
OBJECTIVE 1:

After reviewing the results, describe how the objective was validated by this experiment?

List the second Learning Objective:
OBJECTIVE 2:

After reviewing the results, describe how the objective was validated by this experiment?

List the third Learning Objective:
OBJECTIVE 3:

After reviewing the results, describe how the objective was validated by this experiment?

Conclusion:

If required, attach to this form: ☐ Answers to Questions, ☐ Tables, and ☐ Graphs.

CHAPTER 2

EXPERIMENT 2-1

RESISTANCE MEASUREMENTS

OBJECTIVES

At the completion of this experiment, you will be able to:
- Read nonlinear meter scales.
- Calibrate and/or operate an ohmmeter.
- Measure resistance in ohms.

SUGGESTED READING

Chapters 1 and 2, *Basic Electronics,* Grob/Schultz, Tenth Edition

INTRODUCTION

This experiment will familiarize you with resistance measurements and the ohmmeter. Ohmmeters measure resistance in *ohms*, which is the opposition to current flow in a circuit. An ohmmeter works by sending a known value of current through a component and measuring the amount of current that returns. The difference is then used to determine the resistance. For example, if all the current sent is returned, the ohmmeter reading will be zero ohms (0 Ω) or no resistance. Or, if no current returns, the reading will be very large (millions of ohms) or infinity. Anything in between is determined by the internal circuitry and the applied voltage.

An ohmmeter can measure amounts of resistance from zero ohms, such as a short piece of copper wire (a *short circuit*) to infinite ohms, such as air (an *open circuit*). But, most of the time, you will use the ohmmeter to measure the value of resistors in ohms.

Types of Ohmmeters

For practical purposes, no single piece of equipment is dedicated to measuring ohms. Some older labs (schools) may use dedicated ohmmeters, but they are inadequate for use in the field and not found in modern test and production facilities. In general, ohmmeters are a functional part of multipurpose meters, which often also measure voltage and current. In fact, the only single-function ohmmeter that is practical is a continuity checker, which is used simply to verify that a circuit is closed (continuous) or open (no continuity).

Both digital and analog meters are used, with the digital meter currently the meter of choice. Analog meters use a meter movement and have nonlinear scales which are used only for resistance measurements—this is because the concept of infinite ohms cannot be fit onto a readable scale. In addition, all multifunction meters require that you adjust the function switch for volts, amperes, or ohms, and you may also be required to adjust the range. With the range properly set, you simply multiply the scale reading by the range setting. Ohmmeters can be part of the following types of multipurpose meters.

VOM (volt-ohm-milliammeter): This multifunction meter measures voltage, current, or resistance. As an ohmmeter, it can be digital or analog. Most versions use a magnetic meter movement with a needle or pointer and a nonlinear scale for resistance measurements. Other scales appear on the meter for voltage and current. Many VOMs combine function and range switches. The VOM is usually battery operated and portable. It also can usually measure low values of current, and both dc and ac voltages. See Fig. 2-1.1.

VTVM (vacuum tube voltmeter): This multifunction meter is based on vacuum tube technology and requires ac power to operate the vacuum tube inside. VTVMs usually have separate switches for function and range. For years the VTVM was considered the best bench tool for technicians because it measured both ac and dc voltages. However, it cannot measure current and is rarely used in modern workplaces since it has been replaced by digital meters that perform the same tasks. Like the VOM, the VTVM must be calibrated before measuring resistance. It also can measure power (in watts). See Fig. 2-1.2.

EQUIPMENT

Ohmmeter: DMM, VOM, or VTVM

COMPONENTS

30 resistors, all 0.25 W unless indicated otherwise:

(1) 10 Ω   (1) 4.7 kΩ
(1) 56 Ω   (1) 5.6 kΩ (0.5 W or less)
(1) 100 Ω   (3) 10 kΩ (0.5 W or less)

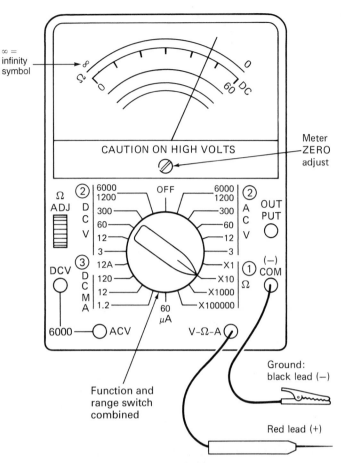

Fig. 2-1.1 VOM. Linear and nonlinear scale (ohms). A generic VOM measures resistance in ohms (Ω), ac and dc voltage, and current in amperes (A). The VOM shown is set for resistance: $R \times 10$.

(1) 220 Ω	(1) 22 kΩ (0.5 W or less)
(1) 390 Ω	(1) 33 kΩ (0.5 W or less)
(1) 470 Ω	(2) 47 kΩ (0.5 W or less)
(1) 680 Ω	(1) 68 kΩ (0.5 W or less)
(1) 820 Ω	(1) 86 kΩ (0.5 W or less)
(1) 1 kΩ	(2) 100 kΩ (0.5 W or less)
(1) 1.2 kΩ	(1) 220 kΩ (0.5 W or less)
(1) 1.5 kΩ	(1) 470 kΩ (0.5 W or less)
(1) 2.2 kΩ	(1) 1.2 MΩ (0.5 W or less)
(1) 3.3 kΩ	(1) 3.3 MΩ (0.5 W or less)

USING A VOM

Function/Range Switch: This rotary switch should be in the OFF position when not in use; otherwise, the VOM's battery may become depleted. Not only does this switch allow you to choose the desired functions (ohmmeter, voltmeter, ammeter), but it also indicates the range. VOMs often seem more difficult to use than VTVMs because they have so many switches and inputs located in a small area.

Ohms Adjust: This is the zero adjustment for measuring resistance (ohms). Use it to line up the needle directly on the zero line when using the VOM as an ohmmeter when the leads are shorted.

COM: This is the common or negative input. That is, it is the place where the black ground lead is plugged into the meter. Remember that VOMs, like most electronic instruments, have two inputs: a negative and a positive.

VΩA: This is the positive input. Plug the other lead (red, if available) into this jack. Note that the words *jack, input, terminal,* and *plug-in* are often used interchangeably. You will become used to this terminology as you continue. This plug-in is used whenever measuring volts, ohms, or amperes (dc milliamperes).

Other Inputs: Do not be concerned with the other inputs at this time. You will not be using them. However, notice that the zero adjust is not marked on the VOM itself. This is because it is not adjusted (balanced) like the VTVM. If your VOM needle is not on the zero line in the dc volts function, adjust it here with a screwdriver.

Scales: The VOM has several scales. The nonlinear scale at the top is used for resistance measurements in ohms. The linear scale below is usually used for dc voltage. And the milliamperes scale (current measurements) may be another scale altogether, depending upon the manufacturer.

Function Switch: This switch turns the VTVM on. Actually, it is a rotary switch and should be in the OFF position when not in use. However, VTVMs require some warm-up time and are often left on during lab hours. The function switch allows you to choose the function you want from the VTVM. Thus, setting the function switch turns the VTVM into either an ohmmeter or a voltmeter (AC or +DC or −DC). You will probably use it as either an ohmmeter or a +dc voltmeter during your beginning studies. (See Fig. 2-1.2.)

Range Switch: This switch determines the scale that will be used. Notice that Fig. 2-1.2 shows eight different range settings. Each range setting corresponds to the function switch setting. For example, "15 V, $R \times 100$" can be used for either a voltage or a resistance (ohms) measurement. In this case, if the function switch is on OHMS, the range would automatically be $R \times 100$, not 15 V.

Scales (Ohms and DC/RMS): The OHMS scale, marked Ω for resistance, is on top. It is a nonlinear scale (not equally divided). It goes from zero to infinity (∞). The scales below it are used for voltage and power measurements. DC and RMS are voltage measurements.

Zero Adjustment: This knob is a potentiometer that allows you to position the needle directly on the zero line before making a measurement.

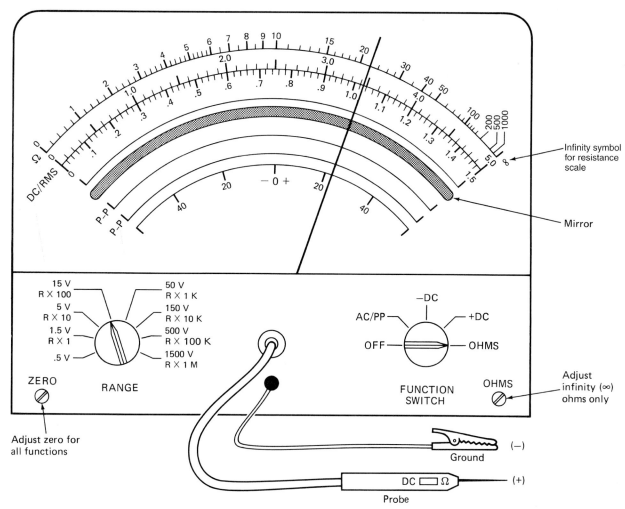

Fig. 2-1.2 VTVM (typical generic type). The top scale is nonlinear (0 to infinity), used for resistance (Ω). The remaining scales are linear (0 to 5, 0 to 1.5, etc.), used for voltage. Note that most VTVMs have a mirror strip between scales to prevent parallax errors. These visual errors are similar to those made by passengers in a car trying to read the speedometer from an angle. To avoid parallax errors, line up the needle with the mirror reflection and with the division line on the scale.

Ohms Adjustment: This knob, also a potentiometer, allows you to line up the needle on the other end of the scale, on the infinity line. It is only used for resistance measurements (ohmmeter function).

Ground Lead: This lead is usually connected first. It is the negative (−), or ground, side. First, connect this ground strap or lead to one end of the component you are measuring. Then, make connection with the probe (+).

Probe: Often called the *positive lead*, this pointed probe has a switch on it for measuring either ohms or dc voltage. After the ground lead is connected, it is easy to use this probe to make contact with the component you are measuring.

DVM or DMM (digital voltmeter or digital multimeter): This is the most common type of multimeter for field use. It can measure resistance, voltage, and current. It is usually small, portable (battery powered), and accurate. Of course, digital meters have an LED (light-emitting diode) display with the values displayed in digits. As with all meters, you must adjust the function and, sometimes, the range. The digital meter is the meter of choice for today's technician. See Fig. 2-1.3.

PROCEDURE

1. Reading Nonlinear Resistance Scales: Refer to Fig. 2-1.4. It shows the type of scales found on a VTVM. The top scale is nonlinear (uneven) and has an R label to indicate that it is used for resistance measurements only. The other scales below are used for voltage and power measurements. The scale goes from 0 (zero) to 1000, with infinity as the last increment or point on the scale. Notice also that the increment spacing is greater between

EXPERIMENT 2-1

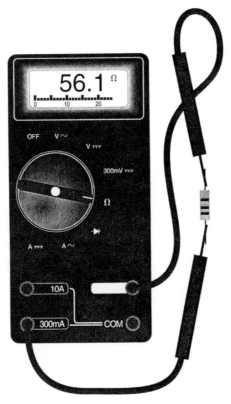

Fig. 2-1.3 DMM.

some numbers. This is due to the way the vacuum tube operates. For that reason, VTVM measurements are most accurate when the range is set so that the needle reading is near the center of the scale. There are 12 measurement readings in this figure. The first four readings have the values (answers in ohms) given where the range is shown multiplied by the needle position. Determine the best answers for readings 5 through 12, and be sure to record the values in Table 2-1.1 also.

2. Calibrate the Ohmmeter:

DMM: Digital Meter. Most handheld digital meters are easy to operate. First, become familiar with the meter and check with the instructor or the manual if you have any general questions. Digital meters normally require no calibration procedure. However, you should check to see whether the meter measures zero and infinite ohms correctly. Set the function to OHMS. With the leads shorted (connected), verify the reading of zero ohms. Next, disconnect the leads and verify infinite or many megohms. Try this in other ranges if applicable.

VTVM or VOM: Be sure that you are familiar with the meter operation as described in the previous sections. Refer to the manual for more information. When you are ready, use the following procedure to calibrate the meter so that it will give the correct resistance measurements.

Turn on the VTVM by setting the function switch to the OHMS position. Notice that the needle will go toward the infinity (∞) line. This is because infinite resistance is the starting point; between the ground lead and the positive lead (probe), there is only air, or infinite resistance.

Switch the probe to the OHMS position.

After allowing the VTVM to warm up for 1 or 2 min, connect both leads together, resulting in a short circuit, or zero resistance. The needle should now go in the other direction, toward zero.

Keeping both leads connected (short-circuited), use the ZERO adjust to align the needle (pointer) with the zero line. Be sure to use the mirror for proper alignment and reduction of parallax error. If the pointer, the zero line, and the mirror reflection are one, you have zeroed the ohmmeter correctly.

Disconnect the two leads. The needle will move toward the infinity line. Now, use the OHMS adjust and align the pointer with the infinity line. Alignment of the pointer, the infinity line, and the mirror reflection of the needle (pointer) will eliminate parallax errors.

Repeat the zero adjustment once more (short-circuit the leads together), and the meter should now be calibrated.

3. Resistor Measurements:

Set the range for R × 10 if using a VOM/VTVM; DMM not required.

Connect the two leads across the 100 Ω resistor as shown in Fig. 2-1.5. It does not matter which side of the resistor you use; you may use either side. Also, never measure resistance when voltage is applied. You may damage the meter if you do.

Note: When measuring resistance, either end of the resistor can be connected. The resistor's polarity (+ or −) is due to the current passing through the resistor.

You should see the needle resting on or near the number 10 for DVM/VOM; DMM should read 100.

For VTVM/VOM: Multiply the number that the needle is indicating by the range setting. For example, if the needle is resting exactly between 9 and 10, at 9.5, multiply 9.5 × 10. The 10 is for R × 10. The result is 95 Ω. Do not be concerned if your measurement is not exactly 100 Ω. Remember that there may be differences due to manufacturing tolerances. Even a 100 Ω resistor with 10 percent tolerance (silver, fourth band) may be 90 Ω and still be good.

For all: Measure all the resistors listed in the component section. Write the measured value next to the nominal (color band) value in Table 2-1.2.

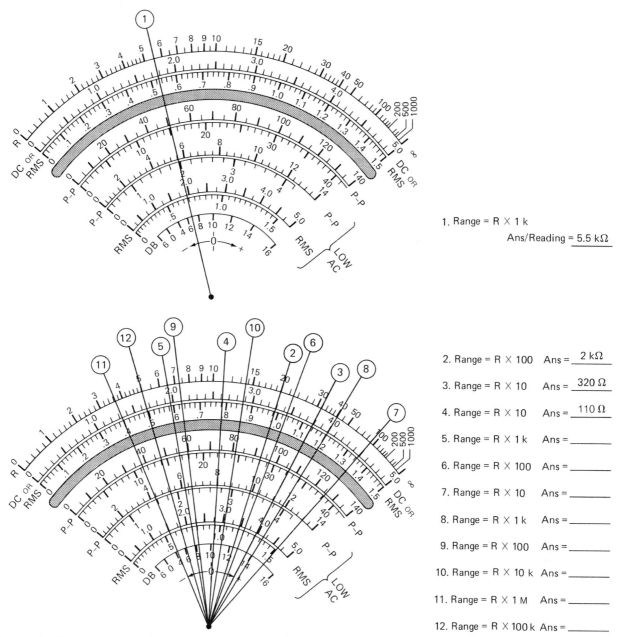

1. Range = R × 1 k
   Ans/Reading = 5.5 kΩ

2. Range = R × 100   Ans = 2 kΩ
3. Range = R × 10   Ans = 320 Ω
4. Range = R × 10   Ans = 110 Ω
5. Range = R × 1 k   Ans = _____
6. Range = R × 100   Ans = _____
7. Range = R × 10   Ans = _____
8. Range = R × 1 k   Ans = _____
9. Range = R × 100   Ans = _____
10. Range = R × 10 k Ans = _____
11. Range = R × 1 M   Ans = _____
12. Range = R × 100 k Ans = _____

Fig. 2-1.4 Ohmmeter and voltmeter scales (VTVM type). Record your answers to items 5 to 12 in Table 2-1.1.

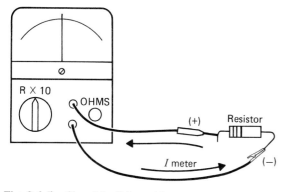

Fig. 2-1.5 Circuit building aid.

For VTVM/VOM: Always attempt to change ranges so that your measurements fall within the middle of the scale.

Readjust the OHMS (infinity) and ZERO controls each time you change ranges.

Remember that you cannot be entirely exact when reading scales. Do not be overly concerned if your values are not perfect.

NAME	DATE

QUESTIONS FOR EXPERIMENT 2-1

Answer TRUE (T) or FALSE (F) to the following:

_____ 1. It is always necessary to allow VTVMs to warm up prior to use.

_____ 2. The VTVM can measure current.

_____ 3. An ohmmeter will show zero ohms when the leads are not connected together (open circuit).

_____ 4. Linear scales are used for resistance measurements.

_____ 5. It is necessary to adjust infinity (∞) and zero ohms whenever changing ranges on an ohmmeter.

_____ 6. A continuity check gives the value in ohms.

_____ 7. Shorting the leads together on an ohmmeter results in zero ohms.

_____ 8. Parallax is an error resulting from reading meter scales from an angular view.

_____ 9. An ohmmeter cannot be damaged by measuring voltage.

_____ 10. The range switch is only used for voltage measurements.

Short Answer:

_____ 11. Refer to Appendix C and write an explanation of the differences between 4-band and 5-band resistors.

TABLES FOR EXPERIMENT 2-1

TABLE 2–1.1 Answers to Fig. 2-1.4

No.	Range	Reading
5	R × 1 k	_____
6	R × 100	_____
7	R × 10	_____
8	R × 1 k	_____
9	R × 100	_____
10	R × 10 k	_____
11	R × 1 M	_____
12	R × 100 k	_____

TABLE 2–1.2 Resistance Measurements Using the Ohmmeter

Nominal Value, Ω	Measured Value, Ω	Nominal Value, Ω	Measured Value, Ω
_____	_____	_____	_____
_____	_____	_____	_____
_____	_____	_____	_____
_____	_____	_____	_____
_____	_____	_____	_____
_____	_____	_____	_____
_____	_____	_____	_____
_____	_____	_____	_____
_____	_____	_____	_____
_____	_____	_____	_____
_____	_____	_____	_____
_____	_____	_____	_____
_____	_____	_____	_____
_____	_____	_____	_____
_____	_____	_____	_____

EXPERIMENT RESULTS REPORT FORM

Experiment No: _____   Name: _____
                                 Date: _____
Experiment Title: _____   Class: _____
_____   Instr: _____

Explain the purpose of the experiment:

List the first Learning Objective:
OBJECTIVE 1:

After reviewing the results, describe how the objective was validated by this experiment?

List the second Learning Objective:
OBJECTIVE 2:

After reviewing the results, describe how the objective was validated by this experiment?

List the third Learning Objective:
OBJECTIVE 3:

After reviewing the results, describe how the objective was validated by this experiment?

Conclusion:

If required, attach to this form: ☐ Answers to Questions, ☐ Tables, and ☐ Graphs.

CHAPTER 2

EXPERIMENT 2-2

RESISTOR V & I MEASUREMENTS

OBJECTIVES

At the completion of this experiment, you will be able to:
- Measure voltage with a voltmeter.
- Measure current with an ammeter.
- Use a dc power supply.

SUGGESTED READING

Chapters 1 and 2, *Basic Electronics*, Grob/Schultz, Tenth Edition

INTRODUCTION

Ammeters measure the amount of current (electron flow) in a circuit. Current is measured in units called *amperes*, or *amps*. Ammeters are almost always used in basic electronics lab courses and in physics lab courses because they are excellent for teaching and verifying the flow of current. But they are rarely used by technicians for repair work because ammeters usually measure only direct current (dc) and are not useful where alternating current (ac) is found. In this experiment, you will be inserting the ammeter in simple circuits and measuring small and safe amounts of dc current.

Voltmeters are the most useful tool for testing and troubleshooting dc circuits. They measure the amount of electrical force (charge or a potential difference between two points). The measurement unit is called the *volt*. Wherever there is current, there is also voltage. Most voltmeters measure dc voltage, and many also measure ac voltage, which is covered later. For this experiment, you will be measuring only dc voltage.

Common Values of Current and Voltage

The *milliampere* is a common value of current. The milliampere is one-thousandth of an amp, and that is enough current to do a lot of useful work on a circuit board. Throughout this course, you will mostly be measuring milliamperes and sometimes *microamperes*. In general, any value of current greater than a few amperes is considered high and can present a danger or hazard. Current in the area of 10 amperes and above is high enough to be deadly.

The *millivolt* is also a common value of voltage in dc electronics. It is one-thousandth of a volt. Values of hundreds of millivolts are common on circuit boards, as are values of voltage below 10 or 20 volts (V). Voltage is somewhat different from current, because large amounts of voltage can exist without a lot of current and may not be dangerous if it cannot supply current. However, it is always best to be cautious and safe.

Types of Ammeters and Voltmeters

The most commonly used voltmeters are digital handheld meters with Liquid Crystal displays, often referred to as *DVMs* (*digital voltmeters*) or *DMMs* (*digital multimeters*). However, older meters such as the *VOM* (*volt-ohm-milliammeter*) and the *VTVM* (*vacuum tube voltmeter*) are still used in many places today, especially schools. And some schools may use dedicated single-function ammeters and voltmeters which may have been built especially as teaching tools. For information about the DVM/DMM, VOM, and VTVM, refer to Experiment 2-1, on resistance measurements, where the meters were shown.

DC Power Supply

A source of dc (direct current) power is essential for activating any circuit. In fact, a battery is the most common source of dc power but it is not a variable type of supply, and it will run out of power if not recharged. For lab work, a variable and constant source of dc power is essential. For that reason, your lab is equipped with a dc power supply that may be capable of supplying 10 or more volts and is also variable. See Fig. 2-2.1 for a typical dc power supply.

The dc power supply is similar to a battery but is adjustable. Most bench supplies can maintain a constant voltage and, sometimes, a constant current. Such power supplies usually have their ranges or maximum values listed on the face, for example, 0 to 10 V or 0 to

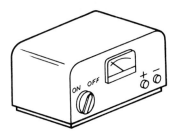

Fig. 2-2.1 DC power supply.

30 V. Also, bench supplies can often be adjusted to values less than a volt, including millivolts.

Because every lab has a different type and manufacturer of power supplies, it is necessary to learn how to safely operate the power supply in your situation. One way to do this is to read the manual or simply to look at the controls. The simplest of power supplies may have only positive and negative terminals, a display, and a knob to adjust the voltage. However, there are sophisticated supplies that have both positive and negative voltages, constant or variable current adjustments, and even swept outputs. For this experiment you will need a supply that adjusts between 0 and 10 V.

EQUIPMENT

Voltmeter
Ammeter
DC power supply

COMPONENTS

Resistors (all 0.25 W unless indicated otherwise):

(1) 100 Ω
(1) 1 kΩ
(1) 10 kΩ
(1) 330 Ω
(1) 560 Ω

PROCEDURE

1. Reading simple (linear) voltmeter scales: Refer to Figure 2-2.2. There are eight typical voltmeter scales shown (*a* to *h*). Notice that each meter face has two scales. Also, notice that the scales are linear, that is, equally divided. Try to read these scales. Remember that it is the range switch that determines which scale is used. For example, meter A is on the 10 V range. Therefore the reading is 6.5 V. If meter A were on the 100 V range, the reading would be 65 V. Be sure you understand that the range indicates the greatest value you can measure on the scale. Fill in the voltage readings for meters *b* through *h*.

2. Setting the Voltmeter

DMM: Digital Meters. Set the function switch to the appropriate VOLTS position. You will be measuring less than 10 V throughout this experiment. Be sure that the leads are properly connected. Digital meters are usually easy to operate and require only a minimal amount of adjustment.

VTVM or VOM: Turn on the meter. Set the function switch to positive (+) dc volts. If you have a probe (VTVM only), set it to dc volts. Adjust the "ZERO" control to be sure that you start with the needle aligned with the zero volts line.

3. Setting the power supply: Locate the power supply and locate its terminals (+) and (−). These are similar to the ends of a battery. If you are using a power supply with a variable knob, turn it counterclockwise to start at zero. Also, if the power supply has any settings, be sure to set it for 10 V or less with minimal current.

Turn on the power supply.

4. Measuring power supply voltage: Connect the voltmeter directly to the power supply as shown in Fig. 2-2.3 (schematic) or Fig. 2-2.4 (circuit-building aid). First, connect the negative lead or ground side to the negative output of the power supply. This is standard practice. Then connect the positive lead (probe) to the positive (+) side of the power supply.

Note: When a *voltmeter* is shown connected in a schematic, the circled symbol Ⓥ is used.

Adjust the power supply to read 3 V by slowly increasing the voltage control knob. If you are using a digital supply, adjust it in a similar manner. When your voltmeter reads 3 V, stop. Even if the power supply display is a little more or less than the meter reading, it does not matter. The voltmeter is making the measurement, and that is the data you want.

Try increasing and decreasing the power supply in minimal amounts between 0 and 5 V, watching the voltmeter. If you are using a meter movement (needle), be sure never to allow the needle to bump up to full-scale deflection. This may ruin the meter movement or bend the needle pointer. For example, if you are going to measure 10 V, set the meter to a higher scale, such as 20 V.

Obtain any battery less than 9 V and try measuring it with the voltmeter. Because a voltmeter has a very high resistance (millions of ohms), it cannot drain the battery of its current.

5. Measuring resistor voltage: Refer to the circuit in Fig. 2-2.5, where a 1 kΩ resistor is used in the path of current flow. Notice that the leads of the voltmeter are connected *across* the resistor like a

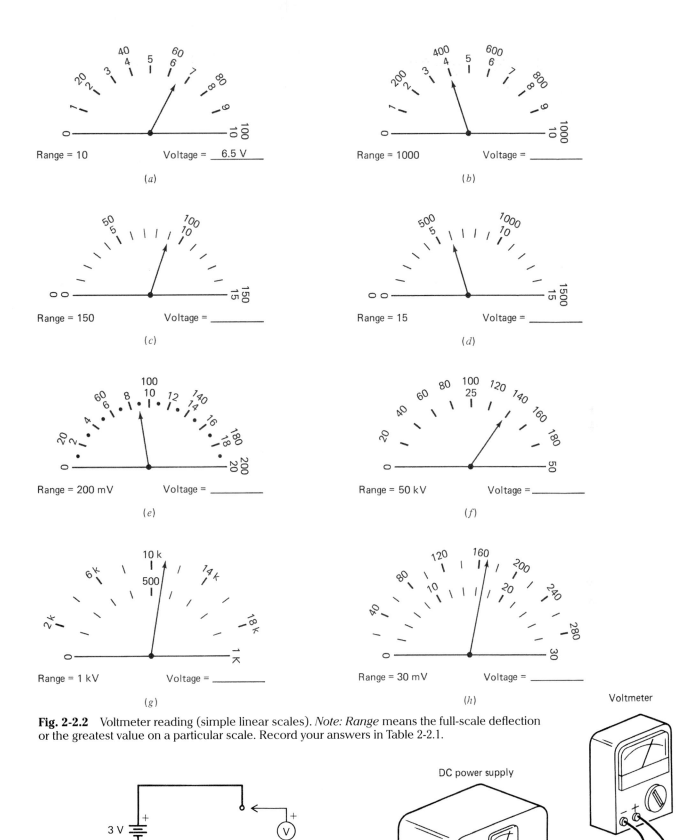

Fig. 2-2.2 Voltmeter reading (simple linear scales). *Note: Range* means the full-scale deflection or the greatest value on a particular scale. Record your answers in Table 2-2.1.

Fig. 2-2.3 Simple schematic of a dc voltage measurement. (A schematic is a type of electronic road map or blueprint.) The 3 V on the left means that the power supply is set at 3 V. The V in a circle on the right indicates a voltmeter (VTVM or VOM).

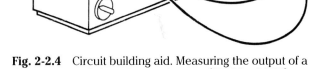

Fig. 2-2.4 Circuit building aid. Measuring the output of a dc power supply by using a voltmeter (VTVM or VOM).

EXPERIMENT 2-2

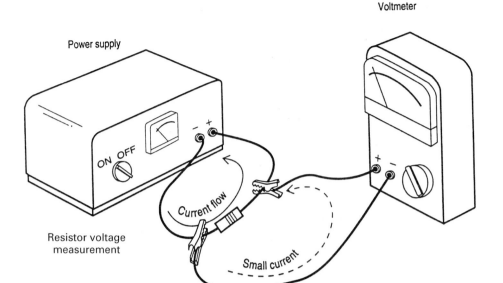

Fig. 2-2.5 Resistor voltage measurement.

bridge. This allows a very small (microampere) amount of current to travel into the highly resistive voltmeter to be measured with little or no effect on the rest of the circuit.

Connect the circuit of Fig. 2-2.5 with 3-V output from the power supply. Record the voltage reading in Table 2-2.2.

Replace the 1 kΩ resistor with a 100 Ω resistor without adjusting the power supply and record the voltage reading on the voltmeter.

Replace the 100 Ω resistor with a 10 kΩ resistor without adjusting the power supply and record the voltage reading on the voltmeter.

6. Measuring current with an ammeter: You have already learned how to use an ohmmeter to measure resistance and a voltmeter to measure voltage. For both those meters, it was necessary to place the two (+ and −) leads *across* the component you measured. However, an ammeter is actually connected *in* the circuit—in the path of current (electron) flow. Therefore, the leads are used to connect the ammeter into a circuit only after you break the circuit open. Refer to Fig. 2-2.6.

Notice that the range/function switch is set to 120 mA (full-scale deflection). Also notice that the electron flow (current) leaves the negative side of the battery, flows through the resistor, and enters the negative side of the ammeter. Then it exits the ammeter and returns to the positive side of the battery.

Let us simplify Fig. 2-2.6 by replacing the ammeter with the symbol A. Now, the same circuit of Fig. 2-2.6 can be seen in the schematic of Fig. 2-2.7.

Here, it should be understood that to connect the ammeter, the circuit was broken between the positive side of the battery and the resistor. Also, to determine where the polarities are, notice that wherever electron flow enters a component, that side of the component is then considered negative. Where electron flow leaves the component, closer to the positive side of the battery source, it is considered positive. Thus, the resistor and ammeter, in Fig. 2-2.7, are shown where electron flow determines polarity. Of course, electron

Fig. 2-2.6 Ammeter connected in a circuit to measure current.

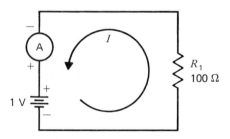

Fig. 2-2.7 Schematic representation of Fig. 2-2.6.

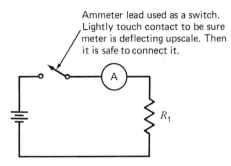

Fig. 2-2.8 Ammeter in circuit.

flow must enter the common, or negative, side of the ammeter and exit from the positive side.

Be sure that power is turned off before making any connections and then turned on again.

Finally, an easy way to protect an ammeter, or any meter, is to allow one lead (positive, for example) to act as a switch, as shown in Fig. 2-2.8. Here, the positive lead of the ammeter can be touched lightly to the connection so that you can be sure the proper polarity (meter goes upscale) is there. Lightly tap or touch the positive lead to the place you are going to connect it. Watch the meter needle; it should go upscale without pegging (attempting to go out of range).

- Connect the circuit of Fig. 2-2.7, which is a 100 Ω resistor in the path of current flow with the ammeter but do not turn the power supply on yet.
- Slowly adjust the power supply voltage for 1 V across the resistor. You can use a voltmeter to measure the voltage across the resistor and adjust to 1 V if necessary. Note the value of current on the ammeter and record it in Table 2-2.3.
- Replace the 100 Ω resistor with a 330 Ω resistor. The voltage should still be 1 V across the resistor. Note the value of current on the ammeter and record it in Table 2-2.3.
- Replace the 330 Ω resistor with a 560 Ω resistor. The voltage should still be 1 V across the resistor. Note the value of current on the ammeter and record it in Table 2-2.3.
- Turn off the power supply and disconnect the circuit.

QUESTIONS FOR EXPERIMENT 2-2

In your own words, describe the different ways an ohmmeter, an ammeter, and a voltmeter are connected to make measurements.

TABLES FOR EXPERIMENT 2-2

TABLE 2–2.1
Answers to Fig. 2-2.2

Scale	Reading, V
a	6.5 V
b	
c	
d	
e	
f	
g	
h	

TABLE 2–2.2 Voltage Measurements

Resistor Value	Voltage
1 kΩ	
100 Ω	
10 kΩ	

TABLE 2–2.3 Current Measurements

Resistor Value	Voltage	Current, I
100 Ω	1 V	
330 Ω	1 V	
560 Ω	1 V	

EXPERIMENT RESULTS REPORT FORM

Experiment No: _____   Name: _____
                                 Date: _____
Experiment Title: _____   Class: _____
_____   Instr: _____

Explain the purpose of the experiment:

List the first Learning Objective:
OBJECTIVE 1:

After reviewing the results, describe how the objective was validated by this experiment?

List the second Learning Objective:
OBJECTIVE 2:

After reviewing the results, describe how the objective was validated by this experiment?

List the third Learning Objective:
OBJECTIVE 3:

After reviewing the results, describe how the objective was validated by this experiment?

Conclusion:

If required, attach to this form: ☐ Answers to Questions, ☐ Tables, and ☐ Graphs.

CHAPTER 3

EXPERIMENT 3-1

OHM'S LAW

OBJECTIVES

At the completion of this experiment, you will be able to Validate the Ohm's law expression, where:

- $V = I \times R$
- $I = \dfrac{V}{R}$
- $R = \dfrac{V}{I}$

SUGGESTED READING

Chapter 3, *Basic Electronics,* Grob/Schultz, Tenth Edition

INTRODUCTION

Ohm's law is the most widely used principle in the study of basic electronics. In 1828, Georg Simon Ohm experimentally determined that the amount of current *I* in a circuit depended upon the amount of resistance *R* and the amount of voltage *V*. If any two of the factors *V, I,* or *R* are known, the third factor can be determined by calculation: $I = V/R$, $V = IR$, and $R = V/I$. Also, the amount of electric power, measured in watts, can be determined indirectly by using Ohm's law:

$$P = \dfrac{V^2}{R} \quad \text{or} \quad P = I^2R$$

Ohm's law is usually represented by the equation $I = V/R$. Because it is a mathematical representation of a physical occurrence, it is important to remember that the relationship between the three factors (*I, V,* and *R*) can also be expressed as follows:

Current is directly proportional to voltage if the resistance does not vary.

Current is inversely proportional to resistance if the voltage does not vary.

This experiment provides data that will validate the relationships between current, voltage, and resistance.

EQUIPMENT

Ohmmeter
Voltmeter
Ammeter
Power supply
Connecting leads
Circuit board

COMPONENTS

Resistors (all 0.25 W):

(1) 100 Ω   (1) 820 Ω
(1) 330 Ω   (1)   1 kΩ
(1) 560 Ω

PROCEDURE

1. Connect the circuit of Fig. 3-1.1*a* with the power supply turned off. The polarity of the power supply, the voltmeter, and the ammeter must be correct in order to avoid damage to the equipment.

Note: Use the circuit building aid of Fig. 3-1.1*b* if you have trouble.

2. Refer to Table 3-1.1. Calculate the current for the first voltage setting (1.5 V, 100 Ω nominal value). Record the value in Table 3-1.1 under the heading Calculated Current.

3. Check your circuit connection by tracing the path of electron flow. Be sure the ammeter reading is correct for the value of calculated current. Also, be sure the voltmeter range is correct.

4. Before turning the power supply on, turn the voltage adjust knob to zero. Now, turn the power supply on. Use the voltmeter to monitor the power supply and adjust for 1.5 V applied voltage.

Note: The ammeter can also be monitored. Unless you are using a digital meter, the pointer or needle should deflect upscale.

5. Read the value of measured current and record the results in Table 3-1.1.

6. Repeat procedures 2 to 5 for each value of applied voltage listed in Table 3-1.1.

Note: It is not necessary to turn the power supply off unless you are disconnecting the circuit in order to change meter ranges.

7. After recording all the measured and calculated values of current for Table 3-1.1, turn off the power

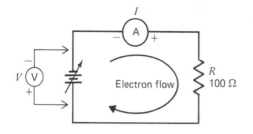

Ammeter: _____

Voltmeter: _____

Ohmmeter: _____

Show the meter number or model used for measuring values in both circuits.

(a)

(b)

Fig. 3-1.1 (a) Ohm's law schematic. (b) Circuit building aid. (Note that arrows indicate electron flow or current path.)

supply. Be sure that the voltage adjust is set to zero. Disconnect the circuit. The calculated values of power can be done later. Indicate which formula you used.

8. Refer to Table 3-1.2. Use an ohmmeter to measure and record the values of each resistor listed in Table 3-1.2.

Note: Remember to adjust the ohmmeter for zero and infinity, depending upon the type you are using.

9. Connect the circuit of Fig. 3-1.2. Keep the power supply turned off and begin with the first value of resistance in Table 3-1.2, $R = 1000\ \Omega$ (1 kΩ).

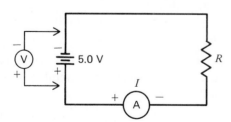

Fig. 3-1.2 Ohm's law circuit.

Note: This is the same basic circuit as Fig. 3-1.1. However, the ammeter is located in a different part of the circuit's current path. Use the same precautions as you did with Fig. 3-1.1.

10. Refer to Table 3-1.2. Calculate the current for the value of resistance listed in Table 3-1.2 (5.0 V, 1 kΩ nominal value). Record the calculated value in Table 3-1.2 under the heading Calculated Current.

11. Check your circuit. Be sure that the ammeter range is correct for the value of calculated current. Also, be sure that the voltmeter range is correct.

12. Turn on the power supply and adjust for 5.0 V applied voltage. Monitor the meters as you did for the circuit of Fig. 3-1.1.

13. Read the value of measured current and record the results in Table 3-1.2.

14. Turn the power supply off. Replace the resistor with the next value of resistance listed in Table 3-1.2.

15. Repeat steps 10 to 14 for each value of resistance listed in Table 3-1.2.

16. Turn the power supply off. Disconnect the circuit.

QUESTIONS FOR EXPERIMENT 3-1

Choose the correct answer.

___ 1. If the circuit of Fig. 3-1.1 had 10 kΩ of resistance, the amount of applied voltage necessary to produce 1 mA would be:
A. 1000 V   B. 100 V   C. 10 V   D. 1 V

___ 2. In the circuit of Fig. 3-1.1, if the applied voltage was increased, the amount of power would:
A. Decrease   B. Increase   C. Stay the same

___ 3. Compared to a voltmeter, the ammeter in Fig. 3-1.1 is:
A. Drawn differently and connected differently   B. Drawn the same and connected differently   C. Drawn differently and connected the same   D. Drawn the same and connected the same

___ 4. Referring to the circuit of Fig. 3-1.2, if the voltage was doubled for each step but the resistance was halved, the current (Table 3-1.2) would:
A. Increase by twice as much   B. Decrease by one-half   C. Decrease by one-fourth   D. Increase by four times as much

___ 5. According to Ohm's law and the data gathered in the experiment:
A. The more resistance, the more current with constant voltage   B. The more current, the less voltage with constant resistance   C. The less resistance, the more current with constant voltage   D. The less current, the more voltage with constant resistance

___ 6. If the terminals (negative and positive) of the power supply in Fig. 3-1.2 were reversed, it would be necessary to:
A. Reverse the terminals of the ammeter and the resistor   B. Reverse the terminals of the voltmeter and the ammeter   C. Reverse the terminals of the voltmeter and the resistor   D. Reverse the terminals of the resistor only

___ 7. Because the ammeter is connected in the same path as the resistor, would you expect the ammeter's resistance to be:
A. Very large   B. Very small   C. Medium value

___ 8. To obtain a current value of 30 mA, the amount of voltage and resistance necessary would be:
A. $V = 100$ V, $R = 333$ Ω   B. $V = 15$ V, $R = 400$ Ω   C. $V = 3$ V, $R = 100$ Ω   D. A, B, and C

TABLES FOR EXPERIMENT 3-1

TABLE 3–1.1 Ohm's Law

Applied Voltage, V	Nominal Resistance, Ω	Measured Current, mA	Calculated Current,* mA	Calculated Power†
1.5	100	_____	_____	_____
2.5	100	_____	_____	_____
3.5	100	_____	_____	_____
4.5	100	_____	_____	_____
6.0	100	_____	_____	_____
7.0	100	_____	_____	_____
8.0	100	_____	_____	_____
9.0	100	_____	_____	_____

*Formula: Applied voltage/nominal resistance

TABLE 3–1.2 Ohm's Law

Applied Voltage, V	Nominal Resistance, Ω	Measured Resistance, Ω	Measured Current, mA	Calculated Current,* mA
5.0	1000	_____	_____	_____
5.0	820	_____	_____	_____
5.0	560	_____	_____	_____
5.0	330	_____	_____	_____
5.0	100	_____	_____	_____

*Formula: Applied voltage/measured resistance

EXPERIMENT RESULTS REPORT FORM

Experiment No: _____ Name: _____
                                        Date: _____
Experiment Title: _____ Class: _____
_____ Instr: _____

Explain the purpose of the experiment:

List the first Learning Objective:
OBJECTIVE 1:

After reviewing the results, describe how the objective was validated by this experiment?

List the second Learning Objective:
OBJECTIVE 2:

After reviewing the results, describe how the objective was validated by this experiment?

List the third Learning Objective:
OBJECTIVE 3:

After reviewing the results, describe how the objective was validated by this experiment?

Conclusion:

If required, attach to this form: ☐ Answers to Questions, ☐ Tables, and ☐ Graphs.

EXPERIMENT 3-2

APPLYING OHM'S LAW

OBJECTIVES

At the completion of this experiment, you will be able to:
- Calculate V, I, or R from measured values.
- Determine the difference between calculated and measured power.
- Design a simple circuit using Ohm's law.

SUGGESTED READING

Chapter 3, *Basic Electronics*, Grob/Schultz, Tenth Edition

INTRODUCTION

As you already learned in the previous experiment, Ohm's law proves that there is a proportional relationship between voltage, current, and resistance. This law forms the basis for much of the troubleshooting work that technicians do every day. Regardless of the complexity of a circuit, engineers and technicians still use Ohm's law to determine voltage, current, or resistance. This experiment will reinforce your understanding of Ohm's law and show you how to apply it to simple circuits.

Ohm's Law Review

To review, Ohm's law states that:

$V = I/R$ Voltage equals current multiplied by resistance

$I = V/R$ Current equals voltage divided by resistance

$R = V/I$ Resistance equals voltage divided by current

The relationship between V and I is *directly proportional* because as one factor is increased, the other factor is increased at the same rate, but only if R does not change. But the relationship between R and I is *inversely proportional* because as one factor is increased, the other factor is decreased at the same rate, but only if V does not change (fixed value). For example, if the resistance R is increased two times, the current I will decrease by one-half.

In Fig. 3-2.1, two circuits have the same value of current although the values of V and R are different for each circuit.

Ohm's law can also be used to calculate power using these formulas:

$P = V \times I$
$P = I^2 R$
$P = V^2/R$

These relationships are used to calculate the amount of power (in watts) dissipated in a circuit. This is important for protecting circuits from overheating and in determining whether a component is operating within its allowed range or wattage specification. In general, whenever current travels through a resistance, power is used, resulting in heat or light being created. For example, when current passes through a resistor, friction causes heat to be released or dissipated. If too much current passes through a resistor, exceeding its power rating, it may burn or melt the resistor. But most circuits are designed with fuses to prevent excessive current from causing too much damage. By applying Ohm's law, it is easy to determine the levels of current in a circuit.

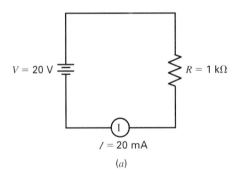

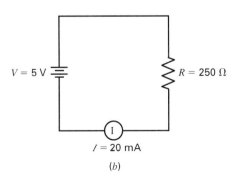

Fig. 3-2.1 Two circuits with the same current but different values of V and R.

EQUIPMENT

DC power supply
Ohmmeter
Ammeter
Voltmeter

COMPONENTS

Resistors: (1) 330 Ω, (1) 470 Ω, and (1) 1 kΩ or less for design (see procedure step 11)

PROCEDURE

1. Measure the 330 Ω and 470 Ω resistors to be sure that they are in tolerance, and record their values in Table 3-2.1. Be sure to calibrate (adjust) the ohmmeter for infinity and zero, unless you are using a digital meter. If any resistor is out of tolerance, replace it with a good one.

2. Refer to the circuit schematic of Fig. 3-2.2a. Use Ohm's law to calculate the current in the circuit. Record the value in Table 3-2.1.

3. Connect the circuit of Fig. 3-2.2a.

4. Measure the voltage across the resistor and the current in the circuit. Record both values, V and I, in Table 3-2.1.

5. Calculate the power dissipated, using the nominal values of resistance and voltage and the formula $P = V^2/R$. Record the calculated power dissipation in Table 3-2.1.

6. Calculate the power dissipated in the resistor, using the measured values of I and R only and the formula $P = I^2/R$.

7. Connect the circuit of Fig. 3-2.2b where $R = 330\ \Omega$ and there is also an ammeter connected in the circuit.

8. Carefully adjust the power supply voltage so that the current in the circuit in Fig. 3-2.2b is approximately equal to the current in the circuit in Fig. 3-2.2a. Record the value of voltage in Table 3-2.1.

9. Repeat steps 4 through 6 above for the circuit in Fig. 3-2.2b.

10. **Percentage of error in power measurments:** Calculate the percentage of error between the calculated and measured values of power in Table 3-2.1. To do this, find the difference between the two values and divide that difference by the calculated value: % error = difference between measured and calculated value ÷ calculated value. It does not matter whether the measured or the calculated value is greater; simply find the difference between them and divide by the calculated value. Record the percentage of error in Table 3-2.1.

11. Design a circuit that has the approximate value of current (±10 percent) as the circuits of Fig. 3-2.2a and b but with the following three specification restrictions: (1) voltage must be between 1 and 15 V; (2) resistance must be less than 1 kΩ; and (3) the resistor used must be a standard value available in your lab (listed in Appendix I).

12. On a separate sheet of paper, draw the schematic for the circuit, showing the design values V, I, and R. Be sure to label all components carefully and show the meter connections. Also, show the design calculations (Ohm's law calculations) you used and be sure to indicate the calculated values of V, I, and R.

13. Measure the resistor you have chosen and record the value in Table 3-2.2.

14. Repeat steps 4 through 6 above for the circuit you designed. Record all the values in Table 3-2.2.

Note: You may want to record the values on a separate sheet of paper before recording them in the table in case the design is invalid.

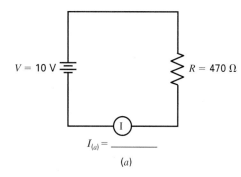

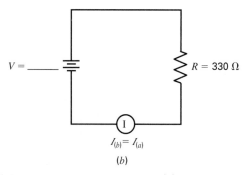

Fig. 3-2.2 Ohm's law test circuits. (a) Given values. (b) Unknown values.

NAME	DATE

QUESTIONS FOR EXPERIMENT 3-2

1. In the circuits of Fig. 3-2.2a and b, what aspects of Ohm's law were validated?

2. Describe any difference between the power calculations using the measured values and the calculated values. Refer to the percentage of error values.

3. Explain why you think an ammeter can be connected in the circuit and not affect the flow of current.

4. Explain why the voltage measured across the resistor is the same as the applied voltage.

5. In the circuit you designed, explain how you could reduce the dissipated power by one-half.

TABLES FOR EXPERIMENT 3-2

TABLE 3–2.1 Approximate Values Are Acceptable

Circuit	Measured Value, R	Measured Value, V	Measured Value, I	Calculated Value, I	Calculated Value, $P = V^2/R$	Measure Value, $P = I^2R$	% Error, P
a (470 Ω)	_____	_____	_____	_____	_____	_____	_____
b (330 Ω)	_____	_____	_____	_____	_____	_____	_____

TABLE 3–2.2

Measured Value, R	Measured Value, V	Measured Value, I	Calculated Value, $P = V^2/R$*	Measured Value, $P = I^2R$	% Error P
_____	_____	_____	_____	_____	_____

*Refers to calculated values on separate sheet.

EXPERIMENT RESULTS REPORT FORM

Experiment No: _____   Name: _____
                                  Date: _____
Experiment Title: _____   Class: _____
_____   Instr: _____

Explain the purpose of the experiment:

List the first Learning Objective:
OBJECTIVE 1:

After reviewing the results, describe how the objective was validated by this experiment?

List the second Learning Objective:
OBJECTIVE 2:

After reviewing the results, describe how the objective was validated by this experiment?

List the third Learning Objective:
OBJECTIVE 3:

After reviewing the results, describe how the objective was validated by this experiment?

Conclusion:

If required, attach to this form: ☐ Answers to Questions, ☐ Tables, and ☐ Graphs.

CHAPTER 4

EXPERIMENT 4-1

SERIES CIRCUITS

OBJECTIVES

At the completion of this experiment, you will be able to:
- Recognize the basic characteristics of series circuits.
- Compare the mathematical relationships existing in a series circuit.
- Compare the mathematical calculations to the measured values of a series circuit.

SUGGESTED READING

Chapter 4, *Basic Electronics,* Grob/Schultz, Tenth Edition

INTRODUCTION

An electric circuit is a complete path through which electrons can flow from the negative terminal of the voltage source, through the connecting wires or conductors, through the load or loads, and back to the positive terminal of the voltage source.

If the circuit is arranged so that the electrons have only one possible path, the circuit is called a *series circuit*. Therefore, a series circuit is defined as a circuit that contains only one path for current flow. Figure 4-1.1 shows a series circuit with several resistors.

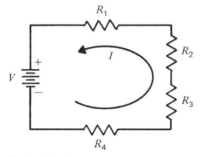

Fig. 4-1.1 Series circuit.

One of the most important aspects of a series circuit is its relationship to current. Current in a series circuit is determined by Ohm's law, which defines a proportional relation between voltage and the total circuit resistance. This relationship between total voltage and total circuit resistance results in the current being the same value throughout the entire circuit. In other words, the measured current will be the same value at any point in a series circuit.

Figure 4-1.2 shows a series circuit consisting of a battery and two resistors. The battery is labeled V_t and provides the total voltage across the circuit. The resistors are labeled R_1 and R_2. The resistance values are $R_1 = 100\ \Omega$ and $R_2 = 2.7\ k\Omega$. The power supply has been adjusted to a level of 20 V.

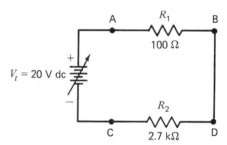

Fig. 4-1.2 Series circuit for analysis.

Figure 4-1.2 shows several points labeled A, B, C, and D. These are the points at which current could be measured. To measure the current at these points, it is necessary to break into the circuit and insert an ammeter in series. Remember, in a series circuit the amount of current measured will be the same at each of these points.

The total current, labeled I_T, flowing throughout the series circuit depends upon two factors: the total resistance R_T and the applied total voltage V_T. The applied total voltage was given in Fig. 4-1.2 such that

$$V_T = 20\ V$$

To determine the current, the total resistance R_T must be calculated by

$$R_T = R_1 + R_2$$

The total resistance for the circuit shown in Fig. 4-1.2 is

$$R_T = 100\ \Omega + 2700\ \Omega = 2800\ \Omega$$

For circuits that have three (3) or more series resistors, the above formula may be modified as shown to include the additional resistors:

$$R_T = R_1 + R_2 + R_3 + \cdots$$

When solving for the current I, Ohm's law states that

$$I = \frac{V}{R}$$

EXPERIMENT 4-1 **41**

When solving for current I_T,

$$I_T = \frac{V_T}{R_T}$$
$$= \frac{20 \text{ V}}{2800 \text{ }\Omega}$$
$$= 0.00714 \text{ A}$$
$$= 7.14 \text{ mA}$$

Applied voltage will be dropped proportionally across individual resistances, depending upon their value of resistance. As shown in Fig. 4-1.3, the IR voltage drops can be solved as follows. The sum of the voltage drops will be the total applied voltage. This is stated by

$$V_T = IR_1 + IR_2$$
$$\text{or} \quad V_T = V_{R_1} + V_{R_2}$$

where

$$V_{R_1} = I_T R_1$$
$$= (7.14 \text{ mA})(100 \text{ }\Omega)$$
$$= 0.71 \text{ V}$$

and

$$V_{R_2} = I_T R_2$$
$$= (7.14 \text{ mA})(2700 \text{ }\Omega)$$
$$= 19.27 \text{ V}$$

Therefore,

$$V_T = 0.71 \text{ V} + 19.27 \text{ V}$$
$$= 19.98 \text{ V}$$
$$= 20 \text{ V (approximately)}$$

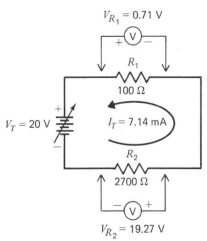

Fig. 4-1.3 Determining the IR (voltage) drops.

EQUIPMENT

Breadboard
DC power supply
Voltmeter
Ammeter
Ohmmeter

COMPONENTS

Resistors (all 0.25 W unless indicated otherwise):

(1) 10 Ω (1) 15 Ω (1) 150 Ω, 1 W

PROCEDURE

1. With an ohmmeter, measure each resistor value for the resistors required for this experiment. Connect the resistors in series and measure the total resistance R_T. Record the results in Table 4-1.1.

2. Using the information in the introduction and Fig. 4-1.4, calculate the total resistance R_T, the series current I_T, and the IR voltage drops across R_1, R_2, and R_3. Record the results in Table 4-1.2.

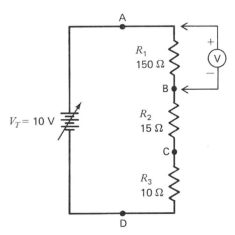

Fig. 4-1.4 Series circuit.

3. Connect the circuit in Fig. 4-1.4 and turn on the dc power supply. Using a voltmeter, adjust the power supply level to 10 V dc.

4. Measure the IR voltage drops across each resistance as indicated in Fig. 4-1.4. For example, the voltmeter is shown connected across R_1 to measure V_{R_1}. Record the results in Table 4-1.2.

5. After recording the measured voltage drops, compare the measured and calculated values. Use the following formula for determining this percentage:

$$\% = \left| \frac{\text{difference between meas. and calc. values}}{\text{calc. values}} \right| \times 100$$

If any error is greater than 20 percent, then repeat calculations or measurements.

6. Measure the current at points A, B, C, and D in Fig. 4-1.4. Record this information in Table 4-1.3.

Note 1: To measure current, you must actually break the circuit at these points and insert your ammeter (or current-measuring resistor, if you are not using an ammeter) in accordance with the information presented in the introduction.

Note 2: Check with your instructor in case you need to leave this circuit connected. When all results have been recorded, call your instructor over to your lab station to verify your results.

QUESTIONS FOR EXPERIMENT 4-1

Choose the correct answer.

_____ 1. In a series circuit, total current I_t is equal to:
   A. $R_T \times I_T$ B. $V_T \times R_T$ C. V_T/R_T D. R_T/V_T

_____ 2. In a series circuit, the current is:
   A. Different through every resistor in series   B. Always the same through every resistor in series   C. Calculated by using Ohm's law as $I = V \times R$ D. Found only by using the voltmeter

_____ 3. The total voltage in a series circuit is:
   A. Equal to total resistance   B. Found by adding the current through each resistor   C. Equal to the sum of the series IR voltage drops   D. Found by using an ohmmeter

_____ 4. In a series circuit with 10 V applied:
   A. The greater the total resistance, the greater the total current   B. The greater the total current, the greater the total resistance   C. The IR voltage drops will each equal 10 V   D. The sum of the IR voltage drops will equal 10 V

_____ 5. When an IR voltage drop exists in a series circuit:
   A. The polarity of the resistor is equal to positive   B. The polarity of the resistor is equal to negative   C. The polarity of the resistor is less than the total current on both sides   D. The polarity of the resistor is positive on one end and negative on the other because of current flowing through it

Write a short answer to the following questions.

6. Does $V_T = I_{R_1} + I_{R_2} + I_{R_3}$ in a series circuit? Explain your answer.

7. Is the measurement current the same at all parts of the series circuit? Explain your answer.

8. Does $R_T = R_1 + R_2 + R_3$ in a series circuit? Explain your answer.

TABLES FOR EXPERIMENT 4-1

TABLE 4–1.1

Nominal Resistance, Ω	Measured, Ω
$R_1 = 150$	_____
$R_2 = 15$	_____
$R_3 = 10$	_____
$R_t = 175$	_____

TABLE 4–1.2

	Calculated	Measured	% Error
R_T	_____	_____	_____
I_T	_____	_____	_____
V_{R_1}	_____	_____	_____
V_{R_2}	_____	_____	_____
V_{R_3}	_____	_____	_____

TABLE 4–1.3

Point	Current, mA
A	_____
B	_____
C	_____
D	_____

EXPERIMENT RESULTS REPORT FORM

Experiment No: _____ Name: _____
                                 Date: _____
Experiment Title: _____ Class: _____
_____   Instr: _____

Explain the purpose of the experiment:

List the first Learning Objective:
OBJECTIVE 1:

After reviewing the results, describe how the objective was validated by this experiment?

List the second Learning Objective:
OBJECTIVE 2:

After reviewing the results, describe how the objective was validated by this experiment?

List the third Learning Objective:
OBJECTIVE 3:

After reviewing the results, describe how the objective was validated by this experiment?

Conclusion:

If required, attach to this form: ☐ Answers to Questions, ☐ Tables, and ☐ Graphs.

CHAPTER 4

EXPERIMENT 4-2

SERIES CIRCUITS—RESISTANCE

OBJECTIVES

At the completion of this experiment, you will be able to:
- Recognize the basic characteristics of series circuits; for example, the total resistance R_T of a series string is equal to the sum of the individual resistances.
- Understand that a combination of series resistances is often called a *string* and that the string resistance equals the sum of the individual resistances.
- Compare the mathematical calculations to the measured values of a series circuit and recognize that the amount of current between two points in a circuit equals the potential difference divided by the resistance between these points.

SUGGESTED READING

Chapter 4, *Basic Electronics*, Grob/Schultz, Tenth Edition

INTRODUCTION

When a series circuit is connected across a voltage source, such as a power supply, as shown in Fig. 4-2.1, the electrons forming the current must pass through all the series resistances. This path is the only way the electrons can return to the power supply. With two or more resistances in the same current path, the total resistance across the voltage source is the total of all individual resistances found in what is referred to as a *series string*. This "total" resistance is referred to as R_T.

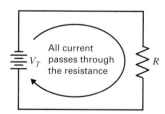

Fig. 4-2.1 Series resistance.

For example, if the circuit in Fig. 4-2.2 is analyzed, the total resistance R_T is the sum of the two individual resistances R_1 and R_2. In this case the total resistance is equal to 1150 Ω (because the string resistance of R_1 (470 Ω) is added to R_2 (680 Ω). The total resistive

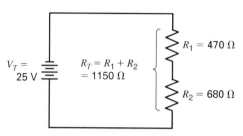

Fig. 4-2.2 R_T equals the sum of R_1 and R_2.

opposition to current flow is the same as if a 1150 Ω resistor were substituted for the two resistors, as shown in Fig. 4-2.3.

According to Ohm's law, the amount of current between two points in a circuit equals the potential difference divided by the resistance between these points. Since the entire string is connected across the voltage source, the current equals the voltage applied across the entire string divided by the total series resistance of the string (between points A and B in Fig. 4-2.3). In this case a power supply applies 25 V across 1150 Ω to produce 21.7 mA. This is the same amount of current flow through R_1 and R_2 shown in Fig. 4-2.2.

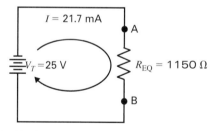

Fig. 4-2.3 Equivalent series resistance.

EQUIPMENT

Breadboard
DC power supply
VOM/DMM
    Voltmeter
    Ammeter
    Ohmmeter

COMPONENTS

All resistors are 0.25 W unless indicated otherwise:

(1) 100 Ω resistor
(1) 150 Ω resistor

(1) 220 Ω resistor
(1) 390 Ω resistor
(1) 560 Ω resistor

PROCEDURE

1. Measure and record each resistor value for the resistors required in this experiment. Record the results in Table 4-2.1.

2. Connect the circuit shown in Fig. 4-2.4. First calculate and then measure the total resistance from point A to point B. Record this information in Table 4-2.2. In addition, record the percentage of error between these values, where

$$\% \text{ error} = \left| \frac{\text{difference between meas. and calc. values}}{\text{calc. values}} \right| \times 100$$

If the error is greater than 10 percent, repeat the calculations and measurements.

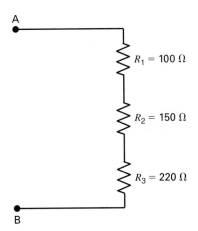

Fig. 4-2.4 Series resistance measurements.

3. Connect the circuit shown in Fig. 4-2.5. Adjust the supply voltage to 15 V and, using a voltmeter, take measurements and perform the necessary calculations required to complete Table 4-2.3.

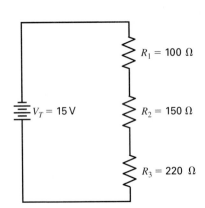

Fig. 4-2.5 Series string resistance.

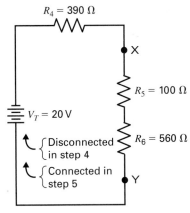

Fig. 4-2.6 Series resistance with an applied voltage source.

4. Connect the circuit as shown in Fig. 4-2.6. Calculate the expected resistance between points X and Y and record the results in Table 4-2.4. With the *power turned off and totally disconnected* from the circuit, measure and record, in Table 4-2.4, the resistance between X and Y.

5. With the ohmmeter disconnected from the circuit, connect the dc power supply to the circuit in Fig. 4-2.6 and adjust to 20 V. With an ammeter, measure the current at point Y and record the result in Table 4-2.4. Break the circuit at this point and insert the ammeter into the circuit as shown in Fig. 4-2.7.

6. Measure the voltage across R_5 and R_6 and record it in Table 4-2.4.

7. Calculate the current for R_5 and R_6 and record the results.

8. Determine and record the percentage of error for current using current at point Y. Record the results.

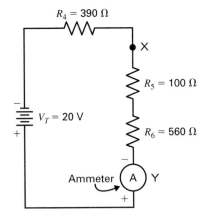

Fig. 4-2.7 Inserting the ammeter into the circuit.

QUESTIONS FOR EXPERIMENT 4-2

1. Write the formula that determines the total resistance R_T that is found in a series resistance string.

2. Write the formula that determines the amount of current flow for the circuit shown in Fig. 4-2.5 if the potential difference is 85 V.

3. In your estimation would $V_T = IR_1 + IR_2 + IR_3 + \cdots$ be true for all series resistive strings? Explain.

4. Do you believe that the measured current is the same at all points of a series resistive string when a potential difference is present across the circuit? Explain.

5. Could you always expect that $R_T = R_1 + R_2 + R_3 + \cdots$ in a series resistive circuit? Explain.

CRITICAL THINKING QUESTIONS

Note: The following questions are designed to help you analyze the previous laboratory experiment in a complete and in-depth fashion. To answer these questions, you should review the related material in Grob/Schultz, *Basic Electronics,* Tenth Edition.

1. Explain where "sources" of error exist in this experiment. How could these sources of error be reduced?

2. The purpose of a series circuit is to connect different components that need the same current. After reviewing your results for this experiment, explain how this purpose supports, or agrees with, your findings.

3. Explain why the current is the same in all parts of a series circuit.

4. Is it the case that the mathematical calculations of the measured values of a series circuit are such that the amount of current between two points in a circuit equals the potential difference divided by the resistance between these points? Explain.

5. Explain what is meant by the term *series string*.

TABLES FOR EXPERIMENT 4-2

TABLE 4–2.1 Individual Resistor Values

Resistors	Measured	% Tolerance
100	_____	_____
150	_____	_____
220	_____	_____
390	_____	_____
560	_____	_____

TABLE 4–2.2 Total Resistance R_T and Percentage of Error

	Calculated	Measured	% Error
Total Resistance	_____	_____	_____

TABLE 4–2.3 Series Circuit Measurements

Resistors	Value	Measured*	Measured Voltage	Calculated Current
R_1	100	_____	_____	_____
R_2	150	_____	_____	_____
R_3	220	_____	_____	_____

*Values from previous table.

TABLE 4–2.4 Series Circuit Measurements and Percentage of Error

Resistance	Expected	Measured	Measured Voltage	Calculated Current	% Error
X to Y	_____	_____			
Current Y		_____			
R_5			_____	_____	_____
R_6			_____	_____	_____

EXPERIMENT RESULTS REPORT FORM

Experiment No: _____   Name: _____
                                 Date: _____
Experiment Title: _____   Class: _____
_____   Instr: _____

Explain the purpose of the experiment:

List the first Learning Objective:
OBJECTIVE 1:

After reviewing the results, describe how the objective was validated by this experiment?

List the second Learning Objective:
OBJECTIVE 2:

After reviewing the results, describe how the objective was validated by this experiment?

List the third Learning Objective:
OBJECTIVE 3:

After reviewing the results, describe how the objective was validated by this experiment?

Conclusion:

If required, attach to this form: ☐ Answers to Questions, ☐ Tables, and ☐ Graphs.

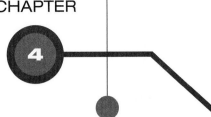

EXPERIMENT 4-3

SERIES CIRCUITS—ANALYSIS

OBJECTIVES

At the completion of this experiment, you will be able to:
- Recognize that, when you know the current I for one component connected into a series circuit, this value is used for the current I in all components, since the current is the same in all parts of a series circuit.
- Determine that, in order to calculate I, the total V_T can be divided by the total R_T, or an individual IR drop can be divided by its resistance.
- Compare the mathematical calculations to the measured values of a series circuit to the extent that, when the individual voltage drops around the series circuit are known, they can be added to equal the applied V_T. Further, you will demonstrate that a known voltage drop can be subtracted from the total V_T to find the remaining voltage drop.

SUGGESTED READING

Chapter 4, *Basic Electronics*, Grob/Schultz, Tenth Edition

INTRODUCTION

It is useful to remember general methods for analyzing series circuits. For example, when analyzing series resistive circuits, if the circuit current for one component is known, you can use this value for currents in all the remaining series components, since the current is the same in all parts of a series circuit.

In order to calculate the circuit current I, the total applied voltage V_T can be divided by the total circuit resistance R_T, or the voltage drops (IR drops) of an individual component can be divided by its resistance R.

When the individual voltage drops around a circuit are known, they can be added to equal the applied voltage V_T. This also means that a known voltage drop can be subtracted from the total applied voltage in order to find the remaining voltage drop.

A common application of series circuits is to use a resistance to drop the voltage from the voltage source to a lower value, as shown in Fig. 4-3.1. The load resistance represented here is a transistor amplifier

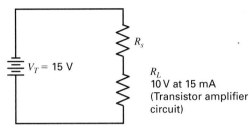

Fig. 4-3.1 Series dropping resistor.

which would normally operate at 10 V_{dc} with a constant dc load current of 15 mA. The load requirements are 10 V_{dc} at 15 mA. The available voltage source can be lowered only to 15 V_{dc} minimum. A suitable series voltage-dropping resistor R_s must be designed so that the voltage-dropping resistor is inserted in series to provide a voltage drop V_s that will make V_L equal to 10 V_{dc}. The required voltage drop across V_s is the difference between V_L and the higher V_T. For example:

$$V_s = V_T - V_L = 15 - 10 = 5 \text{ V}$$

This voltage drop of 5 V must be provided with a current of 15 mA, because the current is the same through R_s and R_L. To calculate R_s:

$$R_s = 5 \text{ V}/15 \text{ mA} = 333.33 \text{ }\Omega$$

EQUIPMENT

Breadboard
DC power supply
VOM/DMM
   Voltmeter
   Ammeter
   Ohmmeter

COMPONENTS

All resistors can be 0.25 W or 0.5 W unless indicated otherwise:

(1) 100 Ω resistor
(1) 150 Ω resistor
(1) 220 Ω resistor
(1) 390 Ω resistor
(1) 560 Ω resistor

PROCEDURE

1. Measure and record each resistor value for the resistors required in this experiment. Record the results in Table 4-3.1, p. 48.

2. Connect the circuit shown in Fig. 4-3.2. First calculate and then measure the total resistance from point A to point B. Record this information in Table 4-3.2. In addition, record the percentage of error between these values, where

$$\% \text{ error} = \left|\frac{\text{difference between meas. and calc. values}}{\text{calc. values}}\right| \times 100$$

If the error is greater than 10 percent, repeat the calculations and measurements.

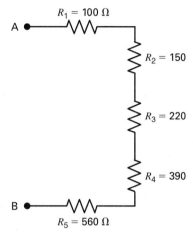

Fig. 4-3.2 Series resistive circuit.

3. Connect the circuit as shown in Fig. 4-3.3 and apply 20 V_{dc}. With the supply voltage on, open the circuit at point A and, using an ammeter, measure and record the current in Table 4-3.3, p. 49. Disconnect the ammeter and reconnect the circuit.

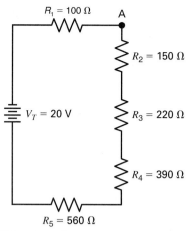

Fig. 4-3.3 Voltage applied to the series resistive circuit.

4. Using a voltmeter, measure the *IR* voltage drop across the resistors R_1, R_2, R_3, and R_4, and record this information in Table 4-3.3.

5. By mathematical calculation, record the current in Table 4-3.3. If the series current *I* is known for one component, this value is the current *I* found in all components in series.

6. Subtract the *IR* voltage drops of R_1, R_2, R_3, and R_4 from the total applied voltage and record the results in Table 4-3.4 as IR_5.

7. Complete Table 4-3.4. Calculated *IR* drops are calculated current times resistance. Use the formula shown in step 2 in determining the percentage of error. If the error is greater than 10 percent, repeat the calculations and measurements.

8. Using a voltmeter, measure and record, in Table 4-3.4, the *IR* voltage drop across R_5. Does this step demonstrate that a known voltage drop can be subtracted from the total V_T to find the remaining voltage drop? Explain in Table 4-3.5.

NAME	DATE

QUESTIONS FOR EXPERIMENT 4-3

1. Explain the significance of the following statement: In a series resistive circuit, in order to calculate I, the total V_T can be divided by the total R_T, or an individual IR drop can be divided by its resistance.

2. Describe what a zero voltage drop could be.

3. Describe the nature of the IR drop and current value of a component which displays the characteristics of a short.

4. In a series circuit, if the IR voltage drop is measured to be the same across two resistors, what can be said about these two resistors?

5. In a series circuit, if the current is measured to be 0 mA and is the same in all resistances, what would you suspect is taking place in this circuit?

CRITICAL THINKING QUESTIONS

Note: The following questions are designed to help you analyze the previous laboratory experiment in a complete and in-depth fashion. To answer these questions, you should review the related material in Grob/Schultz, *Basic Electronics*, Tenth Edition.

1. As described in the introduction of this experiment, a common application of series circuits is to use a resistance to drop the voltage from the voltage source to a lower value. Design a series-dropping resistor R_s so that the load requires 5 V at 50 mA, while the supply voltage is limited to 14 V_{dc}.

2. Determine the resistor's wattage rating for the design problem as described in critical thinking question 1. Using manufacturer's specification sheets and/or supplier catalogs, select a resistor that closely matches your component design. Using this value, and assuming that your design is not a perfect match, how will the voltage drops and circuit current values be affected? Explain.

3. The purpose of a series circuit is to connect different components that need the same current. After reviewing your results for this experiment, explain how this purpose supports, or agrees with, your findings.

4. Explain why the current is the same in all parts of a series circuit.

5. Explain where the sources of errors exist within this experiment. How could these sources of error be reduced?

TABLES FOR EXPERIMENT 4-3

TABLE 4–3.1 Individual Resistor Values

Resistors	Measured	% Tolerance
100		
150		
220		
390		
560		

TABLE 4–3.2 Total Resistance R_T and Percentage of Error

	Calculated	Measured	% Error
Total Resistance			

TABLE 4–3.3 Current and Voltage Measurements
Current = _____ at point A.

Point A Resistors	Value	Measured*	Measured Voltage	Calculated Current
R_1	100	_____	_____	_____
R_2	150	_____	_____	_____
R_3	220	_____	_____	_____
R_4	390	_____	_____	_____

*Values from previous table.

TABLE 4–3.4 The Total Voltage *IR* Drops and Percentage of Error
Voltage Applied = _____

	Measured *IR* Drops*	Calculated *IR* Drops	% Error
IR_1	_____	_____	_____
IR_2	_____	_____	_____
IR_3	_____	_____	_____
IR_4	_____	_____	_____
Sum of *IR* Drops	_____		
V_T − sum = IR_5	_____		

*Values from previous table.

TABLE 4–3.5 Analyzing *IR* Drops

Calculated IR_5*		Explain:
Measured IR_5		

*Use value of IR_5 from Table 4-3.4.

EXPERIMENT RESULTS REPORT FORM

Experiment No: _____   Name: _____
                                          Date: _____
Experiment Title: _____   Class: _____
_____   Instr: _____

Explain the purpose of the experiment:

List the first Learning Objective:
OBJECTIVE 1:

After reviewing the results, describe how the objective was validated by this experiment?

List the second Learning Objective:
OBJECTIVE 2:

After reviewing the results, describe how the objective was validated by this experiment?

List the third Learning Objective:
OBJECTIVE 3:

After reviewing the results, describe how the objective was validated by this experiment?

Conclusion:

If required, attach to this form: ☐ Answers to Questions, ☐ Tables, and ☐ Graphs.

EXPERIMENT 4-4

SERIES CIRCUITS—WITH OPENS

OBJECTIVES

At the completion of this experiment, you will be able to:
- Recognize that an open circuit is a break in the current path. Since the current is the same in all parts of a series circuit, an open in any part results in no current flow for the entire circuit.
- Understand that the resistance of an open circuit path is very high because an insulator such as air takes the place of the conducting path of the circuit.
- Verify that at the point of open in a series circuit, the value of current is practically zero, although the supply voltage is considered maximum.

SUGGESTED READING

Chapter 4, *Basic Electronics*, Grob/Schultz, Tenth Edition

INTRODUCTION

An open circuit is a break in the current path. The resistance of the open path is very high because an insulator, in this case air, takes the place of a conducting part of the circuit. Since the current is the same in all parts of a series circuit, as shown in Fig. 4-4.1a, an open in any part results in no current throughout the entire circuit. This is illustrated in Fig. 4-4.1b. The open which exists between points A and B provides an infinite resistance to the series circuit. Since the open circuit resistance at this point is extremely large when compared to the resistances of R_1 and R_2, and since the supply voltage remains unchanged, the resulting current flow is considered to be zero. The current flow in Fig. 4-4.1b is zero and therefore results in there being no IR voltage drop across any of the series resistances.

It is extremely important to note, however, that even though there is no current flow in an open series circuit and therefore no IR drops, the voltage source still maintains its output voltage and can be a potential shock hazard for the technician working on such open circuits.

EQUIPMENT

Breadboard
DC power supply
VOM/DMM
    Voltmeter
    Ammeter
    Ohmmeter

COMPONENTS

All resistors can be 0.25 W or 0.5 W unless indicated otherwise:

(1) 100 Ω resistor
(1) 150 Ω resistor
(1) 220 Ω resistor
(1) 390 Ω resistor

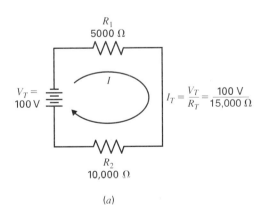
(a)

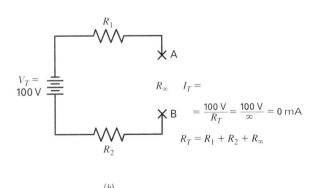
(b)

Fig. 4-4.1 Series resistive circuits. (*a*) Series circuit with a complete path for current flow. (*b*) Series circuit with an open that establishes an infinite series resistance.

EXPERIMENT 4-4   **59**

PROCEDURE

1. Measure and record each resistor value for the resistors required in this experiment. Record the results in Table 4-4.1.

2. Connect the circuit shown in Fig. 4-4.2. First calculate and then measure the total resistance from point A to point B. Record this information in Table 4-4.2. In addition, record the percentage of error between these values, where

$$\% \text{ error} = \left| \frac{\text{difference between meas. and calc. values}}{\text{calc. values}} \right| \times 100$$

If the error is greater than 10 percent, repeat the calculations and measurements.

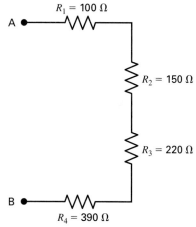

Fig. 4-4.2 Series resistive circuit without a supply voltage.

3. Connect the supply voltage to the circuit as shown in Fig. 4-4.3. Adjust this voltage to 20 V_{dc}.

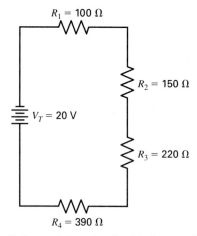

Fig. 4-4.3 Series resistive circuit with the supply voltage connected.

4. Using a voltmeter, measure and record, in Table 4-4.3, the voltage drops across each of the series resistances.

5. Using an ammeter and with the supply voltage on, open the circuit at each of the points, as shown in Fig. 4-4.4, and measure and record the currents in Table 4-4.3.

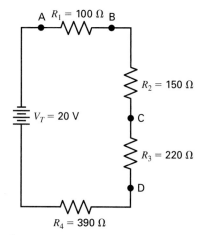

Fig. 4-4.4 Current and voltage measurements in a series resistive circuit.

6. With the supply voltage on, open the circuit as shown in Fig. 4-4.5, and repeat steps 4 and 5. Again, record your measurements in Table 4-4.3.

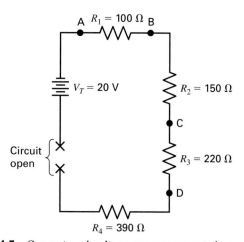

Fig. 4-4.5 Current and voltage measurements in a series resistive open circuit.

QUESTIONS FOR EXPERIMENT 4-4

1. Define, in your own words, what the unique characteristics of a series resistive circuit are.

2. Describe what happens to the current in a series resistive circuit if an open in the circuit path occurs.

3. Describe what happens to the voltage drops in a series resistive circuit if an open in the circuit path occurs.

4. In a series resistive circuit that provides a given amount of current flow, when the resistors are compared why does the resistor with a higher ohmic value also have a larger voltage drop across it? Is this also true for a circuit with an open?

5. Give an application of a series circuit.

CRITICAL THINKING QUESTIONS

Note: The following questions are designed to help you analyze the previous laboratory experiment in a complete and in-depth fashion. To answer these questions, you should review the related material in Grob/Schultz, *Basic Electronics,* Tenth Edition.

1. Explain where the sources of error exist in this experiment. How could these sources of error be reduced?

2. How could the idea of open circuits apply to the 120 V_{ac} voltage from the power line in your home?

3. Explain why the current is the same in all parts of a series circuit. Further, describe why the *IR* drops are zero in an open series resistive circuit.

4. Demonstrate mathematically why the current of an open circuit is practically zero.

5. Why is the source voltage in an open series resistive circuit present with or without current flow in the external circuit?

TABLES FOR EXPERIMENT 4-4

TABLE 4–4.1 Individual Resistor Values

Resistors	Measured	% Tolerance
100	_____	_____
150	_____	_____
220	_____	_____
390	_____	_____

TABLE 4–4.2 Total Resistance R_T and Percentage of Error

	Calculated	Measured	% Error
Total Resistance	_____	_____	_____

TABLE 4-4.3 Current and Voltage Measurements

	Circuit without Open			Circuit with Open		
Resistor Values	Measured Voltage	Circuit Points	Measured Current	Measured Voltage	Circuit Points	Measured Current
R_1	_____	A	_____	_____	A	_____
R_2	_____	B	_____	_____	B	_____
R_3	_____	C	_____	_____	C	_____
R_4	_____	D	_____	_____	D	_____

EXPERIMENT RESULTS REPORT FORM

Experiment No: _____   Name: _____
                                          Date: _____
Experiment Title: _____   Class: _____
_____   Instr: _____

Explain the purpose of the experiment:

List the first Learning Objective:
OBJECTIVE 1:

After reviewing the results, describe how the objective was validated by this experiment?

List the second Learning Objective:
OBJECTIVE 2:

After reviewing the results, describe how the objective was validated by this experiment?

List the third Learning Objective:
OBJECTIVE 3:

After reviewing the results, describe how the objective was validated by this experiment?

Conclusion:

If required, attach to this form: ☐ Answers to Questions, ☐ Tables, and ☐ Graphs.

CHAPTER 4

EXPERIMENT 4-5

SERIES-AIDING AND SERIES-OPPOSING VOLTAGES

OBJECTIVES

At the completion of this experiment, you will be able to:
- Define the terms *series-aiding* and *series-opposing*.
- Construct a series-aiding and a series-opposing circuit.
- Measure current and voltage; anticipate correct polarity connections.

SUGGESTED READING

Chapter 4, *Basic Electronics,* Grob/Schultz, Tenth Edition

INTRODUCTION

In many practical applications, a circuit may contain more than one voltage source. Voltage sources that cause current to flow in the same direction are considered to be *series-aiding,* and their voltages add. Voltage sources that tend to force current in opposite directions are said to be *series-opposing,* and the effective voltage source is the difference between the opposing voltages. When two opposing sources are inserted into a circuit, current flow would be in a direction determined by the larger source. Examples of series-aiding and series-opposing sources are shown in Fig. 4-5.1a and b.

Series-aiding voltages are connected such that the currents of the sources add, as shown in Fig. 4-5.1a. The 10 V of V_1 produces 1 A of current flow through the 10 Ω resistance of R_1. Also, the voltage source V_2, of 5 V, creates 0.5 A of current flowing through the 10 Ω resistance of R_1. The total current would then be additive and be 1.5 A.

When voltage sources are connected in a series-aiding fashion, where the negative terminal of one source is connected to the positive terminal of the next, voltages V_1 and V_2 are added to find the total voltage.

$$V_T = 5\text{ V} + 10\text{ V} = 15\text{ V}$$

The total current can then be determined by

$$I_T = 15\text{ V}/10\text{ Ω} = 1.5\text{ A}$$

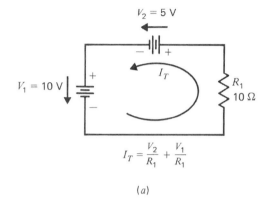

Fig. 4-5.1 (*a*) Series-aiding voltages. (*b*) Series-opposing voltages.

Series-opposing voltages can be subtracted, as shown in Fig. 4-5.1b. The currents that are generated are opposing each other. These voltages can still be algebraically added, keeping in mind their algebraic sign. Here, V_2 is smaller than V_1, and the difference between them is +5 V, resulting in an I_T of 0.5 A.

EQUIPMENT

VOM
Test leads

COMPONENTS

(4) D cells (1.5-V batteries)

PROCEDURE

1. Connect the negative lead of the voltmeter to the negative terminal of the battery, and connect the positive lead of the meter to the positive terminal of the battery. Identify and number each battery 1 through 4. Measure and record in Table 4-5.1 the voltage of each of the dry cells supplied to you.

2. Connect batteries 1 and 2 as series-aiding. Measure and record in Table 4-5.2 their total voltage.

3. Connect batteries 1, 2, and 3 as series-aiding. Measure and record in Table 4-5.2 their total voltage.

4. Repeat this process, measuring and recording the voltages of the four batteries as series-aiding, in Table 4-5.2.

5. Connect two batteries in parallel. (Be sure to connect the negative terminal of one battery to the negative terminal of the other, and connect the positive terminal of one battery to the positive terminal of the other. In this way, we have connected them together.) Measure and record this voltage in Table 4-5.3.

6. Connect three batteries in parallel. Measure and record this voltage in Table 4-5.3. Then connect four batteries in parallel. Measure and record this voltage in Table 4-5.3 also.

7. Connect the batteries in a series-parallel arrangement as shown in Fig. 4-5.2. Measure and record in Table 4-5.4 the voltage from point A to point B.

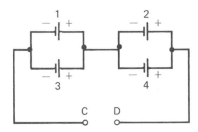

Fig. 4-5.3 Series-aiding circuit.

8. Connect the circuit as shown in Fig. 4-5.3. Measure and record in Table 4-5.4 the voltage from point C to point D.

9. Connect the series-aiding–series-opposing arrangements shown in Fig. 4-5.4a to c. Measure and record in Table 4-5.5 the voltage across each circuit. Before connecting the voltmeter, determine which are the probable positive and negative terminals.

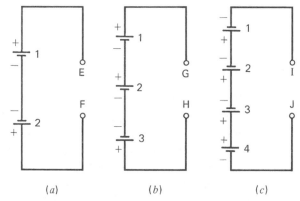

Fig. 4-5.4 (a) Series-opposing circuit. (b) Series-aiding and series-opposing circuit. (c) Series-aiding and series-opposing circuit.

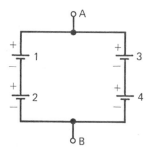

Fig. 4-5.2 Series-parallel batteries.

66 EXPERIMENT 4-5

NAME	DATE

QUESTIONS FOR EXPERIMENT 4-5

1. Name four precautions which must be observed in measuring voltages.

2. What would happen to a dry cell or battery if the positive and negative terminals were short-circuited?

3. What arrangement of six dry cells gives the maximum voltage?

4. Draw a practical arrangement of ten 1.5 V dry cells to give a battery of 7.5 V.

5. Explain the difference in connection between two dry cells connected in series-aiding and in series-opposing.

QUESTIONS FOR EXPERIMENT 4-5

TABLE 4–5.1

Battery	Measured Voltages
1	_____
2	_____
3	_____
4	_____

TABLE 4–5.2

Series-Aiding Battery	Measured Voltages
1 + 2	_____
1 + 2 + 3	_____
1 + 2 + 3 + 4	_____

TABLE 4–5.3

Parallel-Arrangement Battery	Measured Voltages
1 + 2	_____
1 + 2 + 3	_____
1 + 2 + 3 + 4	_____

TABLE 4–5.4

Circuit	Voltages	Terminals
Fig. 4-5.2	_____	A to B
Fig. 4-5.3	_____	C to D

TABLE 4–5.5

Circuit	Voltages	Terminals
Fig. 4-5.4*a*	_____	E to F
Fig. 4-5.4*b*	_____	G to H
Fig. 4-5.4*c*	_____	I to J

EXPERIMENT RESULTS REPORT FORM

Experiment No: _____   Name: _____
                                  Date: _____
Experiment Title: _____   Class: _____
_____   Instr: _____

Explain the purpose of the experiment:

List the first Learning Objective:
OBJECTIVE 1:

After reviewing the results, describe how the objective was validated by this experiment?

List the second Learning Objective:
OBJECTIVE 2:

After reviewing the results, describe how the objective was validated by this experiment?

List the third Learning Objective:
OBJECTIVE 3:

After reviewing the results, describe how the objective was validated by this experiment?

Conclusion:

If required, attach to this form: ☐ Answers to Questions, ☐ Tables, and ☐ Graphs.

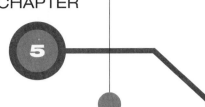

CHAPTER 5

EXPERIMENT 5-1

PARALLEL CIRCUITS

OBJECTIVES

At the completion of this experiment, you will be able to:
- Identify a parallel circuit.
- Accurately measure current in a parallel circuit.
- Use Ohm's law to verify measurements taken in a parallel circuit.

SUGGESTED READING

Chapter 5, *Basic Electronics,* Grob/Schultz, Tenth Edition

INTRODUCTION

A *parallel circuit* is defined as one having more than one current path connected to a common voltage source. Parallel circuits, therefore, must contain two or more load resistances which are not connected in series. Study the parallel circuit shown in Fig. 5-1.1.

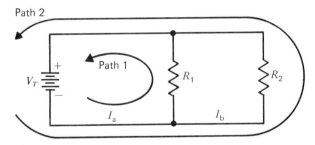

Fig. 5-1.1 Parallel circuit with two branches.

Beginning at the voltage source V_T and tracing counterclockwise around the circuit, two complete and separate paths can be identified in which current can flow. One path is traced from the source through resistance R_1 and back to the source; the other is traced from the source through resistance R_2 and back to the source.

The source voltage in a series circuit divides proportionately across each resistor in the circuit. In a parallel circuit, the same voltage is present across all the resistors of a parallel bank. In other words,

$$V_T = V_{R_1} = V_{R_2} = \cdots$$

The current in a circuit is inversely proportional to the circuit resistance. This fact, obtained from Ohm's law, establishes the relationship upon which the following discussion is developed. A single current flows in a series circuit.

In summary, when two or more electronic components are connected across a single voltage source, they are said to be *in parallel*. The voltage across each component is the same. The current through each component, or branch, is determined by the resistance of that branch and voltage across the bank of branches. Adding a parallel resistance of any value increases the total current. The total resistance of a circuit (R_T) can be found by dividing the total voltage applied (V_T) by the total current (I_T).

When measuring currents and resistances in the following procedure, some important precautions should be observed. When measuring currents, be sure to install the ammeter in a series configuration, with the ammeter connected in series with the individual branch resistances through which the current is flowing. Also, before measuring branch resistances, be sure to turn off all voltage sources.

EQUIPMENT

DC power supply, 0–10 V
Ammeter
Voltmeter
Protoboard or springboard
Test leads

COMPONENTS

Resistors (all 0.25 W):

(2) 1 kΩ   (1) 2.2 kΩ

PROCEDURE

1. Measure the resistance values of R_1, R_2, and R_3 and record in the required locations of Tables 5-1.1 to 5-1.3, where the nominal values are:

$R_1 = 1.0$ kΩ
$R_2 = 1.0$ kΩ
$R_3 = 2.2$ kΩ

2. Connect the circuit as shown with R_1 only in Fig. 5-1.2. Adjust the power supply voltage to 10 V.
3. Using an ammeter, measure and record in Table 5-1.1 the current through points a and b.
4. Using a voltmeter, measure and record in Table 5-1.1 the voltage across R_1.

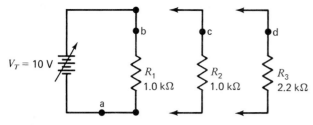

Fig. 5-1.2 Three-branch parallel circuit.

5. To this circuit, add R_2 across (meaning that R_2 is connected in parallel) R_1. With an ammeter, measure and record in Table 5-1.2 the current through points a, b, and c.

6. Using a voltmeter, measure and record in Table 5-1.2 the voltages across R_1 and R_2.
7. Finally, add R_3 across R_2. With an ammeter, measure and record in Table 5-1.3 the current through points a, b, c, and d.
8. Using a voltmeter, measure and record in Table 5-1.3 the voltages across R_1, R_2, and R_3.
9. Calculate the total resistances and the current for each of the three previous circuits and record this information in Tables 5-1.1 to 5-1.3.

NAME _____ DATE _____

QUESTIONS FOR EXPERIMENT 5-1

1. What is a parallel circuit? What circuit characteristics indicate that a parallel circuit condition exists?

2. In the circuit of Fig. 5-1.2, determine the power being dissipated by each resistor using the values of current determined in procedure step 7. Use the formula I^2R = power (watts).

3. Are the voltages the same across each resistor in a parallel circuit?

4. Are the currents the same through each resistor in a parallel circuit?

5. Suppose in procedure step 7 that R_3 developed a short-circuited condition. How would the current flowing through each resistor change? Would the voltage drops across each resistor change? How?

TABLES FOR EXPERIMENT 5-1

TABLE 5–1.1

	Measured	Calculated
R_1	_____	
R_T		_____
V_{R_1}	_____	
I_a	_____	_____
I_b	_____	_____

TABLE 5–1.2

	Measured	Calculated
R_1	_____	
R_2	_____	
R_T		_____
V_{R_1}	_____	
V_{R_2}	_____	
I_a	_____	_____
I_b	_____	_____
I_c	_____	_____

TABLE 5–1.3

	Measured	Calculated
R_1	_____	
R_2	_____	
R_3	_____	
R_T		_____
V_{R_1}	_____	
V_{R_2}	_____	
V_{R_3}	_____	
I_a	_____	_____
I_b	_____	_____
I_c	_____	_____
I_d	_____	_____

EXPERIMENT RESULTS REPORT FORM

Experiment No: _____   Name: _____
                                  Date: _____
Experiment Title: _____ Class: _____
_____   Instr: _____

Explain the purpose of the experiment:

List the first Learning Objective:
OBJECTIVE 1:

After reviewing the results, describe how the objective was validated by this experiment?

List the second Learning Objective:
OBJECTIVE 2:

After reviewing the results, describe how the objective was validated by this experiment?

List the third Learning Objective:
OBJECTIVE 3:

After reviewing the results, describe how the objective was validated by this experiment?

Conclusion:

If required, attach to this form: ☐ Answers to Questions, ☐ Tables, and ☐ Graphs.

CHAPTER 5

EXPERIMENT 5-2

PARALLEL CIRCUITS—RESISTANCE BRANCHES

OBJECTIVES

At the completion of this experiment, you will be able to:
- Recognize that for a parallel resistor circuit each branch current equals the applied voltage divided by the branch resistance.
- Recognize that with a parallel circuit arrangement, any branch that has less resistance allows more current to flow.
- Determine that current can have different values in different parts of parallel circuits.

SUGGESTED READING

Chapter 5, *Basic Electronics*, Grob/Schultz, Tenth Edition

INTRODUCTION

A parallel circuit is formed when two or more components, or resistances, are connected across a voltage source, as shown in Fig. 5-2.1. In this figure, R_1, R_2, R_3, and R_4 are in parallel with each other and the 10-V supply source. This figure can be redrawn, as shown in Fig. 5-2.2, to readily show that each component connection is really equivalent to a direct connection at the terminals of the battery, because the connecting wires exhibit almost no resistance. Since R_1, R_2, R_3, and R_4 are connected directly across the two terminals of the battery, all resistances must have the same potential differences as the battery. It therefore follows that the voltage is the same across each of the components that are connected in a parallel fashion. The parallel circuit arrangement is used, therefore, to connect components that require the same voltage from a common supply voltage.

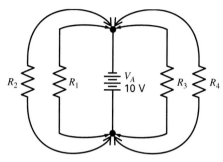

Fig. 5-2.2 Reconfiguration of a parallel circuit with four resistors.

Again (with reference to Fig. 5-2.1), R_1, R_2, R_3, and R_4 can be referred to as *branch* resistances. Although the voltage is the same across each branch resistance, the current divides into branches according to Ohm's law.

In each branch circuit the amount of branch current is equivalent to the applied voltage divided by the individual branch resistor. In other words, when Ohm's law is applied to the parallel circuit, the current equals the voltage applied across the circuit divided by the resistance between the two points at which that voltage is applied. As shown in Fig. 5-2.3, a 15 V source is applied across the 3900 Ω resistor (R_3). This results in a calculated current flow of 0.0038 A

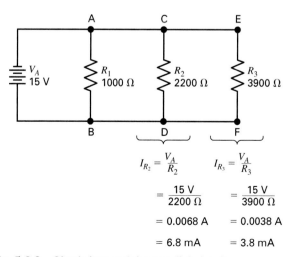

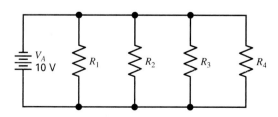

Fig. 5-2.1 A parallel circuit with four resistors.

Fig. 5-2.3 Ohm's law and the parallel circuit.

(3.8 mA) between points E and F. The battery voltage is *also* applied across the parallel resistance of R_2, which applies 15 V across a 2200 Ω resistor. This results in a calculated current of 0.0068 A (6.8 mA) between points C and D. This current provides a different value through R_1 with the same applied voltage, because the resistance is different. Remember that the current in each parallel branch equals the applied voltage divided by each individual branch resistance.

Further, as in a circuit with just one resistance, any branch that has less resistance allows more current to flow. If two branch resistances were equivalent in ohmic value, the two resulting branch currents would be of the same value.

In summary, the current can be different in parallel circuits that have different resistances because the applied voltage is the same across all branches. How much current is in each separate path depends on the amount of resistance in each branch.

EQUIPMENT

Breadboard
DC power supply
VOM/DMM
    Voltmeter
    Ammeter
    Ohmmeter

COMPONENTS

All resistors are 0.25 W unless indicated otherwise:

(1) 390 Ω
(1)   1 kΩ
(1) 1.2 kΩ
(1) 2.2 kΩ
(1) 3.9 kΩ

PROCEDURE

1. Measure and record each resistor value for the resistors required in this experiment. Record the results in Table 5-2.1.

2. Connect the circuit shown in Fig. 5-2.4. First calculate and then measure the total resistance from point A to point B. Record this information in Table 5-2.2. In

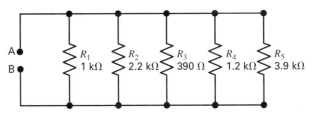

Fig. 5-2.4 Calculate and measure the total circuit resistance.

addition, record the percentage of error between these values, where:

$$\% \text{ error} = \left| \frac{\text{difference between meas. and calc. values}}{\text{calc. values}} \right| \times 100$$

If the error is greater than 10 percent, repeat the calculations and measurements.

3. Connect the circuit shown in Fig. 5-2.5. Adjust the supply voltage to 15 V and, using a voltmeter, take measurements required to complete Table 5-2.3 (Part A).

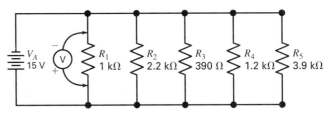

Fig. 5-2.5 Measuring the voltage existing in a parallel branch circuit.

4. Referring to Fig. 5-2.6, adjust the supply voltage to 15 V and, using an ammeter, take the current measurements at points A through F and perform the necessary calculations required to complete Table 5-2.3 (Part B).

5. Using the formula as shown in step 3, record in Table 5-2.3 (Part C) the percentage of error between the calculated and measured currents.

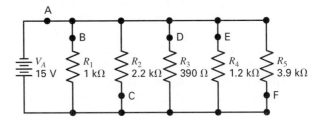

Fig. 5-2.6 Measuring the current existing in a parallel branch circuit.

QUESTIONS FOR EXPERIMENT 5-2

1. Write the formula that determines the total resistance R_T of the circuit depicted in Fig. 5-2.3.

2. Mathematically determine the branch current passing through R_1, as shown in Fig. 5-2.3. Show all calculations.

3. In your estimation would $V_T = IR_1 = IR_2 = IR_3 = \cdots$, be true for all parallel resistive branches? Explain.

4. In a parallel circuit, the amount of current in each separate branch depends upon what factor?

5. Could you always expect that $R_T = R_1 + R_2 + R_3 + \cdots$ in a series resistive circuit? Explain.

CRITICAL THINKING QUESTIONS

Note: The following questions are designed to help you analyze the previous laboratory experiment in a complete and in-depth fashion. To answer these questions, you should review the related material in Grob/Schultz, *Basic Electronics,* Tenth Edition.

1. Explain where the sources of errors exist in this experiment. How could these sources of error be reduced?

2. The purpose of a parallel circuit is to connect different components that need the same voltage. After reviewing your results for this experiment, explain how this purpose supports, or agrees with, your findings. Explain.

3. Explain why the voltage is the same across all branches in a parallel circuit.

4. Explain why, for a parallel resistive circuit, each branch current equals the applied voltage divided by the branch resistance.

5. How is it possible for current to be different in various parts of parallel circuits when the applied voltage is found to be the same across all branches?

TABLES FOR EXPERIMENT 5-2

TABLE 5-2.1 Individual Resistor Values

	Nominal Value	Measured Value
$R_1 =$	1 kΩ	
$R_2 =$	2.2 kΩ	
$R_3 =$	390 Ω	
$R_4 =$	1.2 kΩ	
$R_5 =$	3.9 kΩ	

TABLE 5-2.2 Total Resistance R_T and Percentage of Error

	Nominal Calculated Total Resistance $R_1 \| R_2 \| R_3 \| R_4 \| R_5$	Total Resistance Measured	% Error
$R_T =$			

TABLE 5-2.3 Recording Voltage, Current, and Error Measurements

Part A		Tables 5-2.1 and 5-2.2, Measured Resistance	Part B Calculated, $I = V/R$		Part C Measured, I		% Error
	Measured Voltage						
V_A		R_T	I_T	Point A			
V_{R_1}		R_1	I_{R_1}	Point B			
V_{R_2}		R_2	I_{R_2}	Point C			
V_{R_3}		R_3	I_{R_3}	Point D			
V_{R_4}		R_4	I_{R_4}	Point E			
V_{R_5}		R_5	I_{R_5}	Point F			

EXPERIMENT RESULTS REPORT FORM

Experiment No: _____   Name: _____
                                  Date: _____
Experiment Title: _____   Class: _____
_____   Instr: _____

Explain the purpose of the experiment:

List the first Learning Objective:
OBJECTIVE 1:

After reviewing the results, describe how the objective was validated by this experiment?

List the second Learning Objective:
OBJECTIVE 2:

After reviewing the results, describe how the objective was validated by this experiment?

List the third Learning Objective:
OBJECTIVE 3:

After reviewing the results, describe how the objective was validated by this experiment?

Conclusion:

If required, attach to this form: ☐ Answers to Questions, ☐ Tables, and ☐ Graphs.

EXPERIMENT 5-3

PARALLEL CIRCUITS—ANALYSIS

OBJECTIVES

At the completion of this experiment, you will be able to:

- Recognize that, for a parallel circuit, the main-line total current equals the sum of the individual branch currents.
- Use the formula $I_T = I_1 + I_2 + I_3 + I_4 + \cdots$, applied to any number of parallel branches, regardless of whether the individual branch resistances are equal or unequal.
- Recognize that a circuit's equivalent resistance across a common voltage source is found by dividing the common voltage by the total current of all the branches.

SUGGESTED READING

Chapter 5, *Basic Electronics*, Grob/Schultz, Tenth Edition

INTRODUCTION

Components connected in parallel are usually wired directly across each other so that the entire parallel combination is connected across the voltage source, as shown in Fig. 5-3.1. A combination of parallel branches is often called a *bank*, and a bank may contain two or more parallel resistors. The connecting wires provide no resistance and have almost no effect on circuit operation. One advantage of having circuits wired in parallel is that since only one pair of connecting leads is attached to the voltage source, less wire is used. It is common to find a pair of leads connecting all the branches to the terminals of the voltage source.

The main-line current is also known as the *total circuit current* and is equivalent to the sum of the individual branch currents. In Fig. 5-3.2, the supply voltage is 35 V and each of the three resistors establishes an

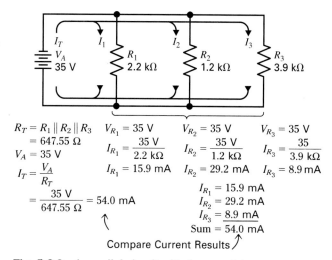

Fig. 5-3.2 A parallel circuit with three resistors.

independent current flow according to Ohm's law. The sum of the three currents is equivalent to the total (main-line) current, as demonstrated by the formula $I_T = I_1 + I_2 + I_3$. This formula applies for any number of parallel branches, whether the resistances are equal or unequal.

The main-line current is equal to the total of the branch currents. All current in the circuit must originate from one side of the voltage source and return to its opposite side in order to form a complete path. Therefore, the main-line current I_T equals the sum of the branch currents.

Analysis of the parallel resistive circuit reveals that the combined equivalent resistance across the supply voltage can be found by Ohm's law, so that the common voltage across the parallel resistances is divided by the total current of all the branches. Again referring to Fig. 5-3.2, note that the parallel equivalent resistance of R_1, R_2, and R_3 is 648 Ω and results in an opposition to the total current flow in the main line.

Two techniques are used to determine the equivalent resistance of a parallel circuit. These are referred to as the *reciprocal resistance formula* and the *total-current method* (see Fig. 5-3.3).

There are unique occasions when the resistance is equal in all branches. At those times, the combined equivalent resistance equals the value of one branch resistance divided by the number of branches.

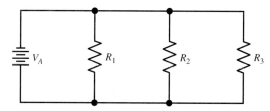

Fig. 5-3.1 A parallel circuit with three resistors.

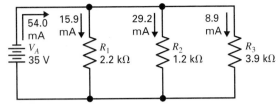

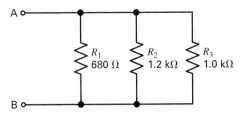

Fig. 5-3.4 Finding the total resistance and percentage of error.

Determining equivalent resistance by the reciprocal resistance formula:

$$\frac{1}{R_{EQ}} = \frac{1}{R_1} + \frac{1}{R_2} + \frac{1}{R_3}$$

$$= \frac{1}{2.2 \text{ k}\Omega} + \frac{1}{1.2 \text{ k}\Omega} + \frac{1}{3.9 \text{ k}\Omega}$$

$$R_{EQ} = 647.55 \text{ }\Omega \cong 648 \text{ }\Omega \text{ (rounded)}$$

Determining equivalent resistance by the total-current method:

$$R_{EQ} = \frac{V_A}{I_T} = \frac{35 \text{ V}}{54.0 \text{ mA}} = 648.1481 \text{ }\Omega \cong 648 \text{ }\Omega \text{ (rounded)}$$

Fig. 5-3.3 Determining an equivalent resistance.

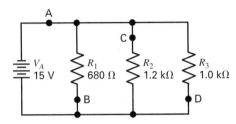

Fig. 5-3.5 Main-line and branch current.

In summary, consider the following when analyzing parallel circuits:

1. When a circuit provides more current with the same applied voltage, the greater value of current corresponds to less resistance because of the Ohm's law inverse relationship between current and resistance.

2. The combination of parallel resistances for a parallel bank is always less than the smallest individual branch resistance.

EQUIPMENT

Breadboard
DC power supply
VOM/DMM
   Voltmeter
   Ammeter
   Ohmmeter

COMPONENTS

All resistors are 0.25 W unless indicated otherwise:

(1) 680 Ω   (1) 1.2 kΩ
(1) 1.0 kΩ   (1) 22 kΩ

PROCEDURE

1. Measure and record each resistor value for the 680 Ω, 1.0 kΩ, and 1.2 kΩ resistors required in the first part of this experiment. Record the results in Table 5-3.1.

2. Connect the circuit shown in Fig. 5-3.4. First calculate and then measure the total resistance from point A to point B. Record this information in Table 5-3.1. In addition, record the percentage of error between these values, where:

$$\% \text{ error} = \left| \frac{\text{difference between meas. and calc. values}}{\text{calc. values}} \right| \times 100$$

If the error is greater than 10 percent, repeat the calculations and measurements.

3. Connect the circuit shown in Fig. 5-3.5. Adjust the supply voltage to 15 V and, using a voltmeter, make voltage measurements across the supply voltage and the individual resistors. Record this information in Table 5-3.2.

4. Again referring to Fig. 5-3.5, while the voltage is connected and turned on, break the circuit at point A and, with an ammeter, measure the current passing through this point. Record this information in Table 5-3.2.

5. Repeat step 4 and make current measurements at points B, C, and D. Record this information in Table 5-3.2.

6. Complete all calculations required in Table 5-3.3. Be certain that you use the measured resistance values from Table 5-3.1 in all calculations. Determine the percentage of error between the calculated and measured currents and record the results in Table 5-3.3.

7. Connect the circuit in Fig. 5-3.6. Measure and record, in Table 5-3.4, each individual resistance value as well as the total resistance from points A and B.

8. Complete and record the calculations required for Table 5-3.4. Explain in the table how the measured and calculated values compare.

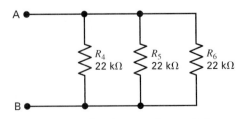

Fig. 5-3.6 Three equal resistances in parallel.

QUESTIONS FOR EXPERIMENT 5-3

1. Why is it true that the combined resistance of a parallel branch circuit equals the value of one branch resistance divided by the number of branches?

2. Define the term *bank* with regard to the parallel resistive circuit.

3. In your estimation would $V_T = I_{R_1} = I_{R_2} = I_{R_3} = \cdots$, be true for all parallel resistive branches? Explain.

4. Does the formula $I_T = I_1 + I_2 + I_3 + I_4 + \cdots$, apply for any number of parallel branches? What if the individual branch resistances were equal or unequal? Explain.

5. What is meant by the *total-current method*?

CRITICAL THINKING QUESTIONS

Note: The following questions are designed to help you analyze the previous laboratory experiment in a complete and in-depth fashion. To answer these questions, you should review the related material in Grob/Schultz, *Basic Electronics*, Tenth Edition.

1. Explain where the sources of errors exist in this experiment. How could these sources of error be reduced?

2. One purpose of this experiment is to validate that the formula for the total current I_T in the main line of a parallel resistive circuit is $I_T = I_1 + I_2 + I_3 + I_4 + \cdots$. After reviewing your results for this experiment, explain how this purpose supports, or agrees with, your findings.

3. Explain why the voltage is the same across all branches in a parallel circuit.

4. Describe the two methods of analyzing main-line current flow as demonstrated in Fig. 5-3.2.

5. Explain what is meant by the statement "When a circuit provides more current with the same applied voltage, the greater value of current corresponds to less resistance because of the Ohm's law inverse relationship between current and resistance." Show an example of this statement.

TABLES FOR EXPERIMENT 5-3

TABLE 5–3.1 Individual Resistor Values

	Nominal Value	Measured Value	% Error
$R_1 =$	680 Ω		
$R_2 =$	1.2 kΩ		
$R_3 =$	1.0 kΩ		
		Step 1	Step 2

		Sum Measured	% Error
Sum =	302.7 Ω		
		Step 2	

TABLE 5–3.2 Voltage Measurements

V_A	_____	I_A	_____
V_{R_1}	_____	I_B	_____
V_{R_2}	_____	I_C	_____
V_{R_3}	_____	I_D	_____
Step 3		Step 4	

TABLE 5–3.3 Comparison of Calculated and Measured Currents

Voltage Measurements from Table 5-3.2		Resistance Measurements from Table 5-3.1		Calculated Currents Columns 1 and 2		Measured Currents from Table 5-3.2		Current % Error
V_{R_1}	_____	R_1	_____	I_1	_____	I_1	_____	_____
V_{R_2}	_____	R_2	_____	I_2	_____	I_2	_____	_____
V_{R_3}	_____	R_3	_____	I_3	_____	I_3	_____	_____

TABLE 5–3.4 Tabulations for Three Equal Resistances in Parallel

	Nominal Values	Measured Values	Calculated by Reciprocal Method	Calculated by Other Method Described in Introduction
$R_4 =$	22 kΩ	_____		
$R_5 =$	22 kΩ	_____		
$R_6 =$	22 kΩ	_____		
		$R_T =$ _____	$R_T =$ _____	$R_T =$ _____

EXPERIMENT RESULTS REPORT FORM

Experiment No: _____   Name: _____
                                  Date: _____
Experiment Title: _____   Class: _____
_____   Instr: _____

Explain the purpose of the experiment:

List the first Learning Objective:
OBJECTIVE 1:

After reviewing the results, describe how the objective was validated by this experiment?

List the second Learning Objective:
OBJECTIVE 2:

After reviewing the results, describe how the objective was validated by this experiment?

List the third Learning Objective:
OBJECTIVE 3:

After reviewing the results, describe how the objective was validated by this experiment?

Conclusion:

If required, attach to this form: ☐ Answers to Questions, ☐ Tables, and ☐ Graphs.

EXPERIMENT 5-4

PARALLEL CIRCUITS—OPENS AND SHORTS

OBJECTIVES

At the completion of this experiment, you will be able to:
- Validate that an open in a parallel circuit results in an infinite resistance that allows no current to flow, whereas the effect of a short circuit is excessive current flow.
- Recognize that short-circuited components are typically not damaged, since they do not have greater than normal current passing through them.
- Recognize that the amount of current resulting from a short circuit is limited by the small resistance of the wire conductors.

SUGGESTED READING

Chapter 5, *Basic Electronics*, Grob/Schultz, Tenth Edition

INTRODUCTION

Parallel Open Circuits: An open in any circuit creates an infinite resistance that permits no current to flow. However, in parallel circuits there is a difference between an open circuit in the main line and an open circuit in a parallel branch, as shown in Fig. 5-4.1. An open circuit in the main line prevents any electrons from flowing in the main line or in any branch. An open in a branch results in no current flow in that particular branch. The current in all the other parallel branches remains the same, because each is connected directly to the voltage source. However, the total main-line current will change, since the total resistance of the circuit changes because one of the branches is open.

One advantage of wiring circuits in parallel is that an open in one component will open only that branch, allowing the other parallel branches to operate with normal voltages and currents.

Parallel Short Circuits: A short circuit has almost a zero resistance level. Its effect, therefore, is to allow excessive current to flow in the shorted circuit, as shown in Fig. 5-4.2. If a wire conductor were connected across a component, the effective resistance would be essentially zero and the current would be extremely high. This high amount of current is limited by the current-carrying capability of the wire.

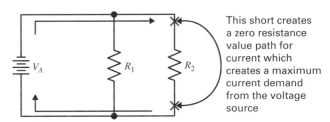

Fig. 5-4.2 A short circuit and excessive current flow.

Typically, in circuits which develop a shorted component, the voltage source cannot usually maintain its voltage level while at the same time providing the excessive current demand to the shorted component. The amount of current flowing can be dangerously high to the extent that wire can become hot enough to burn insulation and other components. To prevent the damaging effects of excessively high current flow, fuses are installed. A *fuse* is a component that intentionally burns open when there is too high a level of current demand from the supply voltage.

Short-circuited components, such as resistors, have no current passing through them. This is because the short circuit is a parallel path with almost no resistance. When this is the case, all current flows in this path, as shown in Fig. 5-4.3. In this figure the short

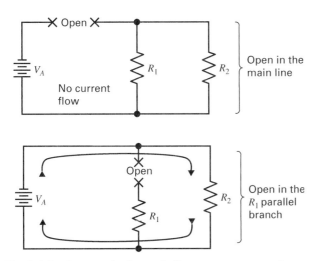

Fig. 5-4.1 An open in the main line versus an open in a branch.

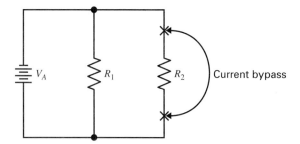

Fig. 5-4.3 A shorted component's current is bypassed.

bypasses the parallel resistances, and these resistances are referred to as being *shorted out*.

When a parallel branch circuit is analyzed, if any individual resistor or component is shorted, all resistive branches are also shorted. This idea can be extended to a short existing across the voltage source; in this case the entire circuit is shorted as well.

Short-circuited components are not usually damaged, since the current passing through the actual component is less than it is under normal operating conditions. If the short circuit does not damage the voltage source and the associated wiring for the circuit, the components will function again when the short is removed and the circuit is restored to normal.

EQUIPMENT

Breadboard
DC power supply
VOM/DMM
    Voltmeter
    Ammeter
    Ohmmeter

COMPONENTS

All resistors are 0.25 W unless indicated otherwise:

(1) 1 kΩ
(1) 1.5 kΩ
(1) 2.2 kΩ
(1) 3.9 kΩ
(1) 5.6 kΩ

For Critical Thinking Question 1:

(1) Clear glass functioning fuse (any current rating)
(1) Clear glass nonfunctioning fuse (any current rating)

PROCEDURE

1. Measure and record each resistor value for the resistors required in this experiment. Record the results in Table 5-4.1.

2. Connect the circuit shown in Fig. 5-4.4. First calculate and then measure the total resistance from point A to point B. Record the results in Table 5-4.2.

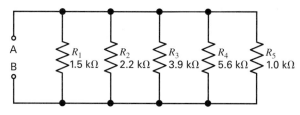

Fig. 5-4.4 Determining the total resistance.

In addition, record the percentage of error between these values, where:

$$\% \text{ error} = \left| \frac{\text{difference between meas. and calc. values}}{\text{calc. values}} \right| \times 100$$

If the error is greater than 10 percent, repeat the calculations and measurements.

Short Circuit

3. Connect the circuit shown in Fig. 5-4.5, but *do not* turn on the power supply. When constructing this circuit, notice that this configuration uses a resistor in series with the supply voltage. This is done so that there will always be a reference level of current being maintained from the power supply. For example, if in this circuit a short exists across the entire parallel branch arrangement, this series resistance will maintain a minimal circuit resistance and a maximum circuit current. This is necessary because if the entire parallel branch arrangement were shorted and there were no series resistance, a theoretically infinite amount of current could be demanded from the power supply. Actually, the power supply that you will be using cannot provide an infinite amount of current, and if a short were provided across the supply, the internal circuitry would be damaged or a fuse opened or a circuit breaker tripped (opened).

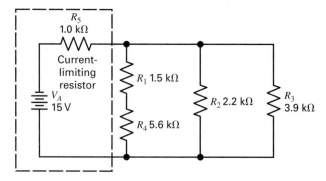

Fig. 5-4.5 Parallel shorts.

4. Turn on the power supply and adjust to 15 V$_{dc}$. Measure and record the voltage drops and currents as needed to complete Table 5-4.3.

5. Using a clip lead of a small piece of wire, place a short across resistance (R_1, R_2, R_3, and R_4) one at a time, and retake voltage and current measurements necessary to complete Table 5-4.4. Be sure to answer the questions in Table 5-4.4.

Open Circuit

6. Connect the circuit shown in Fig. 5-4.6, but *do not* turn on the power supply. Refer to step 3 and notice that, when the *previous* circuit was constructed, this configuration used a resistor in series with the supply voltage. This was done so that there would always be a reference level of current maintained from the power supply. In the analysis of open circuits it is *not* necessary to include a series resistor in the test circuit. This is because, when an open occurs in a parallel resistive circuit, the parallel branch resistance becomes practically infinite. If the open occurs in the main line of the power supply, the total resistance becomes infinite as well. The addition of infinite resistances does not harm the power supply. Because of Ohm's law, an infinitely large resistance creates an infinitely small

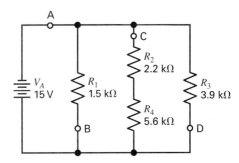

Fig. 5-4.6 Parallel opens.

current in relation to the constant voltage provided by the power supply.

7. Turn on the power supply and adjust it to 15 V_{dc}. Measure and record the voltage drops and currents as needed to complete Table 5-4.5.

8. Open the circuit at the identified points (A, B, C, and D). Retake voltage and current measurements and complete Table 5-4.6. Be sure to answer the question that appears in Table 5-4.6.

NAME	DATE

QUESTIONS FOR EXPERIMENT 5-4

1. What is the result of an open circuit? Explain what happens to the levels of voltage and current in a basic parallel branch circuit.

2. How will an open in a parallel branch affect the main-line current?

3. What is the main advantage of wiring a circuit in a parallel fashion?

4. Why does a short display practically no resistance? Provide a schematic drawing explaining your answer.

5. What limits the current resulting from a short?

CRITICAL THINKING QUESTIONS

Note: The following questions are designed to help you analyze the previous laboratory experiment in a complete and in-depth fashion. To answer these questions, you should review the related material in Grob/Schultz, *Basic Electronics,* Tenth Edition.

1. Ask your laboratory instructor for two fuses, one that functions and one that does not. After studying these components, describe how they are similar and how they are different. Explain in your own words how you think these components operate.

2. Explain in your own words why short-circuited components are rarely damaged.

3. If any individual resistor is shorted in a parallel circuit, why are all other resistive branches shorted as well?

4. Explain what is meant by the term *bypass* with reference to a shorted circuit.

5. When a parallel circuit develops a shorted component, the voltage source cannot usually maintain its voltage level. Explain.

TABLES FOR EXPERIMENT 5-4

TABLE 5–4.1 Individual Resistor Values

		Nominal Resistance	Measured Resistance
Step 1	R_1	1500 Ω	_____
	R_2	2200 Ω	_____
	R_3	3900 Ω	_____
	R_4	5600 Ω	_____
	R_5	1000 Ω	_____

TABLE 5–4.2 Total Resistance R_T and Percentage of Error

		Calculated	Measured	% Error
Step 2	R_T	_____	_____	_____

TABLE 5–4.3 Voltage Drops and Current Measurements

	Measured R Values from Table 5-4.1	Measured Voltage Drop	Current Measured through Resistor
R_1	_____	_____	_____
R_2	_____	_____	_____
R_3	_____	_____	_____
R_4	_____	_____	_____

TABLE 5-4.4 Voltage Drops and Current Measurements with Shorted Components

With Only R₁ Shorted		With Only R₂ Shorted		With Only R₃ Shorted		With Only R₄ Shorted	
V_{R_1} _____	I_1 _____	V_{R_1} _____	I_1 _____	V_{R_1} _____	I_1 _____	V_{R_1} _____	I_1 _____
V_{R_2} _____	I_2 _____	V_{R_2} _____	I_2 _____	V_{R_2} _____	I_2 _____	V_{R_2} _____	I_2 _____
V_{R_3} _____	I_3 _____	V_{R_3} _____	I_3 _____	V_{R_3} _____	I_3 _____	V_{R_3} _____	I_3 _____
V_{R_4} _____	I_4 _____	V_{R_4} _____	I_4 _____	V_{R_4} _____	I_4 _____	V_{R_4} _____	I_4 _____

How do the above measurements reinforce the characteristics of a short?

Verify by calculations the expected voltages and currents for this circuit when R_2 is shorted.

TABLE 5-4.5 Voltage Drops and Current Measurements

	Measured R Values from Table 5-4.1	Measured Voltage Drop	Current Measured through Resistor
R_1	_____	_____	_____
R_2	_____	_____	_____
R_3	_____	_____	_____
R_4	_____	_____	_____

TABLE 5-4.6 Voltage Drops and Current Measurements with Opens

Point A Open		Point B Open		Point C Open		Point D Open	
V_{R_1} _____	I_1 _____	V_{R_1} _____	I_1 _____	V_{R_1} _____	I_1 _____	V_{R_1} _____	I_1 _____
V_{R_2} _____	I_2 _____	V_{R_2} _____	I_2 _____	V_{R_2} _____	I_2 _____	V_{R_2} _____	I_2 _____
V_{R_3} _____	I_3 _____	V_{R_3} _____	I_3 _____	V_{R_3} _____	I_3 _____	V_{R_3} _____	I_3 _____
V_{R_4} _____	I_4 _____	V_{R_4} _____	I_4 _____	V_{R_4} _____	I_4 _____	V_{R_4} _____	I_4 _____

How do the above measurements validate the characteristics of an open circuit?

EXPERIMENT RESULTS REPORT FORM

Experiment No: _____ Name: _____
                                Date: _____
Experiment Title: _____ Class: _____
_____ Instr: _____

Explain the purpose of the experiment:

List the first Learning Objective:
OBJECTIVE 1:

After reviewing the results, describe how the objective was validated by this experiment?

List the second Learning Objective:
OBJECTIVE 2:

After reviewing the results, describe how the objective was validated by this experiment?

List the third Learning Objective:
OBJECTIVE 3:

After reviewing the results, describe how the objective was validated by this experiment?

Conclusion:

If required, attach to this form: ☐ Answers to Questions, ☐ Tables, and ☐ Graphs.

CHAPTER 6

EXPERIMENT 6-1

SERIES-PARALLEL CIRCUITS

OBJECTIVES

At the completion of this experiment, you will be able to:
- Identify a series-parallel circuit.
- Accurately measure voltages and current present in a series-parallel circuit.
- Verify measurements with calculated values.

SUGGESTED READING

Chapter 6, *Basic Electronics*, Grob/Schultz, Tenth Edition

INTRODUCTION

Consider the circuit shown in Fig. 6-1.1. Resistors connected together as shown are often called a *resistor network*. In this network, resistors R_2 and R_3 are considered to be in a parallel arrangement. Also, this parallel arrangement is in series with R_1. To calculate the total resistance R_T, the equivalent resistance of R_2 and R_3 must be determined, where

$$R_{eq} = \frac{R_2 \times R_3}{R_2 + R_3}$$
$$= 150 \ \Omega \text{ (approx.)}$$

Therefore, the equivalent resistance of this parallel arrangement is approximately equal to 150 Ω.

The total resistance is now easily found by adding the two resistances R_1 and R_{eq}, where

$$R_T = R_1 + R_{eq}$$

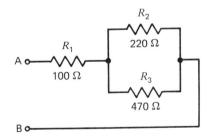

Fig. 6-1.1 Series-parallel circuit.

EQUIPMENT

Ohmmeter	Voltmeter
DC power supply	Protoboard or springboard
Ammeter	Leads

COMPONENTS

Resistors (all 0.25 W unless indicated otherwise):

(1) 68 Ω	(2) 150 Ω
(3) 100 Ω	(1) 220 Ω
(1) 120 Ω	(1) 470 Ω
(1) 150 Ω/1 W	

PROCEDURE

1. Measure and record the actual resistance values shown in Table 6-1.1.
2. Connect the circuit shown in Fig. 6-1.2, where $V = 10 \text{ V}$, $R_1 = 100 \ \Omega$, $R_2 = 220 \ \Omega$, and $R_3 = 470 \ \Omega$.

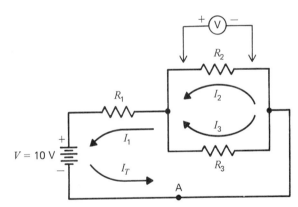

Fig. 6-1.2 Series-parallel circuit.

3. Calculate, measure, and record V_1, V_2, V_3, I_1, I_2, I_3, and I_T (at point A) in Table 6-1.2.
4. Measure and record in Table 6-1.2 the total applied voltage.
5. With the voltage source removed, measure, calculate, and record R_T in Table 6-1.2.
6. Calculate the percentage of error for Table 6-1.2.
7. Construct the circuit shown in Fig. 6-1.3, where

$V_T = 10 \text{ V}$	$R_4 = 120 \ \Omega$
$R_1 = 150 \ \Omega/1 \text{ W}$	$R_5 = 150 \ \Omega$
$R_2 = 100 \ \Omega$	$R_6 = 100 \ \Omega$
$R_3 = 100 \ \Omega$	$R_7 = 68 \ \Omega$

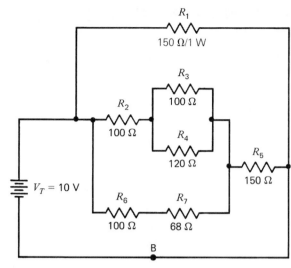

Fig. 6-1.3 Series-parallel circuit.

8. Calculate, measure, and record V_1, V_2, V_3, V_4, V_5, V_6, V_7, I_1, I_2, I_3, I_4, I_5, I_6, I_7, and I_T (at point B) in Table 6-1.3.

9. Measure and record in Table 6-1.3 the total applied voltage V_T.

10. With the voltage source removed, measure, calculate, and record R_T in Table 6-1.3.

11. Calculate the percentage of error for Table 6-1.3.

NAME _____ DATE _____

QUESTIONS FOR EXPERIMENT 6-1

1. In your own words, explain what a series-parallel circuit is.

2. In the circuit shown in Fig. 6-1.2, determine the power being dissipated by each resistor. Use the formula $P/I \cdot V = $ power (watts).

3. In Fig. 6-1.2, are the voltages the same across each resistor?

4. In Fig. 6-1.2, are the currents the same through each resistor?

5. Suppose in Fig. 6-1.3 that R_3 developed a short-circuited condition. How would the current flowing through each resistor change? Would the voltage drop across each resistor change? How?

TABLES FOR EXPERIMENT 6-1

TABLE 6–1.1 (values for Fig. 6-1.3)

Nominal Resistance	Measured Resistance
$R_1 = 150 \, \Omega$	_____
$R_2 = 100 \, \Omega$	_____
$R_3 = 100 \, \Omega$	_____
$R_4 = 120 \, \Omega$	_____
$R_5 = 150 \, \Omega$	_____
$R_6 = 100 \, \Omega$	_____
$R_7 = 68 \, \Omega$	_____

TABLE 6–1.2

	Calculated	Measured	% Error
V_1			
V_2			
V_3			
V_T			
I_1			
I_2			
I_3			
I_T			
R_T			

TABLE 6–1.3

	Calculated	Measured	% Error
V_1			
V_2			
V_3			
V_4			
V_5			
V_6			
V_7			
V_T			
I_1			
I_2			
I_3			
I_4			
I_5			
I_6			
I_7			
I_T			
R_T			

EXPERIMENT RESULTS REPORT FORM

Experiment No: _____   Name: _____
                                          Date: _____
Experiment Title: _____   Class: _____
_____   Instr: _____

Explain the purpose of the experiment:

List the first Learning Objective:
OBJECTIVE 1:

After reviewing the results, describe how the objective was validated by this experiment?

List the second Learning Objective:
OBJECTIVE 2:

After reviewing the results, describe how the objective was validated by this experiment?

List the third Learning Objective:
OBJECTIVE 3:

After reviewing the results, describe how the objective was validated by this experiment?

Conclusion:

If required, attach to this form: ☐ Answers to Questions, ☐ Tables, and ☐ Graphs.

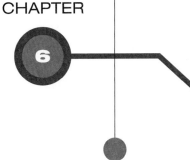

CHAPTER 6

EXPERIMENT 6-2

SERIES-PARALLEL CIRCUITS— RESISTANCE

OBJECTIVES

At the completion of this experiment, you will be able to:
- Recognize the basic characteristics of series-parallel resistive circuits.
- Confirm that, within a series-parallel circuit, main-line currents are the same through series resistances.
- Confirm that voltage drops across parallel resistive components in a series-parallel circuit are equivalent.

SUGGESTED READING

Chapter 6, *Basic Electronics,* Grob/Schultz, Tenth Edition

INTRODUCTION

All components in a parallel configuration with the voltage supply have the same voltage drop shown in Fig. 6-2.1. However, if there is a need to have a voltage less than what is created across one of the parallel resistance branches, an additional resistor can be placed in series within one of the branches, as shown in Fig. 6-2.2.

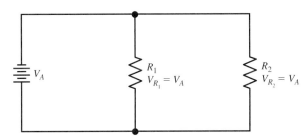

Fig. 6-2.1 Parallel circuit.

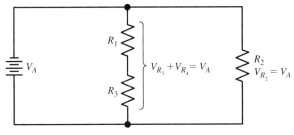

Fig. 6-2.2 Adding a series resistor to a parallel circuit.

Components in series have the same level of current passing through them. If there is a need to have a component with the same voltage level, another component can be placed across (or in parallel with) one of the series resistors.

Series-parallel circuits are used when it is necessary to provide different amounts of voltage and current from components that use only one source of supply voltage.

Finding R_T: Fig. 6-2.3 depicts a series-parallel circuit. In order to determine the total resistance (R_T) of this combination, it is necessary to follow the current from the negative side of the supply voltage, through the resistor network and back to the power supply.

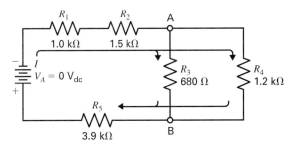

Fig. 6-2.3 Analysis of a series-parallel circuit.

When this series-parallel circuit is analyzed in this fashion, it is seen that all the current is passing through both R_1 and R_2, and because of this, R_1 and R_2 are seen in series with each other. From this series arrangement, the current then divides at junction point A. At this point, some of the current will pass through R_3, and the remainder through R_4. The amount of current that will pass will depend upon Ohm's law, and since at this junction point there is current division, resistors R_3 and R_4 are in parallel with each other. The divided currents will reunite at junction joint B and pass through R_5. In this case, all the current will pass through resistor R_5, and because of this fact, R_5 is seen to be in series with the parallel combination of R_3 and R_4, and then in series with R_1 and R_2. Expressed as a formula, total resistance R_T is:

$$R_T = R_1 + R_2 + R_3 \parallel R_4 + R_5$$
$$= 1 \text{ k}\Omega + 1.5 \text{ k}\Omega + \frac{1}{1/680 \text{ }\Omega + 1/1.2 \text{ k}\Omega} + 3.9 \text{ k}\Omega$$

$$= 1\text{ k}\Omega + 1.5\text{ k}\Omega + 434\text{ }\Omega + 3.9\text{ k}\Omega$$
$$= 6834\text{ }\Omega$$

Total Current: The total current can be calculated if the level of the supply voltage is known. If the power supply in Fig. 6-2.3 were adjusted to 50 V$_{dc}$, the current could be calculated by Ohm's law to be approximately 7.3 mA. There is 7.3 mA of main-line current present, as determined by:

$$R_T = 6834\text{ }\Omega$$
$$V_T = 50\text{ V}_{dc}$$
$$I_T = \frac{V_T}{R_T}$$
$$= \frac{50\text{ V}}{6834\text{ }\Omega}$$
$$\approx 0.0073\text{ A} = 7.3\text{ mA}$$

Current and Voltage Calculations: Now that the mainline current is known, it is seen that it passes through R_1 and R_2. The current passing through these resistors creates a voltage drop of 7.3 V and 10.95 V, respectively. This is determined as:

$$I_T = 7.3\text{ mA} = 0.0073\text{ A}$$
$$I_{R_1} = 0.0073\text{ A}$$
$$I_{R_2} = 0.0073\text{ A}$$
$$V_{R_1} = I_T \times R_1$$
$$= 0.0073\text{ A} \times 1000\text{ }\Omega$$
$$= 7.3\text{ V}$$
$$V_{R_2} = I_T \times R_2$$
$$= 0.0073\text{ A} \times 1500\text{ }\Omega$$
$$= 10.95\text{ V}$$

When the 7.3 mA enters junction A, the current divides through R_4 and R_5. Since R_4 and R_5 are in parallel, the voltage drop is the same across each of these parallel resistors. From the above, the equivalent resistance R_{eq} of R_3 and R_4 was determined to be 434 Ω. The total main-line current of 7.3 mA passes through this equivalent resistance and creates a voltage drop of 3.17 V. Again, since R_3 and R_4 are in parallel, the voltage drop across R_3 is 3.17 V and the voltage drop across R_4 is the same (3.17 V). These calculations are as follows:

$$R_{eq(R_3\|R_4)} = 434\text{ }\Omega$$
$$I_T = 7.3\text{ mA} = 0.0073\text{ A}$$
$$V_{eq(R_3\|R_4)} = I_T \times R_{eq(R_3\|R_4)}$$
$$= 0.0073\text{ A} \times 434\text{ }\Omega$$
$$= 3.1628\text{ V} \cong 3.17\text{ V}$$
$$V_{R_3} \cong 3.17\text{ V}$$
$$V_{R_4} \cong 3.17\text{ V}$$
$$I_3 = \frac{V_{R_3}}{R_3}$$
$$= \frac{3.17\text{ V}}{680\text{ }\Omega}$$
$$= 0.0047\text{ A} \cong 4.7\text{ mA}$$
$$I_4 = \frac{V_{R_4}}{R_4}$$
$$= \frac{3.17\text{ V}}{1.2\text{ k}\Omega}$$
$$= 0.0026\text{ A} \cong 2.6\text{ mA}$$

At junction B the two currents from R_3 and R_4 are reunited, and this 7.3 mA passes through R_5. The voltage drop on R_5 is 28.6 V. This is determined as follows:

$$I_3 + I_4 = 0.0047\text{ A} + 0.0026\text{ A}$$
$$I_5 = 0.0073\text{ A} = 7.3\text{ mA}$$
$$V_{R_5} = I_5 \times R_5$$
$$= 0.0073\text{ A} \times 3900\text{ }\Omega$$
$$= 28.47\text{ V}$$

Note that:

$$V_T \cong V_{R_1} + V_{R_2} + V_{eq(R_3\|R_4)} + V_{R_5}$$
$$50\text{ V} \cong 50\text{ V}$$

Power demands of each component can now also be determined, since the resistance values, voltage drops, and currents are all known. The wattage dissipation needs of the components are determined as follows:

$$P_{R_1} = I_1 \times V_{R_1}$$
$$= 0.0073\text{ A} \times 7.3\text{ V}$$
$$= 0.0533\text{ W} = 53.3\text{ mW}$$
$$P_{R_2} = I_2 \times V_{R_2}$$
$$= 0.0073\text{ A} \times 10.95\text{ V}$$
$$= 0.0799\text{ W} = 79.9\text{ mW}$$
$$P_{R_3} = I_3 \times V_{R_3}$$
$$= 0.0047\text{ A} \times 3.17\text{ V}$$
$$= 0.0149\text{ W} = 14.9\text{ mW}$$
$$P_{R_4} = I_4 \times V_{R_4}$$
$$= 0.0026\text{ A} \times 3.17\text{ V}$$
$$= 0.0082\text{ W} = 8.2\text{ mW}$$
$$P_{R_5} = I_5 \times V_{R_5}$$
$$= 0.0073\text{ A} \times 28.47\text{ V}$$
$$= 0.2080\text{ W} \approx 208\text{ mW}$$
$$P_T = P_1 + P_2 + P_3 + P_4 + P_5$$
$$= 0.0533\text{ W} + 0.0799\text{ W} + 0.0149\text{ W} +$$
$$0.0082\text{ W} + 0.2078\text{ W}$$
$$= 0.3641\text{ W} \approx 364.1\text{ mW}$$

Review Fig. 6-2.4 to determine all current paths and resistor voltage drops as calculated above.

In Summary: There can be any number of parallel strings and more than two series resistances in a string. Still, Ohm's law can be used in the same way for the series and parallel parts of the circuit. The series parts have the same current; the parallel parts have the same voltage. Remember that, to find the total current V/R, the total resistance must include all the resistances present across the two terminals of applied voltage.

EQUIPMENT

Breadboard
DC power supply

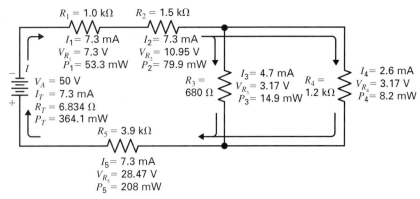

Fig. 6-2.4 Voltage drops, currents, and power dissipations of a series-parallel circuit.

VOM/DMM
  Voltmeter
  Ammeter
  Ohmmeter

COMPONENTS

All resistors are 0.25 W unless indicated otherwise:

(1) 100 Ω
(1) 150 Ω
(1) 220 Ω
(1) 390 Ω
(1) 560 Ω

PROCEDURE

1. Measure and record each resistor value for the resistors required in this experiment. Record the results in Table 6-2.1.
2. Connect the circuit shown in Fig. 6-2.5. First calculate and then measure the total resistance from point A to point B. Record this information in Table 6-2.2. In addition, record the percentage of error between these values, where:

$$\% \text{ error} = \left| \frac{\text{difference between meas. and calc. values}}{\text{calc. values}} \right| \times 100$$

If the error is greater than 10 percent, repeat the calculations and measurements.
3. Connect the circuit shown in Fig. 6-2.6. Perform the necessary calculations as required in Table 6-2.2.
4. For the circuit shown in Fig. 6-2.6, adjust the supply voltage to 15 V and, using a voltmeter, take voltage measurements as required in Table 6-2.2.
5. Again, for the circuit shown in Fig. 6-2.6, and with the power supply adjusted to 15 V, use an ammeter and, by breaking the current path at the identified points, measure the current and record your results in Table 6-2.2.
6. Using the formula described in step 2, calculate the percentages of error between your calculations and voltage and current measurements, as required in Table 6-2.2.

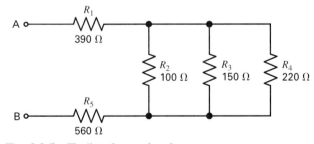

Fig. 6-2.5 Finding the total resistance.

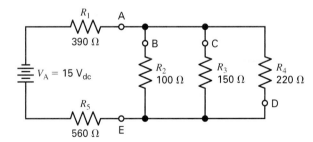

Fig. 6-2.6 Series-parallel circuit under analysis.

EXPERIMENT 6-2

QUESTIONS FOR EXPERIMENT 6-2

1. Write the formula that correctly determines the total resistance of the circuit in Fig. 6-2.6.

2. With reference to Fig. 6-2.6, which resistors are parallel to, or in series with, R_3? Why?

3. If in Fig. 6-2.1, the voltage across R_1 was determined to be 20 V, what would the circuit supply voltage be? Explain.

4. If in Fig. 6-2.4, the current measured through R_5 was 20 mA, what would be the current through R_1? Why?

5. Write the formula that would give the equivalent resistance of five 22 kΩ resistors connected in parallel. Applying your formula, determine this resistive value.

CRITICAL THINKING QUESTIONS

Note: The following questions are designed to help you analyze the previous laboratory experiment in a complete and in-depth fashion. To answer these questions, you should review the related material in Grob/Schultz, *Basic Electronics,* Tenth Edition.

1. In analyzing series-parallel circuits, it is believed that with a parallel string across the main line, the branch currents and the total current can be found without knowing the total resistance. Explain how this principle can be true.

2. In analyzing series-parallel circuits, it is believed that when parallel strings have series resistances in the main line, the total resistance must be calculated to find the total current, assuming no branch currents are known. Explain how this principle can be true.

3. In analyzing series-parallel circuits, if the source voltage is applied across the total resistance of the entire circuit, a total current will be produced that will be found in the main line. Explain how this principle can be true.

4. In analyzing series-parallel circuits, it is believed that any individual series resistance has its own voltage drop that is less than the total applied voltage. Explain how this principle can be true.

5. In analyzing series-parallel circuits, it is believed that any individual branch current must be less than the total main-line current. Explain how this principle can be true.

TABLES FOR EXPERIMENT 6-2

TABLE 6–2.1 Individual Resistor Values

	Nominal Resistance	Measured Value
R_1	390 Ω	
R_2	100 Ω	
R_3	150 Ω	
R_4	220 Ω	
R_5	560 Ω	

TABLE 6–2.2 Measurements and Calculations for Resistance, Voltage, and Current

		Calculated	Measured	% Error
Step 2	R_T	_____	_____	_____
Step 4	V_{R_1}	_____	_____	_____
	V_{R_2}	_____	_____	_____
	V_{R_3}	_____	_____	_____
	V_{R_4}	_____	_____	_____
	V_{R_5}	_____	_____	_____
Step 5	I_1 (Point A)	_____	_____	_____
	I_2 (Point B)	_____	_____	_____
	I_3 (Point C)	_____	_____	_____
	I_4 (Point D)	_____	_____	_____
	I_5 (Point E)	_____	_____	_____

Step 6

EXPERIMENT RESULTS REPORT FORM

Experiment No: _____

Experiment Title: _____
_____

Name: _____
Date: _____
Class: _____
Instr: _____

Explain the purpose of the experiment:

List the first Learning Objective:
OBJECTIVE 1:

After reviewing the results, describe how the objective was validated by this experiment?

List the second Learning Objective:
OBJECTIVE 2:

After reviewing the results, describe how the objective was validated by this experiment?

List the third Learning Objective:
OBJECTIVE 3:

After reviewing the results, describe how the objective was validated by this experiment?

Conclusion:

If required, attach to this form: ☐ Answers to Questions, ☐ Tables, and ☐ Graphs.

EXPERIMENT 6-3

SERIES-PARALLEL CIRCUITS—ANALYSIS

OBJECTIVES

At the completion of this experiment, you will be able to:

- Recognize the basic characteristics of series-parallel resistive banks that are in series with other components.
- Determine that, within a series-parallel circuit, main-line currents are the same through series resistances.
- Confirm that voltage drops across parallel resistive components in a series-parallel circuit are equivalent.

SUGGESTED READING

Chapter 6, *Basic Electronics*, Grob/Schultz, Tenth Edition

INTRODUCTION

Another way to analyze complex series-parallel circuits is to consider resistance banks in series with other resistors. Figure 6-3.1 depicts such a circuit. Here R_1 is in series with a parallel combination of R_2 and R_3. From this point R_4 is seen to be in a series relationship with the parallel combination of R_5 and R_6, leaving R_7 as a series component. If the main-line current from the supply voltage is followed throughout this circuit, we can see that currents are equivalent through series components and that the voltage drops across parallel components are also equivalent. This is demonstrated by the following calculations.

$$R_T = R_1 + R_2 \| R_3 + R_4 + R_5 \| R_6 + R_7$$
$$= 1 \text{ k}\Omega + 5.6 \text{ k}\Omega \| 6.8 \text{ k}\Omega + 3.9 \text{ k}\Omega +$$
$$\qquad 10 \text{ k}\Omega \| 2.4 \text{ k}\Omega + 2.2 \text{ k}\Omega$$
$$= 1 \text{ k}\Omega + 3070.97 \text{ }\Omega + 3.9 \text{ k}\Omega +$$
$$\qquad 1935.48 \text{ }\Omega + 2.2 \text{ k}\Omega$$
$$= 12{,}106.45 \text{ }\Omega$$

Figure 6-3.2 shows the equivalent total resistance for Fig. 6-3.1. This total resistance, with an applied voltage of 100 V_{dc}, will create a current of 8.3 mA, as shown below:

$$I_T = V_A/R_T$$
$$= 80 \text{ V}/12{,}106 \text{ }\Omega$$
$$= 0.0033 \text{ A} = 8.3 \text{ mA}$$

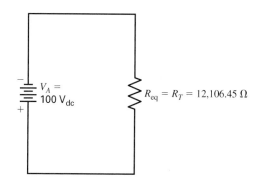

Fig. 6-3.2 The equivalent total resistance.

The values from all calculations for the circuit shown in Fig. 6-3.1 are shown in Fig. 6-3.3.

In summary, there can be more than two parallel resistances in a bank and any number of banks in series. Still, Ohm's law can be applied in the same way to the series and parallel parts of the circuit. The

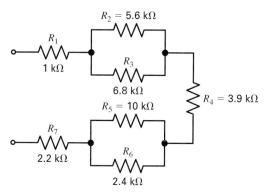

Fig. 6-3.1 Series-parallel circuit for analysis.

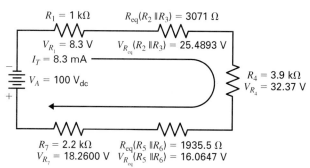

Fig. 6-3.3 Determining the equivalent resistance of a series-parallel circuit.

general procedures for circuits of this type is to find the equivalent resistance of each bank and then add all the series resistances.

Again, as in the previous experiment, the most important facts to know are which components are in series with each other and which parts of the circuit are in parallel.

EQUIPMENT

Breadboard
DC power supply
VOM/DMM
   Voltmeter
   Ammeter
   Ohmmeter

COMPONENTS

All resistors are 0.25 W unless indicated otherwise:

   (1) 100 Ω   (1) 390 Ω
   (1) 150 Ω   (1) 560 Ω
   (1) 220 Ω   (1) 820 Ω

PROCEDURE

1. Measure and record each resistor value for the resistors required in this experiment. Record the results in Table 6-3.1.

2. Connect the circuit shown in Fig. 6-3.4. First, calculate and then measure the total resistance from point A to point B. Record this information in Table 6-3.2. In addition, record the percentage of error between these values, where:

$$\% \text{ error} = \left| \frac{\text{difference between meas. and calc. values}}{\text{calc. values}} \right| \times 100$$

If the error is greater than 10 percent, repeat the calculations and measurements.

3. Connect the circuit shown in Fig. 6-3.5. Adjust the supply voltage to 15 V. Using a voltmeter, take the required measurements to complete Table 6-3.3.

4. With 15 V applied to the circuit, as shown in Fig. 6-3.5, determine with an ammeter the current flowing at each identified point and record this measurement in Table 6-3.3.

5. Perform all calculations required to complete Table 6-3.3.

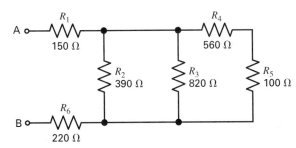

Fig. 6-3.4 Determining the equivalent resistance of the complex circuit.

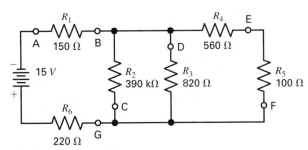

Fig. 6-3.5 Complex series-parallel circuit analysis.

QUESTIONS FOR EXPERIMENT 6-3

1. Draw a diagram showing three resistors in a bank that is in series with three resistors.

2. From Fig. 6-3.4, describe which resistors are in a parallel circuit arrangement.

3. From your readings in Grob/Schultz, *Basic Electronics,* Tenth Edition, describe three of the six characteristics of a parallel resistor circuit.

4. From your readings in Grob/Schultz, *Basic Electronics,* Tenth Edition, describe three of the five characteristics of a series resistor circuit.

5. In Fig. 6-3.5, explain how you would know which resistors are in parallel with each other and which are in series.

CRITICAL THINKING QUESTIONS

Note: The following questions are designed to help you analyze the previous laboratory experiment in a complete and in-depth fashion. To answer these questions, you should review the related material in Grob/Schultz, *Basic Electronics,* Tenth Edition.

1. Determine all circuit calculations (voltage drops and currents) for the circuit shown in Fig. 6-3.5, where the applied voltage is changed to 15 V_{dc}. Determine the power dissipation required for each resistor. If the voltage were increased to 15 V_{dc}, would the 0.25 W power rating of the resistors recommended for use in this experiment be adequate?

2. Determine the total resistance R_T for Fig. 6-3.3. Show all calculations.

3. Determine the power dissipation required for each resistor shown in Fig. 6-3.3. Show all calculations.

4. Determine the total resistance R_T for Fig. 6-3.5. Show all calculations.

5. Determine the power dissipation required for each resistor shown in Fig. 6-3.5. Show all calculations.

TABLES FOR EXPERIMENT 6-3

TABLE 6–3.1 Individual Resistor Values

		Nominal Values	Measured Values
	R_1	150 Ω	_____
	R_2	390 Ω	_____
Step 1	R_3	820 Ω	_____
	R_4	560 Ω	_____
	R_5	100 Ω	_____
	R_6	220 Ω	_____

TABLE 6–3.2 Total Resistance R_T and Percentage of Error

		Measured	Calculated	% Error
Step 2	R_T	_____	_____	_____

TABLE 6–3.3   Voltage, Current Measurements, and Percentage of Error

		Measured	Calculated	% Error
Step 3	V_{R_1}	_____	_____	_____
	V_{R_2}	_____	_____	_____
	V_{R_3}	_____	_____	_____
	V_{R_4}	_____	_____	_____
	V_{R_5}	_____	_____	_____
	V_{R_6}	_____	_____	_____
Step 4	I_T (Point A)	_____	_____	_____
	I_{R_1} (Point B)	_____	_____	_____
	I_{R_2} (Point C)	_____	_____	_____
	I_{R_3} (Point D)	_____	_____	_____
	I_{R_4} (Point E)	_____	_____	_____
	I_{R_5} (Point F)	_____	_____	_____
	I_{R_6} (Point G)	_____	_____	_____

Step 5

EXPERIMENT RESULTS REPORT FORM

Experiment No: _____ Name: _____
                                        Date: _____
Experiment Title: _____ Class: _____
_____   Instr: _____

Explain the purpose of the experiment:

List the first Learning Objective:
OBJECTIVE 1:

After reviewing the results, describe how the objective was validated by this experiment?

List the second Learning Objective:
OBJECTIVE 2:

After reviewing the results, describe how the objective was validated by this experiment?

List the third Learning Objective:
OBJECTIVE 3:

After reviewing the results, describe how the objective was validated by this experiment?

Conclusion:

If required, attach to this form: ☐ Answers to Questions, ☐ Tables, and ☐ Graphs.

EXPERIMENT 6-4

SERIES-PARALLEL CIRCUITS—OPENS AND SHORTS

OBJECTIVES

At the completion of this experiment, you will be able to:
- Confirm the basic characteristics of series-parallel resistive circuits.
- Identify the characteristics and effects of opens in a series-parallel resistive circuit.
- Identify the characteristics and effects of shorts in a series-parallel resistive circuit.

SUGGESTED READING

Chapter 6, *Basic Electronics*, Grob/Schultz, Tenth Edition

INTRODUCTION

The purpose of this experiment is to learn about the characteristics of electrical opens and shorts and how they affect currents and voltages. In this experiment we will see how a series-parallel circuit is affected.

The Short Circuit: A short circuit has practically zero resistance. Its effect, therefore, is to allow excessive current to flow. An open circuit has the opposite effect, because an open circuit has infinitely high resistance with practically zero current. Furthermore, in series-parallel circuits, an open or short circuit in one path changes the circuit for the other resistances. For example, in Fig. 6-4.1, the series-parallel circuit becomes a series circuit with only R_1 when there is a short circuit between points A and B.

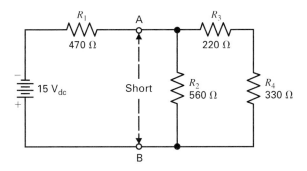

Fig. 6-4.1 Analysis of the short circuit.

Studying the Circuit without a Short: If the circuit shown in Fig. 6-4.1 does not have a short, R_T can be calculated so that R_3 and R_4 are in series and have an equivalent resistance of 550 Ω. This 550 Ω equivalent resistance is in parallel with 560 Ω R_2 values and creates an equivalent resistance of 277.48 Ω. This value is now in series with the 470 Ω value of R_1 and provides an overall total equivalent resistance of 747.48 Ω.

Since R_T is known, the main-line current can be determined from Ohm's law as follows:

$$I_T = V_T/R_T$$
$$= 15\ V_{dc}/747.48\ \Omega$$
$$= 0.0201\ A = 20.1\ mA$$

The total power of the circuit can be calculated as follows:

$$P_T = I_T \times V_T$$
$$= 0.0201\ A \times 15\ V$$
$$= 0.3010\ W = 301\ mW$$

Since the total main-line current I_T is known, the associated voltage drops across each of the resistors can be determined. For example:

$$V_{R_1} = I_T \times R_1$$
$$= 0.0201\ A \times 470\ \Omega$$
$$= 9.45\ V$$

Since the applied voltage is 15 V and 9.45 V was dropped across R_1, V_{R_2} can be determined to be 5.55 V, as follows:

$$V_{R_2} = V_T - V_{R_1}$$
$$= 15\ V - 9.45\ V$$
$$= 5.55\ V$$

Since 5.55 V is dropped across R_2, the current through R_2 can be determined as follows:

$$I_{R_2} = V_{R_2}/R_2$$
$$= 5.55\ V/560\ \Omega$$
$$= 0.0099\ A = 9.9\ mA$$

As you know, currents divide in parallel branches. The main-line current of 20.1 mA divides between the R_2 branch and the $R_3 + R_4$ branch. If 9.9 mA passes through R_2, and the total current I_T is 20.1 mA, the difference of 10.2 mA will pass through the series combination of the $R_3 + R_4$ branch. Since R_3 and R_4 are in series, this 10.2 mA of current will pass through both series resistors, as follows:

$$\begin{aligned} I_{R_3} = I_{R_4} &= I_T - I_{R_2} \\ &= 0.02010 \text{ A} - 0.0099 \text{ A} \\ &= 0.0102 \text{ A} = 10.2 \text{ mA} \end{aligned}$$

If the current through R_3 and R_4 is known to be 10.2 mA, the voltage drop across each can be determined:

$$\begin{aligned} V_{R_3} &= I_{R_3} \times R_3 \\ &= 0.0102 \text{ A} \times 220 \text{ }\Omega \\ &= 2.24 \text{ V} \end{aligned}$$

and

$$\begin{aligned} V_{R_4} &= I_{R_4} \times R_4 \\ &= 0.0102 \text{ A} \times 330 \text{ }\Omega \\ &= 3.36 \text{ V} \end{aligned}$$

Note that the sum of V_{R_3} and V_{R_4} is equivalent to V_{R_2}. This is to be anticipated, since these resistances are in a parallel branch arrangement, and voltage drops are found to be equivalent in parallel branches.

$$V_{R_3} + V_{R_4} = V_{R_2}$$
$$2.24 \text{ V} + 3.36 \text{ V} = 5.60 \text{ V}$$
(approximately equal to 5.55 V)

Power in the circuit can be determined as follows:

$$\begin{aligned} P_{R_1} &= I_{R_1} \times V_{R_1} \\ &= 0.0201 \text{ A} \times 9.45 \text{ V} \\ &= 0.1899 \text{ W} = 189.9 \text{ mW} \end{aligned}$$

$$\begin{aligned} P_{R_2} &= I_{R_2} \times V_{R_2} \\ &= 0.0099 \text{ A} \times 5.55 \text{ V} \\ &= 0.0549 \text{ W} = 54.9 \text{ mW} \end{aligned}$$

$$\begin{aligned} P_{R_3} &= I_{R_3} \times V_{R_3} \\ &= 0.0102 \text{ A} \times 2.24 \text{ V} \\ &= 0.0228 \text{ W} = 22.8 \text{ mW} \end{aligned}$$

$$\begin{aligned} P_{R_4} &= I_{R_4} \times V_{R_4} \\ &= 0.0102 \text{ A} \times 3.36 \text{ V} \\ &= 0.0343 \text{ W} = 34.3 \text{ mW} \end{aligned}$$

Studying the Same Circuit with a Short: If the circuit shown in Fig. 6-4.1 now possesses a short between points A and B, R_T can be calculated so that R_3 and R_4 are *still* in series and have an equivalent resistance of 550 Ω. This 550 Ω equivalent resistance is *still* in parallel with the 560 Ω R_2 value and creates an equivalent resistance of 277.48 Ω. However, this equivalent resistance is seen to be in parallel with a short, which displays practically no resistance (0 Ω). In this case, the equivalent resistance of the parallel branches is now 0 Ω. This ohmic value is in series with the 470 Ω value of R_1 and provides an overall total equivalent resistance of 470 Ω.

Since R_T is known, the main-line current can be determined from Ohm's law as follows.

$$\begin{aligned} I_T &= V_T/R_T \\ &= 15 \text{ V}_{dc}/470 \text{ }\Omega \\ &= 0.0319 \text{ A} = 31.9 \text{ mA} \end{aligned}$$

The total power of the circuit can be calculated as follows:

$$\begin{aligned} P_T &= I_T \times V_T \\ &= 0.0319 \text{ A} \times 15 \text{ V} \\ &= 0.4787 \text{ W} = 478.7 \text{ mW} \end{aligned}$$

Since the total main-line current I_T is known, the associated voltage drops across each of the resistors can be determined. For example:

$$\begin{aligned} V_{R_1} &= I_T \times R_1 \\ &= 0.0319 \text{ A} \times 470 \text{ }\Omega \\ &= 14.99 \text{ V (which in practical terms is 15 V)} \end{aligned}$$

Since the applied voltage is 15 V and 15 V was dropped across R_1, V_{R_2} can be determined to be 0 V, as follows:

$$\begin{aligned} V_{R_2} &= V_T - V_{R_1} \\ &= 15 \text{ V} - 15 \text{ V} \\ &= 0 \text{ V} \end{aligned}$$

Since 0 V is dropped across R_2, the current through R_2 can be determined as follows:

$$\begin{aligned} I_{R_2} &= V_{R_2}/R_2 \\ &= 0 \text{ V}/560 \text{ }\Omega \\ &= 0 \text{ A} = 0 \text{ mA} \end{aligned}$$

$$I_{R_3} = 0 \text{ A} = 0 \text{ mA}$$

$$I_{R_4} = 0 \text{ A} = 0 \text{ mA}$$

$$\begin{aligned} V_{R_3} &= I_{R_3} \times R_3 \\ &= 0 \text{ A} \times 220 \text{ }\Omega \\ &= 0 \text{ V} \end{aligned}$$

$$\begin{aligned} V_{R_4} &= I_{R_4} \times R_4 \\ &= 0 \text{ A} \times 330 \text{ }\Omega \\ &= 0 \text{ V} \end{aligned}$$

$$\begin{aligned} P_{R_1} &= I_{R_1} \times V_{R_1} \\ &= 0.0319 \text{ A} \times 15 \text{ V} \\ &= 0.4787 \text{ W} = 478.7 \text{ mW} \end{aligned}$$

$$\begin{aligned} P_{R_2} &= I_{R_2} \times V_{R_2} \\ &= 0 \text{ A} \times 0 \text{ V} \\ &= 0 \text{ W} \end{aligned}$$

$$\begin{aligned} P_{R_3} &= I_{R_3} \times V_{R_3} \\ &= 0 \text{ A} \times 0 \text{ V} \\ &= 0 \text{ W} \end{aligned}$$

$$\begin{aligned} P_{R_4} &= I_{R_4} \times V_{R_4} \\ &= 0 \text{ A} \times 0 \text{ V} \\ &= 0 \text{ W} \end{aligned}$$

Summary Comparison between the Normal and Short Circuit: It should be seen in this circuit that, when a short was created, the overall resistance of the circuit decreased, main-line current increased, and overall power dissipation increased markedly.

	Normal	Short Circuit
R_T	747.48 Ω	470 Ω
I_T	20.1 mA	31.9 mA
P_T	301 mW	478.7 mW
V_{R_1}	9.45 V	15 V
V_{R_2}	5.55 V	0 V
I_{R_2}	9.9 mA	0 mA
I_{R_3}	10.2 mA	0 mA
I_{R_4}	10.2 mA	0 mA
V_{R_3}	2.24 V	0 V
V_{R_4}	3.36 V	0 V
P_{R_1}	189.9 mW	478.7 mW
P_{R_2}	54.9 mW	0 mW
P_{R_3}	22.8 mW	0 mW
P_{R_4}	34.3 mW	0 mW

The Open Circuit: As an example of an open circuit, the series-parallel circuit in Fig. 6-4.2 becomes a series circuit with just R_1 and R_2 when there is an open circuit between terminals C and D. Since the component values and supply voltages are the same as in the circuit shown in Fig. 6-4.1, we will develop the characteristics of the open in this circuit by concentrating on the component calculated values.

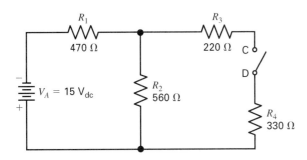

Fig. 6-4.2 Analysis of the open circuit.

When analyzing the effects of this open circuit problem, remember that, in the normal closed circuit, $R_T = 747.48$ Ω, $I_T = 20.1$ mA, $P_T = 301$ mW, $V_{R_1} = 9.45$ V, $V_{R_2} = 5.55$ V, $I_{R_2} = 9.9$ mA, $I_{R_3} = 10.2$ mA, $I_{R_4} = 10.2$ mA, $V_{R_3} = 2.24$ V, $V_{R_4} = 3.36$ V, $P_{R_1} = 189.9$ mW, $P_{R_2} = 54.9$ mW, $P_{R_3} = 22.8$ mW, and $P_{R_4} = 34.3$ mW.

Studying the Effect of an Open Circuit: If the circuit shown in Fig. 6-4.2 is open from points C to D, R_T can be calculated so that the series resistances R_3 and R_4 are open and maintain an infinite resistance. This infinite resistance is in parallel with R_2, and the parallel equivalent resistance can be considered to be equivalent to R_2, or 560 Ω. This value is now in series with the 470 Ω value of R_1 and provides an overall total equivalent resistance of 1030 Ω.

Since R_T is known, the main-line current can be determined from Ohm's law as follows:

$$I_T = V_T/R_T$$
$$= 15\ V_{dc}/1030\ \Omega$$
$$= 0.0146\ A \approx 14.6\ mA$$

The total power of the circuit can be calculated as follows:

$$P_T = I_T \times V_T$$
$$= 0.0146\ A \times 15\ V$$
$$= 0.219\ W = 219\ mW$$

Since the total main-line current I_T is known, the associated voltage drops across each of the resistors can be determined, for example:

$$V_{R_1} = I_T \times R_1$$
$$= 0.0146\ A \times 470\ \Omega$$
$$= 6.8620\ V$$

Since the applied voltage is 15 V and 6.8620 V was dropped across R_1, V_{R_2} can be determined to be 8.1380 V, as follows:

$$V_{R_2} = V_T - V_{R_1}$$
$$= 15\ V - 6.8620\ V$$
$$= 8.1380\ V$$

Since 8.1380 V is dropped across R_2, the current through R_2 can be determined to be:

$$I_{R_2} = V_{R_2}/R_2$$
$$= 8.1380\ V/560\ \Omega$$
$$= 0.0145\ A = 14.5\ mA$$

As you know, currents divide in parallel branches. However, the main-line current of 14.6 mA does not divide between the R_2 branch and the $R_3 + R_4$ branch, since the $R_3 + R_4$ branch is open. It would follow therefore that:

$$I_{R_3} = I_{R_4} = 0\ A$$

If the current is known to be 0 A through R_3 and R_4, the voltage drop across each can be determined, as follows:

$$V_{R_3} = I_{R_3} \times R_3$$
$$= 0\ A \times 220\ \Omega$$
$$= 0\ V$$

and

$$V_{R_4} = I_{R_4} \times R_4$$
$$= 0\ A \times 330\ \Omega$$
$$= 0\ V$$

Power in the circuit can be determined as follows:

$$P_{R_1} = I_{R_1} \times V_{R_1}$$
$$= 0.0146\ A \times 6.8620\ V$$
$$= 0.1002\ W = 100.2\ mW$$

$$P_{R_2} = I_{R_2} \times V_{R_2}$$
$$= 0.0146\ A \times 8.1380\ V$$
$$= 0.1188\ W = 118.8\ mW$$

$$P_{R_3} = I_{R_3} \times V_{R_3}$$
$$= 0\,\text{A} \times 0\,\text{V}$$
$$= 0\,\text{W}$$
$$P_{R_4} = I_{R_4} \times V_{R_4}$$
$$= 0\,\text{A} \times 0\,\text{V}$$
$$= 0\,\text{W}$$

Summary Comparison between the Normal and Short Circuit: It should be seen in this circuit that, when an open was created, the overall resistance of the circuit increased, main-line current decreased, and overall power dissipation changed markedly.

	Normal	Open Circuit
R_T	747.48 Ω	1030 Ω
I_T	20.1 mA	14.6 mA
P_T	301 mW	218.4 mW
V_{R_1}	9.45 V	6.8620 V
V_{R_2}	5.55 V	8.1380 V
I_{R_2}	9.9 mA	14.5 mA
I_{R_3}	10.2 mA	0 mA
I_{R_4}	10.2 mA	0 mA
V_{R_3}	2.24 V	0 V
V_{R_4}	3.36 V	0 V
P_{R_1}	189.9 mW	100.2 mW
P_{R_2}	54.9 mW	118.8 mW
P_{R_3}	22.8 mW	0 mW
P_{R_4}	34.3 mW	0 mW

EQUIPMENT

Breadboard
DC power supply
VOM/DMM
   Voltmeter
   Ammeter
   Ohmmeter

COMPONENTS

All resistors are 0.25 W unless indicated otherwise:

(1) 100 Ω   (1) 560 Ω
(1) 120 Ω   (1) 680 Ω
(1) 470 Ω   (1) 820 Ω

PROCEDURE

1. Measure and record each resistor value for the resistors required in this experiment. Record the results in Table 6-4.1.
2. Connect the circuit shown in Fig. 6-4.3. First calculate and then measure the total resistance from point A to point B. Record this information in Table 6-4.2. In addition, record the percentage of error

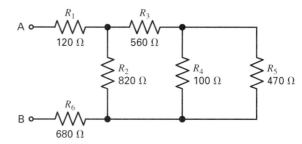

Fig. 6-4.3 Finding the total resistance of a series-parallel circuit.

between these values, where:

$$\% \text{ error} = \left| \frac{\text{difference between meas. and calc. values}}{\text{calc. values}} \right| \times 100$$

If the error is greater than 10 percent, repeat the calculations and measurements.

3. Connect the circuit shown in Fig. 6-4.4. Calculate all voltage drops and component currents, and record them in Table 6-4.3.

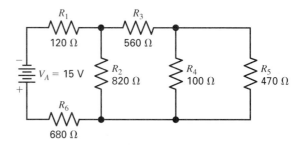

Fig. 6-4.4 Series-parallel circuit under analysis.

4. Adjust the supply voltage to 15 V and, using a voltmeter, take voltage drop measurements across each component. Record your results in Table 6-4.3.

Voltage Measurements—Short Circuit

5. Maintaining a supply voltage of 15 V, use a clip lead to short R_1. Take voltage drop measurements V_{R_1}, V_{R_2}, V_{R_3}, V_{R_4}, V_{R_5}, and V_{R_6}. Record the results in Table 6-4.3.
6. Remove the clip-lead short across R_1.
7. Maintaining a supply voltage of 15 V, use a clip lead to short R_2. Take voltage drop measurements V_{R_1}, V_{R_2}, V_{R_3}, V_{R_4}, V_{R_5}, and V_{R_6}. Record the results in Table 6-4.3.
8. Remove the clip-lead short across R_2.
9. Maintaining a supply voltage of 15 V, use a clip lead to short R_3. Take voltage drop measurements V_{R_1}, V_{R_2}, V_{R_3}, V_{R_4}, V_{R_5}, and V_{R_6}. Record the results in Table 6-4.3.
10. Remove the clip-lead short across R_3.
11. Maintaining a supply voltage of 15 V, use a clip lead to short R_4. Take voltage drop measurements

V_{R_1}, V_{R_2}, V_{R_3}, V_{R_4}, V_{R_5}, and V_{R_6}. Record the results in Table 6-4.3.

12. Remove the clip-lead short across R_4.

13. Maintaining a supply voltage of 15 V, use a clip lead to short R_5. Take voltage drop measurements V_{R_1}, V_{R_2}, V_{R_3}, V_{R_4}, V_{R_5}, and V_{R_6}. Record the results in Table 6-4.3.

14. Remove the clip-lead short across R_5.

15. Maintaining a supply voltage of 15 V, use a clip lead to short R_6. Take voltage drop measurements V_{R_1}, V_{R_2}, V_{R_3}, V_{R_4}, V_{R_5}, and V_{R_6}. Record the results in Table 6-4.3.

16. Remove the clip-lead short across R_6.

Current Measurements—Short Circuit

17. Maintaining a supply voltage of 15 V, use a clip lead to short R_1 from your circuit. With an ammeter, take current measurements of I_{R_2}, I_{R_3}, I_{R_4}, I_{R_5}, and I_{R_6}. Do not take the current measurement through R_1. Record this information in Table 6-4.3.

18. Remove the clip-lead circuit short across R_1.

19. Maintaining a supply voltage of 15 V, use a clip lead to short R_2 from your circuit. With an ammeter, take current measurements of I_{R_1}, I_{R_3}, I_{R_4}, I_{R_5}, and I_{R_6}. Do not take the current measurement through R_2. Record this information in Table 6-4.3.

20. Remove the clip-lead circuit short across R_2.

21. Maintaining a supply voltage of 15 V, use a clip lead to short R_3 from your circuit. With an ammeter, take current measurements of I_{R_1}, I_{R_2}, I_{R_4}, I_{R_5}, and I_{R_6}. Do not take the current measurement through R_3. Record this information in Table 6-4.3.

22. Remove the clip-lead circuit short across R_3.

23. Maintaining a supply voltage of 15 V, use a clip lead to short R_4 from your circuit. With an ammeter, take current measurements of I_{R_1}, I_{R_2}, I_{R_3}, I_{R_5}, and I_{R_6}. Do not take the current measurement through R_4. Record this information in Table 6-4.3.

24. Remove the clip-lead circuit short across R_4.

25. Maintaining a supply voltage of 15 V, use a clip lead to short R_5 from your circuit. With an ammeter, take current measurements of I_{R_1}, I_{R_2}, I_{R_3}, I_{R_4}, and I_{R_6}. Do not take the current measurement through R_5. Record this information in Table 6-4.3.

26. Remove the clip-lead circuit short across R_5.

27. Maintaining a supply voltage of 15 V, use a clip lead to short R_6 from your circuit. With an ammeter, take current measurements of I_{R_1}, I_{R_2}, I_{R_3}, I_{R_4}, and I_{R_5}. Do not take the current measurement through R_6. Record this information in Table 6-4.3.

28. Remove the clip-lead circuit short across R_6.

Voltage Measurements—Open Circuit

29. Maintaining a supply voltage of 15 V, open R_1 by removing it from the circuit. Take voltage drop measurements V_{R_2}, V_{R_3}, V_{R_4}, V_{R_5}, and V_{R_6}. Do not take the voltage drop measurement across R_1. Record these measurements in Table 6-4.4.

30. Remove the open by replacing R_1 into the circuit.

31. Maintaining a supply voltage of 15 V, open R_2 by removing it from the circuit. Take voltage drop measurements V_{R_1}, V_{R_3}, V_{R_4}, V_{R_5}, and V_{R_6}. Do not take the voltage drop measurement across R_2. Record these measurements in Table 6-4.4.

32. Remove the open by replacing R_2 into the circuit.

33. Maintaining a supply voltage of 15 V, open R_3 by removing it from the circuit. Take voltage drop measurements V_{R_1}, V_{R_2}, V_{R_4}, V_{R_5}, and V_{R_6}. Do not take the voltage drop measurement across R_3. Record these measurements in Table 6-4.4.

34. Remove the open by replacing R_3 into the circuit.

35. Maintaining a supply voltage of 15 V, open R_4 by removing it from the circuit. Take voltage drop measurements V_{R_1}, V_{R_2}, V_{R_3}, V_{R_5}, and V_{R_6}. Do not take the voltage drop measurement across R_4. Record these measurements in Table 6-4.4.

36. Remove the open by replacing R_4 into the circuit.

37. Maintaining a supply voltage of 15 V, open R_5 by removing it from the circuit. Take voltage drop measurements V_{R_1}, V_{R_2}, V_{R_3}, V_{R_4}, and V_{R_6}. Do not take the voltage drop measurement across R_5. Record these measurements in Table 6-4.4.

38. Remove the open by replacing R_5 into the circuit.

39. Maintaining a supply voltage of 15 V, open R_6 by removing it from the circuit. Take voltage drop measurements V_{R_1}, V_{R_2}, V_{R_3}, V_{R_4}, and V_{R_5}. Do not take the voltage drop measurement across R_6. Record these measurements in Table 6-4.4.

40. Remove the open by replacing R_6 into the circuit.

Current Measurements—Open Circuit

41. Maintaining a supply voltage of 15 V, open R_1 by removing it from the circuit. Take current measurements I_{R_2}, I_{R_3}, I_{R_4}, I_{R_5}, and I_{R_6}. Do not take the current measurement through R_1. Record these measurements in Table 6-4.4.

42. Remove the open by replacing R_1 into the circuit.

43. Maintaining a supply voltage of 15 V, open R_2 by removing it from the circuit. Take current measurements I_{R_1}, I_{R_3}, I_{R_4}, I_{R_5}, and I_{R_6}. Do not take the current measurement through R_2. Record these measurements in Table 6-4.4.

44. Remove the open by replacing R_2 into the circuit.

45. Maintaining a supply voltage of 15 V, open R_3 by removing it from the circuit. Take current measurements I_{R_1}, I_{R_2}, I_{R_4}, I_{R_5}, and I_{R_6}. Do not take the current measurement through R_3. Record these measurements in Table 6-4.4.

46. Remove the open by replacing R_3 into the circuit.

47. Maintaining a supply voltage of 15 V, open R_4 by removing it from the circuit. Take current measurements I_{R_1}, I_{R_2}, I_{R_3}, I_{R_5}, and I_{R_6}. Do not take the current measurement through R_4. Record these measurements in Table 6-4.4.

48. Remove the open by replacing R_4 into the circuit.

49. Maintaining a supply voltage of 15 V, open R_5 by removing it from the circuit. Take current measurements I_{R_1}, I_{R_2}, I_{R_3}, I_{R_4}, and I_{R_6}. Do not take the current measurement through R_5. Record these measurements in Table 6-4.4.

50. Remove the open by replacing R_5 into the circuit.

51. Maintaining a supply voltage of 15 V, open R_6 by removing it from the circuit. Take current measurements I_{R_1}, I_{R_2}, I_{R_3}, I_{R_4}, and I_{R_5}. Do not take the current measurement through R_6. Record these measurements in Table 6-4.4.

52. Remove the open by replacing R_6 into the circuit.

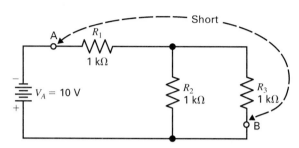

Fig. 6-4.5 Shorted circuit for Question 2.

NAME	DATE

QUESTIONS FOR EXPERIMENT 6-4

1. What effect would a short across a resistor located in series with a parallel resistive branch have on the level of main-line current flow? Explain.

2. What effect would there be (on individual branch circuit levels) of a short from point A to point B within a circuit (opposite in Fig. 6-4.5)? Explain.

3. If in Fig. 6-4.1, R_2 was shorted, what would happen to the current flow and voltage drop of R_3? What would happen to the voltage and current of R_2? Explain.

4. Calculate the needed level of power dissipation for each resistor shown in Fig. 6-4.4. How does power dissipation change for a shorted component? How would it change across an open component? Explain.

5. According to Grob/Schultz, *Basic Electronics,* Tenth Edition, what is the potential difference across an open? Explain.

CRITICAL THINKING QUESTIONS

Note: The following questions are designed to help you analyze the previous laboratory experiment in a complete and in-depth fashion. To answer these questions, you should review the related material in Grob/Schultz, *Basic Electronics,* Tenth Edition.

1. Create a table that compares the calculated values for the normal circuit, short circuit, and open circuit in Figs. 6-4.1 and 6-4.2. Describe the similarities and differences in this table.

2. For the short circuit current analysis in Step 17, the following statement is made: "Maintaining a supply voltage of 15 V, use a clip lead to short R_1 from your circuit. With an ammeter, take current measurements of I_{R_2}, I_{R_3}, I_{R_4}, I_{R_5}, and I_{R_6}. Do not take the current measurement through R_1. Record this information in Table 6-4.3." Why was the current measurement I_{R_1} excluded from the original measurements? What is the anticipated value for I_{R_1}?

3. For the open circuit voltage analysis in Step 29, the following statement is made: "Maintaining a supply voltage of 15 V, open R_1 by removing it from the circuit. Take voltage drop measurements V_{R_2}, V_{R_3}, V_{R_4}, V_{R_5}, and V_{R_6}. Do not take the voltage drop measurement across R_1. Record these measurements in Table 6-4.4." Why was the voltage measurement V_{R_1} excluded from the original measurements? What is the anticipated value for V_{R_1}?

4. For the open circuit current analysis in Step 41, the following statement is made: "Maintaining a supply voltage of 15 V, open R_1 by removing it from the circuit. Take current measurements I_{R_2}, I_{R_3}, I_{R_4}, I_{R_5}, and I_{R_6}. Do not take the current measurement through R_1. Record these measurements in Table 6-4.4." Why was the current measurement I_{R_1} excluded from the original measurements? What is the anticipated value for I_{R_1}?

5. From your results and tables, describe the significance of opens and shorts in analyzing circuit behavior.

TABLES FOR EXPERIMENT 6-4

TABLE 6–4.1   Individual Resistor Values

Resistor	Nominal Value, Ω	Resistive Measurement
R_1	120	_____
R_2	820	_____
R_3	560	_____
R_4	100	_____
R_5	470	_____
R_6	680	_____

TABLE 6–4.2 Total Resistance R_T and Percentage of Error

	Calculated	Measurement	% Error
R_T	_____	_____	_____

TABLE 6–4.3 Short Circuit Calculations and Measurements

	Normal Circuit Calculations	Normal Circuit Measurements	Measurements (with Each Resistor *Shorted* One at a Time)					
			R_1	R_2	R_3	R_4	R_5	R_6
V_{R_1}	_____	_____	_____	_____	_____	_____	_____	_____
V_{R_2}	_____	_____	_____	_____	_____	_____	_____	_____
V_{R_3}	_____	_____	_____	_____	_____	_____	_____	_____
V_{R_4}	_____	_____	_____	_____	_____	_____	_____	_____
V_{R_5}	_____	_____	_____	_____	_____	_____	_____	_____
V_{R_6}	_____	_____	_____	_____	_____	_____	_____	_____
I_{R_1}	_____	_____		_____	_____	_____	_____	_____
I_{R_2}	_____	_____	_____	_____	_____	_____	_____	_____
I_{R_3}	_____	_____	_____	_____	_____	_____	_____	_____
I_{R_4}	_____	_____	_____	_____	_____		_____	_____
I_{R_5}	_____	_____	_____	_____	_____	_____		_____
I_{R_6}	_____	_____	_____	_____	_____	_____	_____	_____

TABLE 6–4.4 Open Circuit Calculations and Measurements

	Measurements (with Each Resistor *Opened* One at a Time)					
	R_1	R_2	R_3	R_4	R_5	R_6
V_{R_1}		_____	_____	_____	_____	_____
V_{R_2}	_____		_____	_____	_____	_____
V_{R_3}	_____	_____		_____	_____	_____
V_{R_4}	_____	_____	_____		_____	_____
V_{R_5}	_____	_____	_____	_____		_____
V_{R_6}	_____	_____	_____	_____	_____	
I_{R_1}		_____	_____	_____	_____	_____
I_{R_2}	_____		_____	_____	_____	_____
I_{R_3}	_____	_____		_____	_____	_____
I_{R_4}	_____	_____	_____		_____	_____
I_{R_5}	_____	_____	_____	_____		_____
I_{R_6}	_____	_____	_____	_____	_____	

EXPERIMENT RESULTS REPORT FORM

Experiment No: _____   Name: _____
                                   Date: _____
Experiment Title: _____   Class: _____
_____   Instr: _____

Explain the purpose of the experiment:

List the first Learning Objective:
OBJECTIVE 1:

After reviewing the results, describe how the objective was validated by this experiment?

List the second Learning Objective:
OBJECTIVE 2:

After reviewing the results, describe how the objective was validated by this experiment?

List the third Learning Objective:
OBJECTIVE 3:

After reviewing the results, describe how the objective was validated by this experiment?

Conclusion:

If required, attach to this form: ☐ Answers to Questions, ☐ Tables, and ☐ Graphs.

EXPERIMENT 6-5

THE WHEATSTONE BRIDGE

OBJECTIVES

At the completion of this experiment, you will:
- Be able to measure current and voltage in a Wheatstone bridge.
- Know how to use a galvanometer.
- Understand how a Wheatstone bridge can be balanced.

SUGGESTED READING

Chapters 5, 6, and 10, *Basic Electronics,* Grob/Schultz, Tenth Edition

INTRODUCTION

Many types of bridge circuits are used in electronics. The bridge circuit in this experiment can become a functional ohmmeter as well as a balanced circuit. Here, the Wheatstone bridge has two input terminals, where the battery (or power supply) terminals are connected to the ratio arms. The ratio arms are the key to understanding how the Wheatstone bridge is balanced. The bridge is balanced when there is an equal division of voltages across the bridge output. This output is a current path across the ratio arms, like a bridge between two series-parallel paths. It does not matter what the ratios are for the bridge to be in a balanced state, because resistances in parallel (two series resistors in this case) have the same voltage across both branches. However, some ratios may create a more sensitive balance than others, depending upon the total current and total resistance of the ratio arms.

If there is an imbalance in the bridge, current will flow through the output path from one ratio arm to the other. In this experiment, a galvanometer is used. It is a dual directional microammeter with the needle zeroed at top dead center.

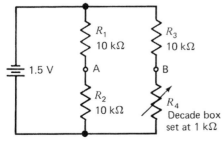

Fig. 6-5.1 Series-parallel circuit for analysis.

It is possible to use a VOM. However, extreme caution will be needed to prevent damage to the meter.

EQUIPMENT

DC power supply, 0–10 V
Decade box
Galvanometer
Test leads
VTVM or DVM

COMPONENTS

(3) 10 kΩ 0.25 W resistors
Plus other desired resistors (all 0.25 W):

(1) 5.6 kΩ   (1) 2.7 kΩ
(1)   1 kΩ   (1) 560 kΩ
(1) 2.2 kΩ

PROCEDURE

1. Connect the circuit of Fig. 6-5.1.
2. Usually, a galvanometer is placed across the output terminals (A and B) to complete the bridge. However, it is better to study the series-parallel aspects of the Wheatstone bridge first to gain a better understanding of how it works. Thus, measure the voltage across points A and B and record the results in Table 6-5.1. Put the ground side of the voltmeter on point B. Also, determine the direction of current flow.
3. Increase the value of R_4 to 5 kΩ and measure the voltage across points A and B and determine current direction. Record the results in Table 6-5.1.
4. Increase the value of R_4 in 1 kΩ steps from 5 to 15 kΩ. Measure the voltage across points A and B at each step, determine current direction, and record the results in Table 6-5.1. This should lead to an understanding of how the value of the ratio arms determines the voltage across the AB output. When finished, turn the power off.

CAUTION: It will be necessary to reverse the voltmeter leads at some point.

5. Insert a galvanometer across points A and B as shown in Fig. 6-5.2. Be sure R_4 is 10 kΩ. The circuit is now a Wheatstone bridge, because there is a true current path across points A and B.

EXPERIMENT 6-5 127

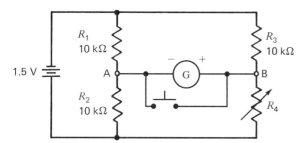

Fig. 6-5.2 Galvanometer circuit.

Note: Many galvanometers have a switch (push button or toggle switch on the front of the meter). Until the switch is closed, current flows through a wire (or short circuit) to protect the meter movement. When the switch is closed, current flows through the meter movement. Be careful not to peg the meter.

6. With equal nominal values of R_1 and R_4, turn the power on and finely adjust R_4 so that no current flows across points A and B. This is the same as 0 V across points A and B. Thus, R_4 (decade box) will be finely adjusted and may be 10.5 kΩ or 11.1 kΩ, etc., unless, of course, you are using precision resistors (1 percent tolerance). The important thing is that both voltage and current at points A and B = 0. If so, the bridge is now balanced. Record the exact value of R_4 (decade box) in Table 6-5.2.

7. Now adjust R_4 so that the needle deflects to the right, approximately halfway to full scale, as shown in Fig. 6-5.3. Measure and record in Table 6-5.2 the voltage across R_4 and R_2. Repeat this procedure by adjusting R_4 with the needle deflecting to the left. Record the value of R_4 and the voltage across R_4 and R_2. Keep in mind the following: the bridge was balanced. Then, the bridge was unbalanced on both ratio arms (left and right). This procedure can be analyzed later to determine how the direction of current flow across the bridge indicates the condition of the bridge. Also record the adjusted value of R_4.

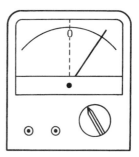

Fig. 6-5.3 Galvanometer.

8. Bring the bridge back to a balanced condition by readjusting R_4 and monitoring the galvanometer. The needle should be dead center—zero current.

9. Turn the power off. Replace R_2 with an unknown resistor with a value between 2 and 8 kΩ. Simply put electrical tape over a resistor. Or, if this is not possible, use a 5.6 kΩ resistor, and pretend you do not know the value.

10. Turn power on. Adjust R_4 until the bridge is balanced. The value of R_4 (decade box) should be the true value of R_2. Repeat steps 9 and 10 with several resistors (1 kΩ, 2.2 kΩ, 2.7 kΩ, and 560 Ω, for example). R_4 should be equal to R_2 (unknown resistor) each time. Every time you adjust R_4, notice the amount of needle deflection. Record the results in a separate table. Be sure to label all values.

QUESTIONS FOR EXPERIMENT 6-5

1. In procedure steps 9 and 10, the Wheatstone bridge was balanced by matching a decade box to the unknown resistor. Thus, the Wheatstone bridge was used as what kind of meter?

2. Explain the difference between a galvanometer and an ohmmeter.

3. In procedure step 5, suppose the schematic of Fig. 6-5.2 showed R_4 as 50 kΩ, and R_1 to R_3 remained at 10 kΩ. With power on, would the needle deflect to the right or the left?

4. Which circuit in Fig. 6-5.4 would be more sensitive (greater needle deflection) when attempting to balance the bridge? Why?

5. Explain any differences and/or similarities between the two circuits in Fig. 6-5.5.

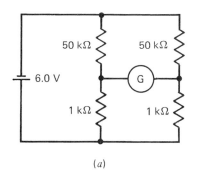

Fig. 6-5.4 Circuits for question 4.

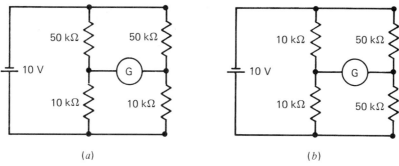

Fig. 6-5.5 Circuits for question 5.

TABLES FOR EXPERIMENT 6-5

TABLE 6–5.1 Data from Fig. 6-5.1

R_4, kΩ	Voltage AB	Current Direction
1	_____	_____
5	_____	_____
6	_____	_____
7	_____	_____
8	_____	_____
9	_____	_____
10	_____	_____
11	_____	_____
12	_____	_____
13	_____	_____
14	_____	_____
15	_____	_____

TABLE 6–5.2 R_1 and $R_3 = 10$ kΩ, R_4 = Decade Box

	Voltage and Current at A and B = 0	Needle Deflection to Right			Needle Deflection to Left		
	R_4, Ω	R_4, Ω	V_{R_4}	V_{R_2}	R_4, Ω	V_{R_4}	V_{R_2}
$R_2 = 10$ kΩ	_____	_____	_____	_____	_____	_____	_____
$R_2 = 5.6$ kΩ	_____	_____	_____	_____	_____	_____	_____
$R_2 = 1$ kΩ	_____	_____	_____	_____	_____	_____	_____
$R_2 = 2.2$ kΩ	_____	_____	_____	_____	_____	_____	_____
$R_2 = 2.7$ kΩ	_____	_____	_____	_____	_____	_____	_____
$R_2 = 560$ Ω	_____	_____	_____	_____	_____	_____	_____

EXPERIMENT RESULTS REPORT FORM

Experiment No: _____ Name: _____
                                        Date: _____
Experiment Title: _____ Class: _____
_____ Instr: _____

Explain the purpose of the experiment:

List the first Learning Objective:
OBJECTIVE 1:

After reviewing the results, describe how the objective was validated by this experiment?

List the second Learning Objective:
OBJECTIVE 2:

After reviewing the results, describe how the objective was validated by this experiment?

List the third Learning Objective:
OBJECTIVE 3:

After reviewing the results, describe how the objective was validated by this experiment?

Conclusion:

If required, attach to this form: ☐ Answers to Questions, ☐ Tables, and ☐ Graphs.

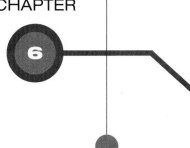

EXPERIMENT 6-6

POSITIVE AND NEGATIVE VOLTAGES TO GROUND

OBJECTIVES

At the completion of this experiment, you will be able to:
- Calculate circuit current and voltage drops found in a voltage-divider circuit.
- Determine the polarity of voltages found in a voltage divider circuit with a common ground.
- Identify voltage divider circuits that display both positive and negative voltages to ground.

SUGGESTED READING

Chapters 6 and 7, *Basic Electronics,* Grob/Schultz, Tenth Edition

INTRODUCTION

In the wiring of practical circuits, one side of the voltage source is usually grounded. In electronic equipment the ground often indicates a metal chassis, which is used as a common return for connections to the voltage source. Where printed-circuit boards are used, usually a common ground path is run around the outside perimeter of the circuit board. In other cases the entire back side of a two-sided board may be used as a common-return path. Note that the chassis ground may or may not be connected to earth ground.

When a circuit has a chassis as a common return, measure the voltages with respect to chassis ground. Consider the voltage divider in Fig. 6-6.1. In Fig. 6-6.1, the circuit shows no ground system. It is instead a closed series circuit. This circuit has an applied power supply voltage V_T of 20 V. To determine the total circuit current, the total circuit resistance R_T must be determined. And R_T can be calculated as

$$R_T = R_1 + R_2 + R_3$$
$$= 4.7 \text{ k}\Omega + 4.7 \text{ k}\Omega + 10 \text{ k}\Omega$$
$$= 19.4 \text{ k}\Omega$$

The circuit current can now be calculated from the Ohm's law relationship:

$$I_T = \frac{V_T}{R_T}$$
$$= \frac{20 \text{ V}}{19.4 \text{ k}\Omega}$$
$$= 1.03 \text{ mA}$$

After the current has been calculated, the individual voltage drops V_1, V_2, and V_3 of the voltage divider can be found from the Ohm's law relationship of

$$V = I \times R$$

For V_1:

$$V_1 = I_T \times R_1$$
$$= 1.03 \text{ mA} \times 4.7 \text{ k}\Omega$$
$$= 4.84 \text{ V} \quad (\text{or } +4.84 \text{ V})$$

For V_2:

$$V_2 = I_T \times R_2$$
$$= 1.03 \text{ mA} \times 4.7 \text{ k}\Omega$$
$$= 4.84 \text{ V} \quad (\text{or } +4.84 \text{ V})$$

For V_3:

$$V_3 = I_T \times R_3$$
$$= 1.03 \text{ mA} \times 10 \text{ k}\Omega$$
$$= 10.31 \text{ V} \quad (\text{or } +10.31 \text{ V})$$

Also the sum of the voltage drops will equal the applied voltage:

$$V_T = V_1 + V_2 + V_3$$
$$= 4.84 \text{ V} + 4.84 \text{ V} + 10.31 \text{ V}$$
$$= 19.99 \text{ V (round to 20 V)}$$

Refer to Fig. 6-6.1 and note that the polarities are included on this schematic. The polarity is determined by how the circuit current and the individual volt-meters

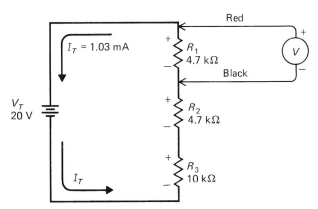

Fig. 6-6.1 Voltage divider circuit without a common ground.

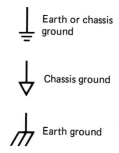

Fig. 6-6.2 Schematic symbols for ground.

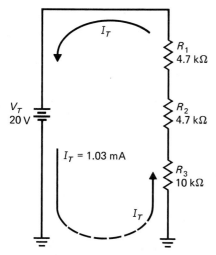

Fig. 6-6.3 Voltage divider circuit using a ground return.

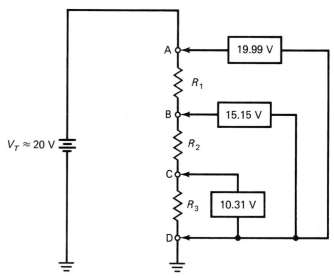

Fig. 6-6.4 Voltage divider circuit where voltages are measured to common ground.

are connected. The polarity of the resistors, which indicates the direction of current flow, and the color of the test leads are also indicated in Fig. 6-6.1.

Figure 6-6.2 shows the schematic symbols for ground. The ground symbol is used in Fig. 6-6.3. Here the same circuit with the same component values and applied voltage is shown. The only addition is the "ground return." Also note that the voltage drops of this circuit are equivalent to those shown in Fig. 6-6.1. The 1.03 mA generated by the battery is pushed out into ground point A and returns to ground point B. This, in effect, means that the ground symbols are connected, perhaps through a cable, foil pattern, or metal chassis.

The circuit shown in Fig. 6-6.4 details the same circuit of Figs. 6-6.3 and 6-6.1. The difference here is that all voltages are taken "with respect to ground." In this case, the voltage from point C to D is +10.31 V. The voltage from point B to D is +15.15 V (where $V_{BD} = VR_2 + VR_3$). The voltage from point A to D is +19.99 V (or 20 V), where $V_{AD} = VR_1 + VR_2 + VR_3$.

The circuit in Fig. 6-6.5 is similar to the one in Fig. 6-6.4. The main difference is that the ground has been moved to point C. This then indicates that all measurements should be taken "with reference" to this point. In this case $V_{AC} = +9.68$ V (where $V_{AC} = VR_1 + VR_2$). The voltage $V_{BC} = +4.84$ V (where $V_{BC} = VR_2$). The voltage $V_{DC} = -10.31$ V (where $V_{DC} = VR_3$). The voltage measured from point D to C results in a negative voltage. This circuit is known as a *positive and negative voltage divider*.

In summary, these dc circuits operate in the same way with or without the ground symbol shown in the schematic. The only factor that changes is the reference point for measuring the voltage.

While this experiment focuses on the use of ground as a reference point, keep in mind that there are several different symbols for ground, as shown in Fig. 6-6.2. The earth ground symbol usually indicates that one side of the power supply is connected to the earth, usually by a metal pipe in the ground (this is the third prong on the ac wall plug). The other two ground symbols are chassis grounds and may or may not be connected to earth (an automobile or an airplane is a good example).

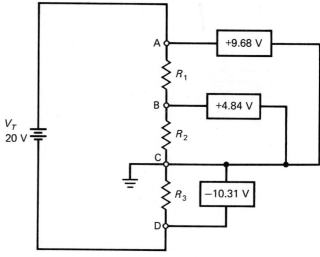

Fig. 6-6.5 Voltage divider displaying both positive and negative voltages to ground.

EQUIPMENT

DC power supply, 0 to 20 V
  Leads
  Breadboard
  Voltmeter

COMPONENTS

Resistors (all 0.25 W):

  (2) 4.7 kΩ
  (1) 10 kΩ

PROCEDURE

1. Connect the circuit of Fig. 6-6.6 to the dc power supply as shown.

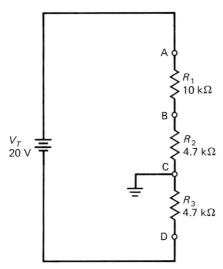

Fig. 6-6.6 Positive and negative voltage divider circuit with ground.

2. With the power supply turned off and disconnected, measure and record in Table 6-6.1 the resistance values of R_1, R_2, and R_3.

3. With the power supply turned off and disconnected, measure and record in Table 6-6.1 the resistive value of R_T.

4. Reconnect the power supply to the circuit shown in Fig. 6-6.6.

5. Calculate and record in Table 6-6.1 the values of R_T, I_T, V_{AC}, V_{BC}, and V_{DC} referenced to ground.

6. Turn on the power supply and adjust its voltage value to 20 V dc.

7. Measure and record in Table 6-6.1: I_T, V_T, V_{AC}, V_{BC}, and V_{DC}. Also note the polarities of V_{AC}, V_{BC}, and V_{DC} in Table 6-6.1.

8. Turn off the power supply, and reverse its polarity.

9. Reconnect the circuit with $V_A = -20$ V, repeat steps 2–7, and record the results in Table 6-6.2.

EXPERIMENT 6-6

NAME	DATE

QUESTIONS FOR EXPERIMENT 6-6

1. Explain the circuit function of a voltage divider.

2. What is the purpose of a circuit ground?

3. Draw an example of a circuit where a voltage is negative with respect to ground.

TABLES FOR EXPERIMENT 6-6

TABLE 6–6.1 (Steps 2–7)

	Measured	Calculated	*Polarity
R_1	————		
R_2	————		
R_3	————		
R_T	————	————	
I_T	————	————	
V_{AC}	————	————	————
V_{BC}	————	————	————
V_{DC}	————	————	————
V_T	————		

TABLE 6–6.2 (Steps 8 and 9)

	Measured	Calculated	*Polarity
R_1	————		
R_2	————		
R_3	————		
R_T	————	————	
I_T	————	————	
V_{AC}	————	————	————
V_{BC}	————	————	————
V_{DC}	————	————	————
V_T	————		

*Note: Polarity with respect to common ground.

EXPERIMENT RESULTS REPORT FORM

Experiment No: _____   Name: _____
                                 Date: _____
Experiment Title: _____ Class: _____
_____   Instr: _____

Explain the purpose of the experiment:

List the first Learning Objective:
OBJECTIVE 1:

After reviewing the results, describe how the objective was validated by this experiment?

List the second Learning Objective:
OBJECTIVE 2:

After reviewing the results, describe how the objective was validated by this experiment?

List the third Learning Objective:
OBJECTIVE 3:

After reviewing the results, describe how the objective was validated by this experiment?

Conclusion:

If required, attach to this form: ☐ Answers to Questions, ☐ Tables, and ☐ Graphs.

CHAPTER 6

EXPERIMENT 6-7

ADDITIONAL SERIES-PARALLEL CIRCUITS

OBJECTIVES

At the completion of this experiment, you will be able to:
- Redraw and simplify a series-parallel circuit.
- Write an equation for the total resistance of a series-parallel circuit.
- Determine which resistances have the greatest effect on R_T.

SUGGESTED READING

Chapter 6, *Basic Electronics*, Grob/Schultz, Tenth Edition

INTRODUCTION

This experiment is similar to Experiment 6-1 on series-parallel circuits. The same concepts discussed in that experiment apply here. The following points summarize the concepts of a series-parallel circuit:
- The R_T of two parallel branches equals the product divided by the sum.

$$R_T = \frac{R_1 \times R_2}{R_1 + R_2}$$

- Adding resistance in series increases the total resistance and decreases the total current.
- Adding resistance in parallel increases the total current and decreases the total resistance.
- The equivalent resistance R_{eq} of a series-parallel network must be determined before the total resistance R_T can be found.

Consider the circuit shown in Fig. 6-7.1. This resistive network has a total resistance of

$$R_T = R_1 \| R_{eq}$$

The two vertical lines in the equation are used as a symbol to indicate resistances in parallel. Thus, $R_T = R_1 \| R_{eq}$ can be literally taken to mean

$$R_T = R_1 \text{ in parallel with } R_{eq}$$

To determine R_T, the circuit must be reduced or redrawn so that all the resistors, except R_1, can be combined into one equivalent resistance R_{eq}. This is because R_1 is in parallel with the equivalent total resistance of R_2 through R_5.

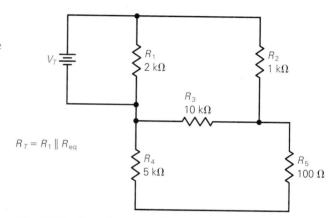

Fig. 6-7.1 Resistive network.

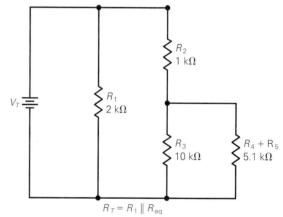

Fig. 6-7.2 Redrawn resistive network.

Although the circuit can be redrawn in more than one way, there is a method that can be used easily: Look for the resistors farthest from the power source, and combine them, working backward toward the source. The result of this method is

$$R_{eq} = [(R_4 + R_5) \| R_3] + R_2$$

The network can be redrawn as shown in Fig. 6-7.2. Notice that the redrawn figure makes it easier to write the equation. In this case, you can begin with the sum of $R_5 + R_4$ in parallel with R_3. This would be

$$R_4 + R_5 = 5.1 \text{ k}\Omega$$

Then, by using the product-over-the-sum method

$$\frac{R_3 \times 5.1 \text{ k}\Omega}{R_3 + 5.1 \text{ k}\Omega} = \frac{10 \text{ k}\Omega \times 5.1 \text{ k}\Omega}{10 \text{ k}\Omega + 5.1 \text{ k}\Omega} = \frac{51.0 \text{ k}\Omega}{15.1 \text{ k}\Omega} = 3.38 \text{ k}\Omega$$

Next, add 3.38 kΩ to 1 kΩ R_2 so that R_{eq} = 4.38 kΩ. Finally,

$$R_T = R_1 \| R_{eq} = \frac{2 \text{ k}\Omega \times 4.38 \text{ k}\Omega}{2 \text{ k}\Omega + 4.38 \text{ k}\Omega} = \frac{8.76 \text{ k}\Omega}{6.38 \text{ k}\Omega} = 1.37 \text{ k}\Omega$$

Not only can you write an equation for R_T by redrawing the circuit, but also you can see how certain resistances have a greater or lesser effect on the network. For example, R_5 has almost no effect on R_T because R_5 is less than 10 percent of the resistance in series and is added to R_4. However, if R_4 were removed from the circuit, then R_5 would have a greater effect on the equivalent resistance.

EQUIPMENT

Ohmmeter
DC power supply
Ammeter
Voltmeter
Protoboard or springboard
Leads

COMPONENTS

Resistors (all 0.25 W):

(2) 4.7 kΩ
(1) 560 Ω
(1) 820 Ω
(2) 10 kΩ
(1) 2.2 kΩ
(1) 1 kΩ

PROCEDURE

1. Measure and record the resistor values shown in Table 6-7.1.
2. Connect the circuit shown in Fig. 6-7.3.
3. Measure and record the voltages around the circuit as shown in Table 6-7.1.

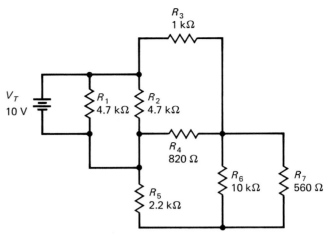

Fig. 6-7.3 Resistive network.

4. Disconnect the power supply V_T, and measure the total circuit resistance. Record the results in Table 6-7.1.
5. Disconnect R_1 and V_T, and measure and record the resistance R_{eq}.
6. Calculate the current through each resistor, using Ohm's law and the measured voltage drops. Record the results in Table 6-7.1 and include I_T, the total circuit current.
7. Remove R_6 from the circuit and measure the total circuit resistance. Record the results in Table 6-7.1.
8. Replace R_2 with a 10 kΩ resistor, and measure the total circuit resistance. Record the results in Table 6-7.1. (R_6 is still removed.)
9. On a separate sheet of 8 × 11 in. paper, redraw the circuit so that only four resistances represent the simplified circuit. Label all resistances so that the combined resistances are easy to identify. For example, one resistance might be $(R_3 + R_4) \| R_2$, etc.
10. Write an equation for the total resistance, and show how to calculate R_T by using R_{eq} and the product-over-the-sum method. Use the same sheet of paper as in step 9 above.

QUESTIONS FOR EXPERIMENT 6-7

1. Which resistor in the circuit of Fig. 6-7.2 has the least effect on R_T and why?

2. What would happen to the circuit of Fig. 6-7.2 if R_1 were decreased to 10 Ω?

3. What would happen to the circuit of Fig. 6-7.2 if R_1 were increased to 10 MΩ?

4. Which resistor in the circuit of Fig. 6-7.2 has the least effect on I_T?

5. Which resistor in the circuit of Fig. 6-7.2 has the greatest effect on I_T?

TABLES FOR EXPERIMENT 6-7

TABLE 6–7.1

Resistance Values Nominal or Calculated	Ω Measured	V Measured	I Calculated
$R_1 = 4.7$ kΩ			
$R_2 = 4.7$ kΩ			
$R_3 = 1$ kΩ			
$R_4 = 820$ Ω			
$R_5 = 2.2$ kΩ			
$R_6 = 10$ kΩ			
$R_7 = 560$ Ω			
$R_T =$ _____	_____	$I_T =$ _____	
$R_{eq} =$ _____	_____		
R_T with R_6 removed	_____		
R_T with $R_2 = 10$ kΩ	_____		

EXPERIMENT RESULTS REPORT FORM

Experiment No: _____   Name: _____
                                        Date: _____
Experiment Title: _____   Class: _____
_____   Instr: _____

Explain the purpose of the experiment:

List the first Learning Objective:
OBJECTIVE 1:

After reviewing the results, describe how the objective was validated by this experiment?

List the second Learning Objective:
OBJECTIVE 2:

After reviewing the results, describe how the objective was validated by this experiment?

List the third Learning Objective:
OBJECTIVE 3:

After reviewing the results, describe how the objective was validated by this experiment?

Conclusion:

If required, attach to this form: ☐ Answers to Questions, ☐ Tables, and ☐ Graphs.

CHAPTER 6

EXPERIMENT 6-8

ADDITIONAL SERIES-PARALLEL OPENS & SHORTS

OBJECTIVES

At the completion of this experiment, you will be able to:
- Determine the changes in circuit current and voltage drops resulting from a short circuit.
- Determine the changes in circuit current and voltage drops resulting from an open circuit.
- Predict changes in circuit current and voltage drops resulting from open and short circuits.

SUGGESTED READING

Chapter 6, *Basic Electronics,* Grob/Schultz, Tenth Edition

INTRODUCTION

A short circuit has practically zero resistance. Its effect, therefore, is to allow excessive current to flow in a circuit, although this is not usually intentional. An open circuit has the opposite effect because an open circuit has infinitely high resistance with practically zero current.

Therefore, if one path in a circuit changes (becomes open or short), the circuit's voltage, resistance, and current in the other paths change as well. For example, the series-parallel circuit shown in Fig. 6-8.1 becomes a series circuit when there is a short across circuit points A and B.

As an example of an open circuit, the series-parallel circuit in Fig. 6-8.2 becomes a series circuit with just R_1 and R_2 when there is an open between points A and B.

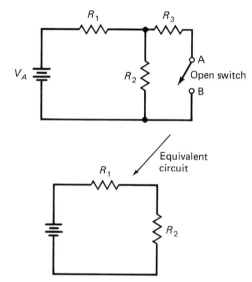

Fig. 6-8.2 Effects of an open in a series-parallel circuit.

The Short Circuit

You can determine the effect of a short in a series-parallel circuit. For example, in the circuit of Fig. 6-8.1, a switch is shown between points A and B. This switch represents a possible flaw in construction or operation of an actual circuit.

Refer to Fig. 6-8.3 where the circuit does not have a short. Specific component values have been assigned

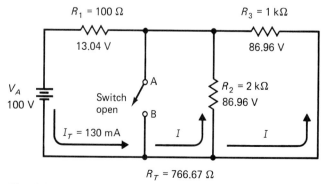

Fig. 6-8.3 Circuit values where no short exists.

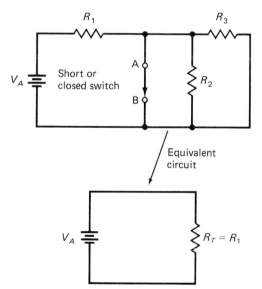

Fig. 6-8.1 Effect of a short in a series-parallel circuit.

to the same circuit as in Fig. 6-8.1. As shown in Fig. 6-8.3, $R_1 = 100\ \Omega$, $R_2 = 1000\ \Omega$, and $R_3 = 2000\ \Omega$. The total resistance R_T can be determined as follows:

$$R_T = R_1 + \frac{R_2 \times R_3}{R_2 + R_3}$$
$$= 766.67\ \Omega$$

Knowing the total resistance is essential in determining the total circuit current. If the applied voltage V_A is 100 V, then

$$I_T = \frac{V_A}{R_T}$$
$$= \frac{100\ \text{V}}{766.67\ \Omega}$$
$$= 0.13\ \text{A} \quad (\text{or } 130\ \text{mA})$$

Another component value that will be interesting to identify is the value of V_1. It can be calculated as

$$V_1 = I_T \times R_1$$
$$= 0.13\ \text{A} \times 100\ \Omega$$
$$= 13.04\ \text{V}$$

The value of V_2 can be found as well:

$$V_2 = V_A - V_1$$
$$= 100\ \text{V} - 13.04\ \text{V}$$
$$= 86.96\ \text{V}$$

The value of V_3 is then

$$V_3 = V_A - V_1$$
$$= 100\ \text{V} - 13.04\ \text{V}$$
$$= 86.96\ \text{V}$$

The voltage values of V_2 and V_3 should be equivalent since R_2 and R_3 form a parallel circuit.

Figure 6-8.3 shows the calculated effect of a short with a series-parallel circuit. If the circuit is shorted from point A to B, then the effects on resistors R_2 and R_3 are eliminated. The elimination of R_2 and R_3 creates predictable changes in circuit current and the voltage drop of R_1. The circuit of Fig. 6-8.4 shows the electrical effects. Since R_2 and R_3 are eliminated, the only effective resistance left in the circuit is R_1.

The total circuit current I_T can then be calculated as

$$I_T = \frac{V_A}{R_1}$$
$$= \frac{100\ \text{V}}{100\ \Omega}$$
$$= 1\ \text{A}$$

The voltage drop across R_1 is then

$$V_1 = I_T \times R_1$$
$$= 1\ \text{A} \times 100\ \Omega$$
$$= 100\ \text{V}$$

The voltage drop across R_2 is thus

$$V_2 = V_A - V_1$$
$$= 100\ \text{V} - 100\ \text{V}$$
$$= 0\ \text{V}$$

The voltage drop across R_3 is

$$V_3 = V_A - V_1$$
$$= 100\ \text{V} - 100\ \text{V}$$
$$= 0\ \text{V}$$

The voltage drops across R_2 and R_3 should equal 0 V since their resistance values are reduced to $0\ \Omega$ because of the short.

In summary, these circuit changes occur in two ways. First, the circuit currents increase significantly. Second, the voltage drop from the increase in current flow also increases.

The Open Circuit

An open circuit provides practically infinite resistance to the applied voltage V_A. Its overall effect on the circuit would be zero or minimal current flow. You can determine the effect of an open path in a series-parallel circuit. For example, in the circuit of Fig. 6-8.5,

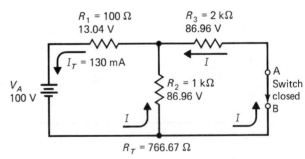

Fig. 6-8.5 Circuit values where no open exists.

component values have been assigned as $R_1 = 100\ \Omega$, $R_2 = 1000\ \Omega$, and $R_3 = 2000\ \Omega$. The total resistance R_T can be determined as follows:

$$R_T = R_1 + \frac{R_2 \times R_3}{R_2 + R_3}$$
$$= 766.67\ \Omega$$

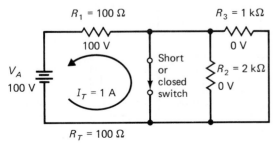

Fig. 6-8.4 Effects of a short.

Knowing the total resistance is essential in determining the total circuit current. If the applied voltage V_A is 100 V, then

$$I_T = \frac{V_A}{R_T}$$
$$= \frac{100 \text{ V}}{766.67 \text{ }\Omega}$$
$$= 0.13 \text{ A} \quad \text{(or 130 mA)}$$

Another component value that will be important to identify is the voltage value of V_1. It can be calculated as

$$V_1 = I_T \times R_1$$
$$= 0.13 \text{ A} \times 100 \text{ }\Omega$$
$$= 13.04 \text{ V}$$

The voltage of V_2 is

$$V_2 = V_A - V_1$$
$$= 100 \text{ V} - 13.04 \text{ V}$$
$$= 86.96 \text{ V}$$

The voltage of V_3 is

$$V_3 = V_A - V_1$$
$$= 100 \text{ V} - 13.04 \text{ V}$$
$$= 86.96 \text{ V}$$

The values are shown in Fig. 6-8.5.

If the circuit is open between points A and B, then R_3 is no longer part of the circuit. The removal of R_3 creates a predictable change in the circuit current and voltage drops of R_1 and R_2.

The circuit of Fig. 6-8.6 shows the overall electrical effects when the switch is opened between points A and B. The circuit current I_T can be calculated after the total resistance R_T is found:

$$R_T = R_1 + R_2$$
$$= 100 \text{ }\Omega + 1000 \text{ }\Omega$$
$$= 1100 \text{ }\Omega$$

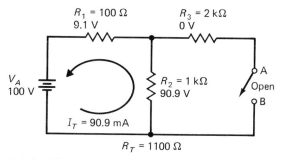

Fig. 6-8.6 Effects of an open.

Then I_T can be calculated as

$$I_T = \frac{V_A}{R_1}$$
$$= \frac{100 \text{ V}}{1100 \text{ }\Omega}$$
$$= 0.0909 \text{ A} \quad \text{(or 90.9 mA)}$$

The voltage drop across R_1 is then

$$V_1 = I_T \times R_1$$
$$= 0.0909 \text{ A} \times 100 \text{ }\Omega$$
$$= 9.09 \text{ or } 9.1 \text{ V}$$

The voltage drop across R_2 is then

$$V_2 = I_T \times R_2$$
$$= 0.0909 \text{ A} \times 1000 \text{ }\Omega$$
$$= 90.9 \text{ V}$$

The voltage value of R_3 is 0 V, due to the open circuit.

EQUIPMENT

Voltmeter
Power supply
Protoboard

COMPONENTS

(1) 150 Ω 1-W resistor
(2) 150 Ω 0.25 W resistors
(1) SPST switch

PROCEDURE

1. Measure and record in Table 6-8.1 the values of the three resistors used in this experiment.
2. Construct the circuit of Fig. 6-8.7, and adjust the voltage of the power supply to 20 V.

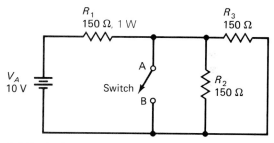

Fig. 6-8.7 Series-parallel circuit.

3. Calculate and record in Table 6-8.1, for the circuit of Fig. 6-8.7, the values of the total resistance R_T; the currents I_{R_1}, I_{R_2}, and I_{R_3}; and the voltages V_{R_1}, V_{R_2}, and V_{R_3}.
4. Measure and record in Table 6-8.1 the measured values of I_{R_1}, I_{R_2}, I_{R_3}, V_{R_1}, V_{R_2}, and V_{R_3}.
5. Connect points A and B.
6. Calculate and record in Table 6-8.1 the values of R_T, I_{R_1}, I_{R_2}, I_{R_3}, V_{R_1}, V_{R_2}, and V_{R_3}.
7. Measure and record in Table 6-8.1 the values of I_{R_1}, I_{R_2}, I_{R_3}, V_{R_1}, V_{R_2}, and V_{R_3}.
8. Turn off the power supply, and disconnect the circuit.
9. Measure and record in Table 6-8.2 the values of the three resistors used in this experiment.

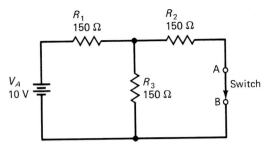

Fig. 6-8.8 Series-parallel circuit.

10. Construct the circuit of Fig. 6-8.8, and adjust the voltage of the power supply to 10 V.

11. Calculate and record in Table 6-8.2, for the circuit of Fig. 6-8.8, the values of the total resistance R_T; currents I_{R_1}, I_{R_2}, and I_{R_3}; and voltages V_{R_1}, V_{R_2}, and V_{R_3}.

12. Measure and record in Table 6-8.2 the measured values of I_{R_1}, I_{R_2}, I_{R_3}, V_{R_1}, V_{R_2}, and V_{R_3}.

13. Disconnect or open points A and B.

14. Calculate and record in Table 6-8.2 the values of R_T, I_{R_1}, I_{R_2}, I_{R_3}, V_{R_1}, V_{R_2}, and V_{R_3}.

15. Measure and record in Table 6-8.2 the values of I_{R_1}, I_{R_2}, I_{R_3}, V_{R_1}, V_{R_2}, and V_{R_3}.

16. Turn off the power supply and disconnect the circuit.

QUESTIONS FOR EXPERIMENT 6-8

1. What are the characteristics of a shorted circuit?

2. What are the characteristics of an open circuit?

3. Compare Tables 6-8.1 and 6-8.2. What are the differences and similarities?

TABLES FOR EXPERIMENT 6-8

TABLE 6–8.1 ($V_A = 10$ V)

Component	Measured	Calculated
Unshorted Circuit, Fig. 6–8.7		
R_1	_____	
R_2	_____	
R_3	_____	
R_T		_____
I_{R_1}	_____	_____
I_{R_2}	_____	_____
I_{R_3}	_____	_____
V_{R_1}	_____	_____
V_{R_2}	_____	_____
V_{R_3}	_____	_____
Shorted Circuit, Fig. 6–8.7		
R_T		_____
I_{R_1}	_____	_____
I_{R_2}	_____	_____
I_{R_3}	_____	_____
V_{R_1}	_____	_____
V_{R_2}	_____	_____
V_{R_3}	_____	_____

TABLE 6–8.2 ($V_A = 10$ V)

Component	Measured	Calculated
Shorted Circuit, Fig. 6–8.8		
R_1	_____	
R_2	_____	
R_3	_____	
R_T		_____
I_{R_1}	_____	_____
I_{R_2}	_____	_____
I_{R_3}	_____	_____
V_{R_1}	_____	_____
V_{R_2}	_____	_____
V_{R_3}	_____	_____
Opened Circuit, Fig. 6–8.8		
R_T		_____
I_{R_1}	_____	_____
I_{R_2}	_____	_____
I_{R_3}	_____	_____
V_{R_1}	_____	_____
V_{R_2}	_____	_____
V_{R_3}	_____	_____

EXPERIMENT RESULTS REPORT FORM

Experiment No: _____ Name: _____
                                      Date: _____
Experiment Title: _____ Class: _____
_____ Instr: _____

Explain the purpose of the experiment:

List the first Learning Objective:
OBJECTIVE 1:

After reviewing the results, describe how the objective was validated by this experiment?

List the second Learning Objective:
OBJECTIVE 2:

After reviewing the results, describe how the objective was validated by this experiment?

List the third Learning Objective:
OBJECTIVE 3:

After reviewing the results, describe how the objective was validated by this experiment?

Conclusion:

If required, attach to this form: ☐ Answers to Questions, ☐ Tables, and ☐ Graphs.

CHAPTER 7

EXPERIMENT 7-1

VOLTAGE DIVIDERS WITH LOADS

OBJECTIVES

At the completion of this experiment, you will be able to:
- Identify a voltage divider circuit.
- Define the purpose for voltage divider circuits.
- Describe voltage divider loading effects.

SUGGESTED READING

Chapter 7, *Basic Electronics,* Grob/Schultz, Tenth Edition

INTRODUCTION

A series circuit is also a voltage divider. That is, the total voltage applied to a series circuit is divided among the series resistors. Because the current is the same value in all parts of a series circuit, voltage is divided among the series resistances in direct proportion to the value of resistance.

For example, imagine four resistors in series. Each resistor is equal in value. Therefore, any one resistor will receive one-fourth the total applied voltage. In other words, the total voltage is divided by 4 across each resistor.

Another way to determine the voltage across any resistor in series, without using a meter, is to add the total resistance, divide that value into any single resistor in series, and multiply by the total voltage. This is the proportional method. For example, see Fig. 7-1.1. There,

$$V_{R_2} = \frac{R_2}{R_T} \times V_T = \frac{2 \text{ k}\Omega}{6 \text{ k}\Omega} \times 6 \text{ V} = 2 \text{ V}$$

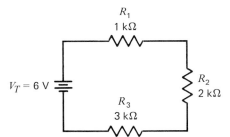

Fig. 7-1.1 Proportional method circuit.

Although these simple voltage dividers are limited in their use, when a load is placed across any series re-sistance, the voltage divider is then extremely useful as a voltage tap. For example, the circuit in Fig. 7-1.2 shows a parallel load current through R_{load}.

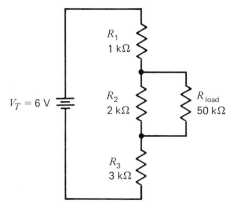

Fig. 7-1.2 Simple voltage divider with parallel load.

This parallel load, if its resistance was more than 10 times the value of R_2, would actually be sharing the *IR* voltage drops across R_2 without changing the division of voltage considerably. For example, if $R_{\text{load}} = 50 \text{ k}\Omega$,

$$\frac{R_2 \times R_{\text{load}}}{R_2 + R_{\text{load}}} = \frac{100 \times 10^6}{50 \times 10^3} = 2 \text{ k}\Omega \text{ (approx.)}$$

Thus, R_{load} could share the same approximate voltage as R_2, but it would have its own current.

While voltage dividers are mainly used to tap off part of a total voltage, it is necessary to remember that the addition of a load will always have some effect upon the circuit current and, many times, the proportional *IR* voltage drops.

EQUIPMENT

DC power supply, 0–10 V
Voltmeter (VTVM, VOM)
Test leads

COMPONENTS

Resistors (all 0.25 W):

(3) 1 kΩ   (1) 47 kΩ
(3) 10 kΩ   (1) 4.7 kΩ
(1) 470 Ω   (1) 100 kΩ

PROCEDURE

1. Connect the circuit of Fig. 7-1.3. Do not connect R_{load}.

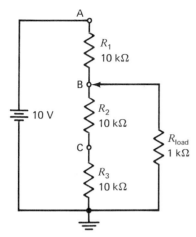

Fig. 7-1.3 Voltage divider.

2. Calculate the *IR* voltage drop across each resistor without the load in the circuit. Use the proportional method. Show the calculations on a separate sheet and record the results in Table 7-1.1. Also, measure total circuit current and record the results in Table 7-1.1.

3. With the load connected to point B, measure the total circuit current, load current, and bleeder current. The bleeder current is the steady drain on the source, the current through R_3. Record the values in Table 7-1.2.

4. Measure the voltages across R_1, R_2, R_3, and R_{load}. Record these values in Table 7-1.2.

5. Connect R_{load} to point A. Repeat steps 3 and 4. Record the results.

6. Connect R_{load} to point C. Repeat steps 3 and 4. Record the results.

7. Change the value of R_{load} to 100 kΩ and repeat steps 3 to 6.

8. Connect the circuit of Fig. 7-1.4. This is a voltage divider with loads.

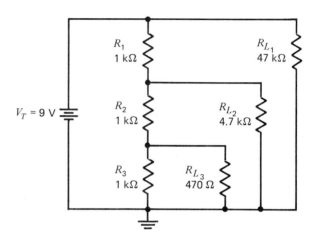

Fig. 7-1.4 Complex voltage divider.

9. Measure and record the *IR* voltage drops across R_1, R_2, R_3, R_{L_1}, R_{L_2}, and R_{L_3}.

10. Calculate the current through each resistor and the total circuit current. Record the results in Table 7-1.3.

NAME	DATE

QUESTIONS FOR EXPERIMENT 7-1

1. Explain what is meant by the term *voltage divider*.

2. Refer to the circuit of Fig. 7-1.4 (loaded voltage divider). Explain what would happen to the total circuit current, voltage, and resistance if R_{L_2} and R_{L_3} were removed.

3. If Fig. 7-1.4 were used to tap the total voltage of 9 V into three equal parts (without the loads), explain why V_{R_1}, V_{R_2}, and V_{R_3} (with respect to ground) would not be 3 V each. Does the value of any one load greatly affect the original unloaded divider?

4. Redraw Fig. 7-1.4 and show the path of current flow by drawing arrows where necessary.

5. What effect would reversing the battery polarity have on the circuit of Fig. 7-1.4?

TABLES FOR EXPERIMENT 7-1

TABLE 7–1.1   No Load Values for Fig. 7-1.3

	Calculated IR Drop	Calculated Current
R_1	_____	
R_2	_____	
R_3	_____	
I_T		_____

TABLE 7-1.2 Measured Values for Fig. 7-1.3, Circuit under Load

$R_L =$ 1 kΩ	Load at Point A		Load at Point B		Load at Point C	
	IR Drops	Current	IR Drops	Current	IR Drops	Current
R_1	_____		_____		_____	
R_2	_____		_____		_____	
R_3	_____		_____		_____	
R_{load}	_____		_____		_____	
I_T		_____		_____		_____
I_B		_____		_____		_____
$R_L =$ 100 kΩ						
R_1	_____		_____		_____	
R_2	_____		_____		_____	
R_3	_____		_____		_____	
R_{load}	_____		_____		_____	
I_T		_____		_____		_____
I_B		_____		_____		_____

TABLE 7-1.3 Circuit Values for Fig. 7-1.4

	IR Drop, Measured	Current, Calculated*
R_1	_____	_____
R_2	_____	_____
R_3	_____	_____
R_{L_1}	_____	_____
R_{L_2}	_____	_____
R_{L_3}	_____	_____
R_T	_____	_____

*V (IR drop measured)/R nominal = calculated current.

EXPERIMENT RESULTS REPORT FORM

Experiment No: _____   Name: _____
                                  Date: _____
Experiment Title: _____ Class: _____
_____   Instr: _____

Explain the purpose of the experiment:

List the first Learning Objective:
OBJECTIVE 1:

After reviewing the results, describe how the objective was validated by this experiment?

List the second Learning Objective:
OBJECTIVE 2:

After reviewing the results, describe how the objective was validated by this experiment?

List the third Learning Objective:
OBJECTIVE 3:

After reviewing the results, describe how the objective was validated by this experiment?

Conclusion:

If required, attach to this form: ☐ Answers to Questions, ☐ Tables, and ☐ Graphs.

EXPERIMENT 7-1

CHAPTER 7

EXPERIMENT 7-2

CURRENT DIVIDERS

OBJECTIVES

At the completion of this experiment, you will be able to:
- Understand how parallel circuits act as current dividers.
- Be familiar with the proportional method for solving branch currents.
- Be familiar with the parallel conductance method for solving branch currents.

SUGGESTED READING

Chapter 7, *Basic Electronics,* Grob/Schultz, Tenth Edition

INTRODUCTION

In the same way that series circuits are also voltage dividers, parallel circuits are also current dividers. That is, the total current is divided among the parallel branches in inverse proportion to the resistance in any branch. Therefore, main-line current increases as branches are added.

To find the branch currents without knowing the total voltage across the bank, a formula can be used. This formula is a proportional method for solving unknown branch currents. For example, the branch currents for Fig. 7-2.1 can be found by using this proportional method. It is based on the fact that currents divide in inverse proportion to their resistances.

Notice two things in Fig. 7-2.1. First, the numerator for each branch resistance is the value of the "opposite" branch resistance. Second, it is only necessary to calculate one branch current and subtract it from the total current. The remainder will be the other branch current.

Another method for determining branch current division is the parallel conductance method. Remember that conductance $G = 1/R$. Note that conductance and current are directly proportional. This is true because the greater the resistance, the less will be the current. With any number of parallel branches, each branch current can be calculated without knowing the voltage across the bank. The formula is

$$I_x = \frac{G_x}{G_T} \times I_T$$

For example, the branch current I_{R_1} in Fig. 7-2.2 can be found as follows:

$$I_{R_1} = \frac{G_1}{G_T} \times I_T$$

$$G_1 = \frac{1}{R_1} = \frac{1}{10\,\Omega} = 0.1\text{ S}$$

Note: Siemens (S) is the reciprocal of ohms.

$$R_T = \frac{1}{1/R_1 + 1/R_2 + 1/R_3}$$
$$= 6.25\,\Omega$$
$$G_T = \frac{1}{R_T} = \frac{1}{6.25\,\Omega} = 0.16\text{ S}$$
$$I_{R_1} = \frac{0.1}{0.16} \times 40\text{ mA}$$
$$= 25\text{ mA}$$

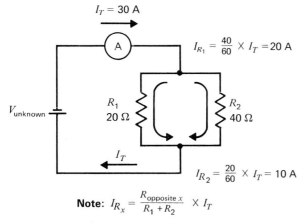

Fig. 7-2.1 Two-branch current divider.

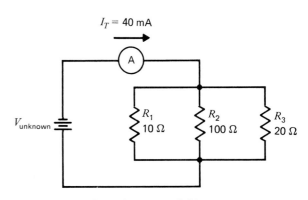

Fig. 7-2.2 Three-branch current divider.

Because $I_{R_1} = 25$ mA, it is easy to see that the remaining currents will be

$$I_{R_2} + I_{R_3} = +40 \text{ mA} - 25 \text{ mA}$$
$$= 15 \text{ mA}$$

Thus, the previous formula (proportional method) can be used to solve for the remaining two branch currents.

Another method is to find the total conductance and use the following formula to solve for each branch current in Fig. 7-2.2:

$$G_1 = \frac{1}{R_1} = \frac{1}{10 \text{ }\Omega} = 0.10 \text{ S}$$
$$G_2 = \frac{1}{R_2} = \frac{1}{100 \text{ }\Omega} = 0.01 \text{ S}$$
$$G_3 = \frac{1}{R_3} = \frac{1}{20 \text{ }\Omega} = 0.05 \text{ S}$$

Therefore, the total conductance is

$$G_1 + G_2 + G_3 = 0.1 + 0.01 + 0.05 = 0.16 \text{ S}$$

To calculate branch currents, use the following formula:

$$I_{R_x} = \frac{G_{R_x}}{G_T} \times I_T$$

In the case of Fig. 7-2.2,

$$I_{R_1} = {}^{10}\!/_{16} \times 40 \text{ mA} = +25.0 \text{ mA}$$
$$I_{R_2} = {}^{1}\!/_{16} \times 40 \text{ mA} = + 2.5 \text{ mA}$$
$$I_{R_3} = {}^{5}\!/_{16} \times 40 \text{ mA} = +12.5 \text{ mA}$$
$$I_T = +40.0 \text{ mA}$$

EQUIPMENT

DC power supply   VTVM/VOM
Ammeter   Leads
Protoboard or springboard

COMPONENTS

Resistors (all 0.25 W unless indicated otherwise):

(1) 100 Ω, 1 W   (1) 560 Ω
(1) 150 Ω   (1) 820 Ω
(1) 390 Ω

PROCEDURE

1. Refer to the circuit of Fig. 7-2.3. Calculate R_T. Record in Table 7-2.1.
2. Assume that $I_T = 100$ mA. Calculate the current through R_1 and R_2 by using the proportional method for two branches in parallel. Show your calculations and record the results in Table 7-2.1.

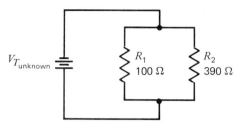

Fig. 7-2.3 Two-branch current divider.

3. Using the values you calculated for I_{R_1} and I_{R_2}, calculate the total voltage across the parallel bank (use Ohm's law). Show your calculations on a separate sheet, and record the value of V_T in Table 7-2.1.
4. Connect the circuit of Fig. 7-2.3. Calculate R_T.
5. Add another resistor, $R_3 = 820$ Ω, to the parallel bank and apply the total voltage you calculated in step 3 to the circuit.
6. Measure the total current and the current through each branch. Record the results in Table 7-2.2.
7. Connect the circuit of Fig. 7-2.4.
8. Adjust the voltage so that total current = 30 mA.
9. Use the parallel conductance formula to determine the branch conductances G_{R_2} to G_{R_5}. Also,

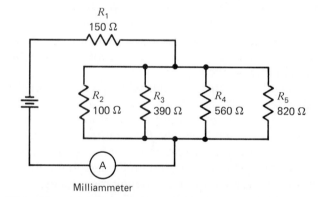

Fig. 7-2.4 Four-branch current divider.

measure the current through each branch and record the results in Table 7-2.3. Show all your calculations and record the results in Table 7-2.3.

Note: Do not measure *any* voltages. Doing so will defeat the purpose of this experiment.

10. Using the values you have determined for the circuit of Fig. 7-2.4, calculate the voltages across R_1 and the bank. Remember, do not measure these voltages. How does the calculated voltage compare to the applied voltage? (Answer in your report.)

QUESTIONS FOR EXPERIMENT 7-2

Answer TRUE (T) or FALSE (F) to the following:

_____ 1. Series circuits divide current, and parallel circuits divide voltages.

_____ 2. Conductance G is the reciprocal of branch current.

_____ 3. Refer to Fig. 7-2.4. If another resistor, $R_6 = 100\ \Omega$, were added in parallel to the bank, the voltage across the bank would increase.

_____ 4. Refer to Fig. 7-2.4. If the series resistor R_1 were short-circuited, the total current would decrease.

_____ 5. Refer to Fig. 7-2.4. If R_3 and R_4 were opened, the voltage across the series resistor R_1 would decrease.

_____ 6. Refer to Fig. 7-2.4. If the total voltage were halved and the total circuit resistance doubled, there would be no effect upon total current.

_____ 7. Refer to Fig. 7-2.4. If R_1 were opened, the total circuit current would increase.

_____ 8. The voltage across any parallel bank is increased as the total conductance of the bank is increased.

_____ 9. The total current in a parallel bank is inversely proportional to its conductance.

_____ 10. The total current in a parallel bank is directly proportional to its total resistance.

TABLES FOR EXPERIMENT 7-2

TABLE 7–2.1

	Nominal R, Ω	Calculated I, mA	Total Calculated V, V
R_1	100		
R_2	390		
R_T			

TABLE 7–2.1

	Nominal R, Ω	Measured I, mA	From Table 7–2.1 V, V
R_1	100		
R_2	390		
R_2	820		
R_T			

TABLE 7–2.3

	Nominal R, Ω	Calculated G, S	Calculated I, mA	Calculated V, V*	Measured I, mA
R_1	150		30.0		30.0
R_2	100				
R_3	390				
R_4	560				
R_5	820				
R_T			30.0		3.0

Note: Calculate R_T.

*$V_x = I$ calculated $\times R$ nominal

EXPERIMENT RESULTS REPORT FORM

Experiment No: _____   Name: _____
                                         Date: _____
Experiment Title: _____   Class: _____
_____   Instr: _____

Explain the purpose of the experiment:

List the first Learning Objective:
OBJECTIVE 1:

After reviewing the results, describe how the objective was validated by this experiment?

List the second Learning Objective:
OBJECTIVE 2:

After reviewing the results, describe how the objective was validated by this experiment?

List the third Learning Objective:
OBJECTIVE 3:

After reviewing the results, describe how the objective was validated by this experiment?

Conclusion:

If required, attach to this form: ☐ Answers to Questions, ☐ Tables, and ☐ Graphs.

CHAPTER 7

EXPERIMENT 7-3

POTENTIOMETERS AND RHEOSTATS AS DIVIDERS

OBJECTIVES

At the completion of this experiment, you will be able to:
- Identify the circuit configuration of a potentiometer.
- Identify the circuit configuration of a rheostat.
- Contrast the differences and similarities of potentiometers and rheostats used in divider circuits.

SUGGESTED READING

Chapters 2 and 7, *Basic Electronics*, Grob/Schultz, Tenth Edition

INTRODUCTION

Potentiometers and rheostats are variable resistances and are used to vary voltage and current in a circuit. A rheostat is a two-terminal device. The potentiometer is a three-terminal device, as shown in Fig. 7-3.1.

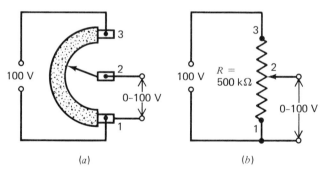

Fig. 7-3.1 Potentiometer connected across voltage source to function as a voltage divider. (*a*) Wiring diagram. (*b*) Schematic diagram.

The maximum resistance is seen between the two end terminals. The middle terminal mechanically adjusts and taps a proportion of this total resistance. A potentiometer can be used as a rheostat by connecting one end terminal to the other, as shown in Fig. 7-3.2.

The primary purpose of a potentiometer (pot) is to tap off a variable voltage from a voltage source, as shown in Fig. 7-3.3.

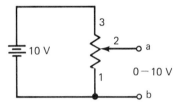

Fig. 7-3.3 Variable voltage source.

As pin 2 is rotated up toward pin 3, the voltage at ab increases until a 10 V level is achieved. If pin 2 is rotated downward toward pin 1, then the voltage present at ab decreases to zero, or approximately zero. The variance of voltages may appear to be presented in a linear or a nonlinear fashion, depending upon the manufacturer's type of potentiometer.

The primary purpose of a rheostat is to vary current though a load. This is accomplished by locating the rheostat in series with the load and source voltage. In this way, the total resistance R_T can be varied and indirectly vary the total current. This circuit configuration is shown in Fig. 7-3.4.

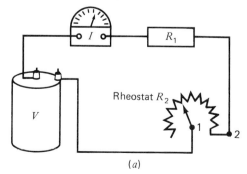
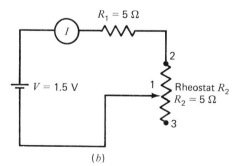

Fig. 7-3.2 Rheostat connected in series circuit to vary the current. (*a*) Wiring diagram with ammeter to measure *I*. (*b*) Schematic diagram.

EXPERIMENT 7-3 163

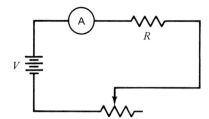

Fig. 7-3.4 Rheostat circuit to vary current.

In summary, rheostats are:

Two-terminal devices
Found in series with loads and voltage sources
Used to vary total current

Potentiometers are:

Three-terminal devices
Found to have end terminals connected across voltage sources
Used to tap off part of the voltage source

EQUIPMENT

DC power supply, 0–10 V
Voltmeter
Ammeter
Ohmmeter
Protoboard
Test leads

COMPONENTS

(1) 100 Ω, 1 W resistor
(1) 1 kΩ, 1 W potentiometer, linear taper

PROCEDURE

1. Connect the circuit shown in Fig. 7-3.5, where $V = 10$ V, $R_1 = 100$ Ω, and R_2 (pot) $= 1$ kΩ.

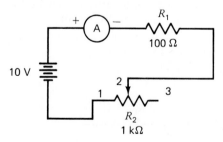

Fig. 7-3.5 Rheostat controlling current.

2. Turn on the power supply and adjust R_2 for a minimum resistance value (maximum I).

3. Remove R_2 from the circuit and connect it to an ohmmeter (points 1 and 2). Set R_2 to 100 Ω and reconnect R_2 to the circuit. Measure the current flow and record in Table 7-3.1.

4. Repeat step 3 in 100-Ω increments (100, 200, 300 Ω, etc.) up to 1000 Ω.

5. Make a graph of resistance (horizontal axis) versus current (vertical axis) from Table 7-3.1.

6. Connect the circuit shown in Fig. 7-3.6, where $V = 10$ V and R_2 (pot) $= 1$ kΩ.

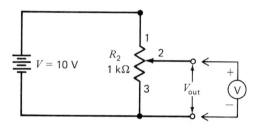

Fig. 7-3.6 Potentiometer voltage divider.

7. Turn on the power supply and adjust R_2 for a minimum resistance value (minimum V).

8. Remove R_2 from the circuit and connect it to an ohmmeter (points 2 and 3). Set R_2 to 100 Ω, and reconnect R_2 to the circuit. Measure the voltage drop across R_2 (pins 2 and 3). Record the results in Table 7-3.2.

9. Repeat step 8 in 100 Ω increments (100, 200, 300 Ω, etc.) up to 1000 Ω.

10. Make a graph of resistance (horizontal axis) versus voltage (vertical axis) from Table 7-3.2.

QUESTIONS FOR EXPERIMENT 7-3

1. How many circuit connections to a potentiometer are needed?

2. How many circuit connections to a rheostat are needed?

3. Determine maximum power consumption from the graphs you completed in steps 5 and 10. What are the actual necessary wattages of R_1 and R_2?

TABLES FOR EXPERIMENT 7-3

TABLE 7–3.1

Resistance, Ω	Measured Current
100	
200	
300	
400	
500	
600	
700	
800	
900	
1000	

TABLE 7–3.2

Resistance, Ω	Measured Voltage
100	
200	
300	
400	
500	
600	
700	
800	
900	
1000	

GRAPH FOR EXPERIMENT 7-3

Graph 7-3.1 Graphing results from table 7-3.2 where resistance is located on the *horizontal axis* versus voltage on the *vertical axis*.

EXPERIMENT RESULTS REPORT FORM

Experiment No: _____   Name: _____
                                 Date: _____
Experiment Title: _____ Class: _____
_____ Instr: _____

Explain the purpose of the experiment:

List the first Learning Objective:
OBJECTIVE 1:

After reviewing the results, describe how the objective was validated by this experiment?

List the second Learning Objective:
OBJECTIVE 2:

After reviewing the results, describe how the objective was validated by this experiment?

List the third Learning Objective:
OBJECTIVE 3:

After reviewing the results, describe how the objective was validated by this experiment?

Conclusion:

If required, attach to this form: ☐ Answers to Questions, ☐ Tables, and ☐ Graphs.

CHAPTER 7

EXPERIMENT 7-4

VOLTAGE DIVIDER DESIGN

OBJECTIVES

At the completion of this experiment, you will be able to:
- Design a voltage divider for given load requirements.
- Understand the concept of negative voltage.
- Understand the concepts of common ground and bleeder current.

SUGGESTED READING

Chapter 7, *Basic Electronics,* Grob/Schultz, Tenth Edition

INTRODUCTION

This experiment requires that you design and test a loaded voltage divider circuit. You can imagine that you are actually designing this divider for the output of a power supply in any piece of electronic equipment.

Often, designing a circuit teaches you more than simply measuring and analyzing data.

In previous experiments, you became familiar with the concepts of voltage and current dividers. You may recall that the loads on a voltage divider are also current dividers because they are parallel circuits.

To review, consider the circuit of Fig. 7-4.1.

1. *Loads*: The loads are resistances that require specific voltages and, often, specific currents. Here, imagine each load is a separate circuit board that requires certain voltages and currents in order to properly perform a function such as amplification.
2. I_B: This is known as *bleeder current*. It is a steady drain on the source and has no load currents passing through it. Typically, bleeder currents are calculated to be 10 percent of the total current.
3. V_T: Total voltage is fixed. It is like a budget that the designer is allowed to work with.
4. *Taps*: These are the places where the loads are connected in order to tap off their specified voltages.
5. R_1, R_2, R_3: These are the voltage divider resistances that you, the designer, will be determining by your calculations. In this case, Fig. 7-4.1 has a given budget of 50 V (total). Your calculations will divide this voltage into the necessary values to supply the loads.
6. *Ground*: This symbol shows a common or earth ground that is connected to one side of the power supply. Remember, ground is a reference point, like zero.

How to Calculate the Voltage Divider

Use a table, as shown in Table 7-4.1, in order to keep your values organized. Find the current in each R.

$$I_{R_1} = I_B = \text{approx 10 percent of } I_T \text{ loads}$$

Thus

$$I_B = 150 \text{ mA} \times 0.1 = 15 \text{ mA}$$

$$I_{\text{Load A}} = 100 \text{ mA}$$
$$I_{\text{Load B}} = 30 \text{ mA}$$
$$I_{\text{Load C}} = 20 \text{ mA}$$
$$I_T = 150 \text{ mA}$$

Knowing the value of $I_B = I_{R_1}, I_{R_2}$ and I_{R_3} are calculated as follows (compare to Fig. 7-4.1):

$$I_{R_2} = I_B + I_{\text{Load C}} = 15 \text{ mA} + 20 \text{ mA} = 35 \text{ mA}$$
$$I_{R_3} = I_{R_2} + I_{\text{Load B}} = 35 \text{ mA} + 30 \text{ mA} = 65 \text{ mA}$$

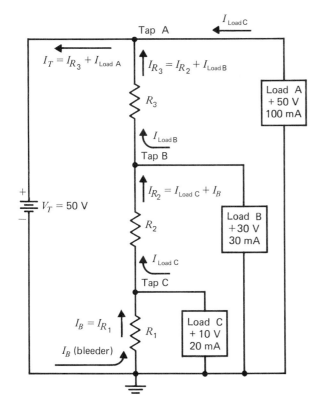

Fig. 7-4.1 Loaded voltage divider.

EXPERIMENT 7-4 **169**

How to Calculate the Voltage Across Each Resistor

The voltages at the taps are also the voltages across the loads with respect to ground. However, the voltages across R_2 and R_3 are not in parallel only with these loads. Therefore, only the voltage across R_1 is the same as $V_{\text{Load C}} = 10$ V. V_{R_2} and V_{R_3} are calculated as follows (compare to Fig. 7-4.1):

$$V_{R_1} = V_{\text{Tap C}} = 10 \text{ V}$$
$$V_{R_2} = V_{\text{Tap B}} - V_{\text{Tap C}} = 30 \text{ V} - 10 \text{ V} = 20 \text{ V}$$
$$V_{R_3} = V_{\text{Tap A}} - V_{\text{Tap B}} = 50 \text{ V} - 30 \text{ V} = 20 \text{ V}$$

How to Calculate Each Resistor

Now that the voltages and currents for each resistor have been determined, it is easy to use Ohm's law to calculate the value of each resistor:

$$R_1 = \frac{V_{R_1}}{I_{R_1}} = \frac{10 \text{ V}}{15 \text{ mA}} = 666.7 \text{ } \Omega$$
$$R_2 = \frac{V_{R_2}}{I_{R_2}} = \frac{20 \text{ V}}{35 \text{ mA}} = 571.4 \text{ } \Omega$$
$$R_3 = \frac{V_{R_3}}{I_{R_3}} = \frac{20 \text{ V}}{65 \text{ mA}} = 307.7 \text{ } \Omega$$

These values of voltage divider resistance should provide the specified load requirements.

EQUIPMENT

DC power supply
Ohmmeter
Ammeter (VOM) optional
Voltmeter (VOM/VTVM)
Protoboard or springboard
Leads

COMPONENTS

Resistors (all 0.25 W):

(1) 150 Ω   (3) 470 Ω

PROCEDURE

1. Design a voltage divider, similar to Fig. 7-4.1, with the following specifications:

$$V_T = 20 \text{ V}$$
Load A = 20 V, 40 mA
Load B = 15 V, 25 mA
Load C = 5 V, 10 mA

2. On a separate sheet, draw the circuit similar to Fig. 7-4.1 and fill in Table 7-4.1 (columns 1, 2, and 4).

3. Verify that your circuit works, by connecting the circuit and measuring the voltages across R_1, R_2, R_3, and each load. Use resistors within 10 percent of the load values. Record the voltages (measured) in Table 7-4.1 (column 3). If any value of voltage falls outside plus or minus 20 percent of your calculated values, redesign the circuit.

4. With your circuit connected, remove earth ground (if attached).

5. Using the circuit of Fig. 7-4.2 as an example, you will move the ground point (common: without an earth ground connection) to point C. Although no actual connection is yet made, point C now becomes the zero reference point.

Note: $V_{R_3} = V_{\text{Point A}} - V_{R_2}$.

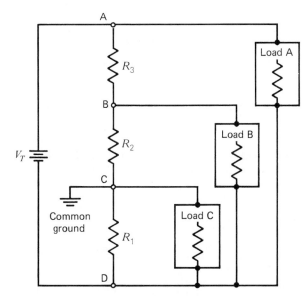

Fig. 7-4.2 Loaded voltage divider. Note that the common ground is shown as an earth (symbol) ground here, but it is not actually a true earth ground.

6. Measure the voltage across each resistor and load with respect to common. This means connect the negative (common) lead of the voltmeter (VTVM or VOM) to point C and measure from there. Note that the voltage across R_1 will now become a negative voltage. Record the results in Table 7-4.2.

7. Move common to point B and measure each voltage. Remember to subtract the voltage across R_2 when you measure V_{R_1}. Record the results in Table 7-4.2.

Note: The loads (A, B, C) are always measured from the original common (point D) in all three circuits. But the divider voltages are measured from points B, C, or D, depending on the configurations.

QUESTIONS FOR EXPERIMENT 7-4

1. Explain the difference between earth ground and a common reference point ground.

2. Explain how a negative voltage can be obtained from a voltage divider. Explain the effects upon I, V, and R, if any.

3. Explain, in your own words, what is meant by *bleeder current*.

4. Explain the difference, if any, between a loaded voltage divider and a series-parallel circuit.

5. Explain how the circuit of Fig. 7-4.1 would be affected if:
   A. R_2 were short-circuited   **B.** R_2 were opened

TABLES FOR EXPERIMENT 7-4

TABLE 7–4.1

Divider	(1) Calculated R Ω	(2) Actual R Value, Ω	(3) Measured V, V	(4) Calculated I, mA
Load A				
Load B				
Load C				
R_1				
R_2				
R_3				

TABLE 7–4.2

Divider	Measured V, V Common = Point C	Measured V, V Common = Point B
Load A		
Load B		
Load C		
R_1		
R_2		
R_3		

EXPERIMENT RESULTS REPORT FORM

Experiment No: _____   Name: _____
                                  Date: _____
Experiment Title: _____   Class: _____
_____   Instr: _____

Explain the purpose of the experiment:

List the first Learning Objective:
OBJECTIVE 1:

After reviewing the results, describe how the objective was validated by this experiment?

List the second Learning Objective:
OBJECTIVE 2:

After reviewing the results, describe how the objective was validated by this experiment?

List the third Learning Objective:
OBJECTIVE 3:

After reviewing the results, describe how the objective was validated by this experiment?

Conclusion:

If required, attach to this form: ☐ Answers to Questions, ☐ Tables, and ☐ Graphs.

EXPERIMENT 7-4

CHAPTER 8

EXPERIMENT 8-1

ANALOG AMMETER DESIGN

OBJECTIVES

At the completion of this experiment, you will be able to:
- Determine the internal resistance of a basic D'Arsonval meter movement.
- Design an ammeter circuit from this meter movement.
- Use this ammeter for actual ammeter measurements.

SUGGESTED READING

Chapter 8, *Basic Electronics*, Grob/Schultz, Tenth Edition

INTRODUCTION

Range of an Ammeter

The small size of the wire with which an ammeter's movable coil is wound places severe limits on the current that may be passed through the coil. Consequently, the basic D'Arsonval movement may be used to indicate or measure only very small currents—for example, microamperes or milliamperes, depending on meter sensitivity.

To measure a larger current, a shunt must be used with the meter. A shunt is a heavy, low-resistance conductor connected across the meter terminals to carry most of the load current. This shunt has the correct amount of resistance to cause only a small part of the total circuit current to flow through the meter coil. The meter current is proportional to the load current. If the shunt is of such a value that the meter is calibrated in milliamperes, the instrument is called a *milliammeter*. If the shunt is of such a value that the meter is calibrated in amperes, it is called an *ammeter*.

A single type of standard meter movement is generally used in all ammeters, no matter what the range of a particular meter. For example, meters with working ranges of 0 to 10 A, 0 to 5 A, or 0 to 1 A all use the same galvanometer movement. The designer of the ammeter calculates the correct shunt resistance required to extend the range of the meter movement to measure any desired amount of current. This shunt is then connected across the meter terminals. Shunts may be located inside the meter case (internal shunt) or somewhere away from the meter (external shunt), with leads going to the meter.

Extending the Range by Use of Shunts

For limited current ranges (below 50 A), internal shunts are most often employed. In this manner, the range of the meter may be easily changed by selecting the correct internal shunt having the necessary current rating. Before the required resistance of the shunt for each range can be calculated, the resistance of the meter movement must be known.

For example, suppose it is desired to use a 100 μA D'Arsonval meter having a resistance of 100 Ω to measure line currents up to 1 A. The meter deflects full scale when the current through the 100 Ω coil is 100 μA. Therefore, the voltage drop across the meter coil is IR, or

$$0.0001 \times 100 = 0.01 \text{ V}$$

Because the shunt and coil are in parallel, the shunt must also have a voltage drop of 0.01 V. The current that flows through the shunt is the difference between the full-scale meter current and the line current. In this case, the meter current is 0.0001 A. This current is negligible compared with the line (shunt) current, so the shunt current is approximately 1 A. The resistance R_S of the shunt is therefore

$$R_S = \frac{V}{I} = \frac{0.01}{1} = 0.01 \text{ }\Omega \text{ (approx.)}$$

and the range of the 100 μA meter has been increased to 1 A by paralleling it with the 0.01 Ω shunt.

The 100 μA instrument may also be converted to a 10 A meter by the use of a proper shunt. For full-scale deflection of the meter, the voltage drop V across the shunt (and across the meter) is still 0.01 V. The meter current is again considered negligible, and the shunt current is now approximately 10 A. The resistance R_S of the shunt is therefore

$$R_S = \frac{V}{I} = \frac{0.01}{10} = 0.001 \text{ }\Omega$$

The same instrument may likewise be converted to a 50 A meter by the use of the proper type of shunt. The current I_S through the shunt is approximately 50 A, and the resistance R_S of the shunt is

$$R_S = \frac{V}{I_S} = \frac{0.01}{50} = 0.0002 \text{ }\Omega$$

EXPERIMENT 8-1 175

EQUIPMENT

DC power supply
Protoboard or springboard
Leads
VTVM or DVM

COMPONENTS

(1) 0 to 1 mA meter movement

Resistors:

    (1) 150 kΩ 0.25 W   Other resistors as calculated.
    (1) 470 Ω 0.25 W

Potentiometers:

    (1) 5 kΩ   (1) 100 kΩ

(1) SPST switch
(1) SPDT switch

PROCEDURE

1. Measure the internal resistance of the meter movement by connecting the circuit shown in Fig. 8-1.1.

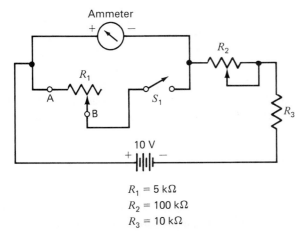

$R_1 = 5\ k\Omega$
$R_2 = 100\ k\Omega$
$R_3 = 10\ k\Omega$

Fig. 8-1.1 Internal resistance measurement.

2. With S_1 open, turn on the power supply and adjust it for 10 V.

3. Adjust R_2 so that the scale upon the meter movement reads at full-scale deflection.

4. Close S_1 and adjust R_1 so that the scale upon the meter movement reads at half-scale deflection. The currents will evenly divide between R_1 and the internal resistance r_m of the meter movement when $R_1 = r_m$.

5. Measure and record in Table 8-1.1 the voltage dropped across V_m.

6. Measure and record r_m in Table 8-1.1 by turning the power supply off, disconnecting R_1 from the circuit, and measuring from point A to B. At this point, $r_m = R_1$.

7. Calculate and record I_m in Table 8-1.1, where

$$I_m = \frac{V_m}{r_m}$$

8. Record in Table 8-1.2 the value of I_m for full-scale deflection, the r_m (meter movement's internal resistance) for the meter movement, and the value V_m needed for full-scale deflection.

9. Construct the following dual-range ammeter in Fig. 8-1.2. Range 1 will measure 30 mA full-scale, and range 2 will measure 100 mA full-scale.

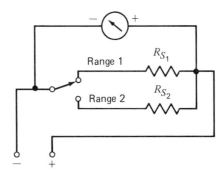

Fig. 8-1.2 Ammeter circuit.

10. Using the following formula, determine the multiplier resistors R_{S_1} and R_{S_2}:

$$R_S = (I_m \times r_m)/I_S$$

For a 30-mA full-scale deflection,

$$R_{S_1} = (I_m \times r_m)/30\ mA$$

Record this value in Table 8-1.2.
For a 100-mA full-scale deflection,

$$R_{S_2} = (I_m \times r_m)/100\ mA$$

Record this value in Table 8-1.2.

11. Connect your ammeter into the circuit configuration shown in Fig. 8-1.3, where I_2 is an ammeter of known accuracy and I_1 is your ammeter design on another meter movement. Complete Table 8-1.3 by turning on and adjusting the power supply in accordance with Table 8-1.3. Record the values of I_1 and I_2 for each power supply setting.

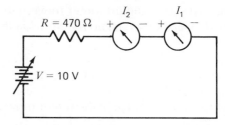

Fig. 8-1.3 Ammeter test setup.

12. Determine the percentage of accuracy for Table 8-1.3.

QUESTIONS FOR EXPERIMENT 8-1

1. Describe how ammeters are connected in a circuit to measure current.

2. Design an ammeter circuit that will measure 1.5 A with a 0- to 100 mA full-scale deflection meter movement.

REPORT

Write a complete report. Discuss the measured and calculated results. Discuss the three most significant aspects of the experiment and write a conclusion.

TABLES FOR EXPERIMENT 8-1

TABLE 8–1.1

r_m	I_m	V_m
_____	_____	_____

TABLE 8–1.2 Meter Movement

I_m, A	r_m, Ω	V_m, V
_____	_____	_____

Shunt R_{S_1}	Shunt R_{S_2}	
_____	_____	_____

TABLE 8–1.3

Range, mA	Voltage Setting	I_1	I_2	% Accuracy
0–30	_____	_____	_____	_____
0–100	_____	_____	_____	_____

EXPERIMENT RESULTS REPORT FORM

Experiment No: _____ Name: _____
                                Date: _____
Experiment Title: _____ Class: _____
_____   Instr: _____

Explain the purpose of the experiment:

List the first Learning Objective:
OBJECTIVE 1:

After reviewing the results, describe how the objective was validated by this experiment?

List the second Learning Objective:
OBJECTIVE 2:

After reviewing the results, describe how the objective was validated by this experiment?

List the third Learning Objective:
OBJECTIVE 3:

After reviewing the results, describe how the objective was validated by this experiment?

Conclusion:

If required, attach to this form: ☐ Answers to Questions, ☐ Tables, and ☐ Graphs.

CHAPTER 8

EXPERIMENT 8-2

ANALOG VOLTMETER DESIGN

OBJECTIVES

At the completion of this experiment, you will be able to:
- Determine the internal resistance of a basic D'Arsonval movement.
- Design a voltmeter from this meter movement.
- Use this voltmeter for actual voltage measurements.

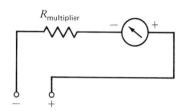

Fig. 8-2.1 Basic voltmeter circuit with $R_{multiplier}$.

SUGGESTED READING

Chapter 8, *Basic Electronics*, Grob/Schultz, Tenth Edition

INTRODUCTION

D'Arsonval Meter

The stationary permanent-magnet moving-coil meter is the basic movement used in most measuring instruments for servicing electric equipment. This type of movement is commonly called the D'Arsonval movement because it was first employed by the Frenchman D'Arsonval in making electrical measurements.

The basic D'Arsonval movement consists of a stationary permanent magnet and a movable coil. When current flows through the coil, the resulting magnetic field reacts with the magnetic field of the permanent magnet and causes the coil to rotate. The greater the amount of current flow through the coil, the stronger the magnetic field produced; the stronger this field, the greater the rotation of the coil. To determine the amount of current flow, a means must be provided to indicate the amount of coil rotation.

Voltmeter

The 100 µA D'Arsonval meter used as the basic meter for the ammeter may also be used to measure voltage if a high resistance is placed in series with the moving coil of the meter. When this is done, the unit containing the resistance is commonly called a *multiplier*. A simplified diagram of a voltmeter is shown in Fig. 8-2.1.

Extending the Range

The value of the necessary series resistance is determined by the current required for full-scale deflection of the meter and by the range of voltage to be measured. Because the current through the meter circuit is directly proportional to the applied voltage, the meter scale can be calibrated directly in volts for a fixed series resistance.

For example, assume that the basic meter (microammeter) is to be made into a voltmeter with a full-scale reading of 1 V. The coil resistance of the basic meter is 100 Ω, and 0.0001 A causes a full-scale deflection. The total resistance R of the meter coil and the series resistance is

$$r_m = \frac{V_m}{I_m}$$
$$= \frac{1}{100\ \mu A}$$
$$= 10{,}000\ \Omega$$

and the series resistance alone is

$$R_1 = 10{,}000 - 100$$
$$= 9900\ \Omega$$

Multirange voltmeters utilize one meter movement with a convenient switching arrangement. A multirange voltmeter with three ranges is shown in Fig. 8-2.2. The total circuit resistance for each of the three ranges, beginning with the 1-V range, is

$$R_1 = \frac{V_m}{I_m} = \frac{1}{100} = 0.01\ M\Omega$$
$$R_2 = \frac{V_m}{I_m} = \frac{100}{100} = 1\ M\Omega$$
$$R_3 = \frac{V_m}{I_m} = \frac{1000}{100} = 10\ M\Omega$$

Voltage-measuring instruments are always connected across (in parallel with) a circuit. If the approximate value of the voltage to be measured is not known, it is best to start with the highest range of the voltmeter and progressively lower the range until a suitable middle third reading is obtained.

EXPERIMENT 8-2 179

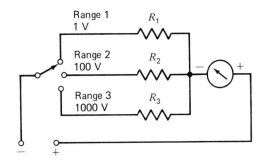

Fig. 8-2.2 Three-range voltmeter circuit.

EQUIPMENT

DC power supply, 10 V
Voltmeter of known accuracy
Protoboard or springboard
Test leads
VTVM or DVM

COMPONENTS

Resistors:

(1) 150 kΩ 0.25 W   Other resistors as calculated.

Potentiometers (linear taper):

(1) 100 kΩ   (1) 5 kΩ

(1) SPST switch
(1) SPDT switch
(1) 50 µA meter movement of unknown resistance

PROCEDURE

1. Measure the internal resistance of the meter movement by connecting the circuit shown in Fig. 8-2.3.

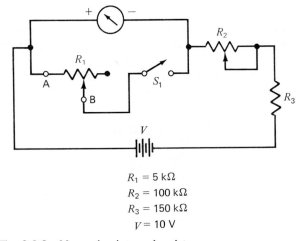

$R_1 = 5\ k\Omega$
$R_2 = 100\ k\Omega$
$R_3 = 150\ k\Omega$
$V = 10\ V$

Fig. 8-2.3 Measuring internal resistance.

2. With S_1 open, turn on the power supply and adjust it for 10 V.

3. Adjust R_2 so that the scale of the meter movement is at full-scale deflection.

4. Close S_1 and adjust R_1 so that the scale upon the movement is at half-scale deflection. The currents will evenly divide between R_1 and the internal resistance r_m of the meter movement when $R_1 = r_m$.

5. Measure and record in Table 8-2.1 the voltage dropped across V_m.

6. Measure and record r_m in Table 8-2.1 by turning the power supply off, disconnecting R_1 from the circuit, and measuring from points A to B. At this point, $r_m = R_1$.

7. Calculate and record I_m in Table 8-2.1, where

$$I_m = \frac{V_m}{r_m}$$

8. Record in Table 8-2.2 the I_m for full-scale deflection, the r_m (meter movement's internal resistance) for the meter movement, and the value of V_m for full-scale deflection.

9. Construct the dual-range voltmeter in Fig. 8-2.4 so that range 1 will measure 5 V full scale and range 2 will measure 10 V full scale. Use the following formula to determine the multiplier resistors R_1 and R_2:

$$R_{\text{multiplier}} = \frac{V_{\text{FS}}}{I_{\text{FS}}} - r_m$$
$$= \frac{V_{\text{intended}}}{I_m} - r_m$$

Note: FS means "full scale."

For a 5 V full-scale deflection,

$$R_1 = \frac{5\ V}{I_m} - r_m$$

For a 10 V full-scale deflection,

$$R_2 = \frac{10\ V}{I_m} - r_m$$

Record these calculated values in Table 8-2.2.

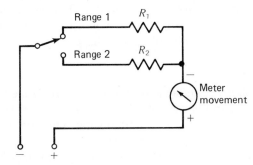

Fig. 8-2.4 Voltmeter dual-range circuit.

10. Connect your voltmeter into the circuit configuration of Fig. 8-2.5, where V_2 is a voltmeter of known accuracy and V_1 is your voltmeter design.

11. Complete Table 8-2.3 by turning on and adjusting the power supply in accordance with Table 8-2.3. Record the values of V_1 and V_2 for each power supply setting.

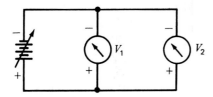

Fig. 8-2.5 Measuring voltages.

12. Determine the percentage of accuracy for Table 8-2.3.

NAME	DATE

QUESTIONS FOR EXPERIMENT 8-2

1. Design a voltmeter that will measure 0 to 30 V dc by using a 100 mA meter movement.

TABLES FOR EXPERIMENT 8-2

TABLE 8-2.1

0–50 μA
$V_m =$ _____
$r_m =$ _____
$I_m =$ _____

TABLE 8-2.2 Meter Movement

$I_m =$ _____

$r_m =$ _____

$V_m =$ _____

$R_1 =$ _____

$R_2 =$ _____

TABLE 8-2.3

Power Supply Voltages	V_1	V_2	% Accuracy
Range 1: 1–5 V			
1 V			
2 V			
3 V			
4 V			
5 V			
Range 2: 1–10 V			
1 V			
2 V			
3 V			
4 V			
5 V			
6 V			
7 V			
8 V			
9 V			
10 V			

EXPERIMENT RESULTS REPORT FORM

Experiment No: _____ Name: _____
                                         Date: _____
Experiment Title: _____ Class: _____
_____ Instr: _____

Explain the purpose of the experiment:

List the first Learning Objective:
OBJECTIVE 1:

After reviewing the results, describe how the objective was validated by this experiment?

List the second Learning Objective:
OBJECTIVE 2:

After reviewing the results, describe how the objective was validated by this experiment?

List the third Learning Objective:
OBJECTIVE 3:

After reviewing the results, describe how the objective was validated by this experiment?

Conclusion:

If required, attach to this form: ☐ Answers to Questions, ☐ Tables, and ☐ Graphs.

CHAPTER 8

EXPERIMENT 8-3

ANALOG OHMMETER DESIGN

OBJECTIVES

At the completion of this experiment, you will be able to:
- Determine the internal resistance of a basic D'Arsonval meter movement.
- Design an ohmmeter from this meter movement.
- Use this ohmmeter for actual ohm measurements.

SUGGESTED READING

Chapter 8, *Basic Electronics*, Grob/Schultz, Tenth Edition

INTRODUCTION

The ohmmeter consists of a dc milliammeter, with a few added features. The added features are:

1. A dc source of potential
2. One or more resistors (one of which is variable)

A simple ohmmeter circuit is shown in Fig. 8-3.1.

The ohmmeter's pointer deflection is controlled by the amount of battery current passing through the moving coil. Before measuring the resistance of an unknown resistor or electric circuit, the test leads of the ohmmeter are first short-circuited together, as shown in Fig. 8-3.1. With the leads short-circuited, the meter is calibrated for proper operation on the selected range. (While the leads are short-circuited, meter current is maximum and the pointer deflects a maximum amount, somewhere near the zero position on the ohms scale.) When the variable resistor is adjusted properly, with the leads short-circuited, the meter pointer will come to rest exactly on the zero graduation. This indicates *zero resistance* between the test leads, which in fact are short-circuited together. The zero readings of series-type ohmmeters are sometimes on the right-hand side of the scale, whereas the zero reading for ammeters and voltmeters is generally to the left-hand side of the scale. When the test leads of an ohmmeter are separated, the meter pointer will return to the left side of the scale, due to the interruption of current and the spring tension acting on the movable-coil assembly.

After the ohmmeter is adjusted for zero reading, it is ready to be connected in a circuit to measure resistance. A typical circuit and ohmmeter arrangement is shown in Fig. 8-3.2.

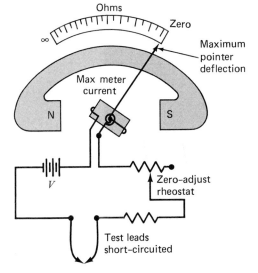

Fig. 8-3.1 Simple series ohmmeter.

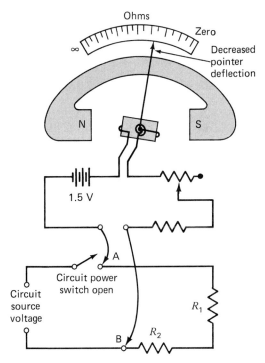

Fig. 8-3.2 Typical ohmmeter arrangement.

The power switch of the circuit to be measured should always be in the off position. This prevents the circuit's source voltage from being applied across the meter, which could cause damage to the meter movement.

The test leads of the ohmmeter are connected across (in parallel with) the circuit to be measured (see Fig. 8-3.2). This causes the current produced by the meter's internal battery to flow through the circuit being tested. Assume that the meter test leads are connected at points A and B of Fig. 8-3.2. The amount of current that flows through the meter coils will depend on the resistance of resistors R_1 and R_2, plus the resistance of the meter. Since the meter has been pre-adjusted (zeroed), the amount of coil movement now depends solely upon the resistance of R_1 and R_2. The inclusion of R_1 and R_2 raised the total series resistance, decreased the current, and thus decreased the pointer deflection. The pointer will now come to rest at a scale figure indicating the combined resistance of R_1 and R_2. If R_1 or R_2, or both, were replaced with a resistor(s) having a larger ohmic value, the current flow in the moving coil of the meter would be decreased still more. The deflection would also be further decreased, and the scale indication would read a still higher circuit resistance. Movement of the moving coil is proportional to the amount of current flow. The scale reading of the meter, in ohms, is inversely proportional to current flow in the moving coil.

EQUIPMENT

DC power supply, 0–10 V
Ohmmeter
Voltmeter
Ammeter
Test leads
VTVM or DVM

COMPONENTS

(1) 0 to 1 mA meter movement
(1) 150 kΩ 0.25 W resistor
(1) 5 kΩ potentiometer, linear taper
(1) 100 kΩ potentiometer, linear taper
(1) 1 MΩ potentiometer, linear taper
(1) 10 kΩ 0.25 W resistor
(1) SPST switch
(1) Decade box

PROCEDURE

1. Measure the internal resistance of the meter movement by connecting the circuit shown in Fig. 8-3.3, where $R_1 = 5$ kΩ, $R_2 = 100$ kΩ, and $R_3 = 10$ kΩ.
2. With S_1 open, turn on the power supply and adjust it for 10 V.
3. Adjust R_2 so that the scale upon the meter movement reads at full-scale deflection.
4. Close S_1 and adjust R_1 so that the scale upon the movement is at half-scale deflection. The currents

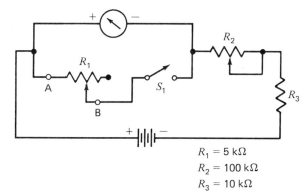

$R_1 = 5$ kΩ
$R_2 = 100$ kΩ
$R_3 = 10$ kΩ

Fig. 8-3.3 Measuring internal resistance.

will evenly divide between R_1 and the internal r_m of the meter movement when $R_1 = r_m$.

5. Measure and record in Table 8-3.1 the voltage dropped across V_m.
6. Measure and record r_m in Table 8-3.1 by turning the power supply off, disconnecting R_1 from the circuit, and measuring from points A to B. At this point, $r_m = R_1$.
7. Calculate and record I_m in Table 8-3.1, where

$$I_m = \frac{V_m}{r_m}$$

8. Record in Table 8-3.1 the r_m for full-scale deflection of a 0 to 1 mA meter movement.
9. Construct the series-type ohmmeter shown in Fig. 8-3.4, where $R_1 = 1.5$ kΩ $- r_m$. (The resistance value of R_1 may have to be created by using a resistance decade box.) R_2 will be used to set the ohmmeter to zero ohms.

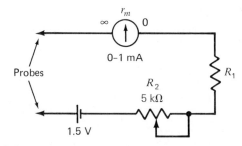

Fig. 8-3.4 Basic series ohmmeter.

10. With the probes not touching, the ohmmeter reads infinity. With the probes touching, adjust R_2 until the meter reads zero, indicating a zero ohms condition (a short circuit).
11. Calibrate this ohmmeter by connecting a 1 MΩ potentiometer across the probes, as shown in Fig. 8-3.5, using a grease pencil to mark the face of the ohmmeter at the infinity value. Complete Table 8-3.2.

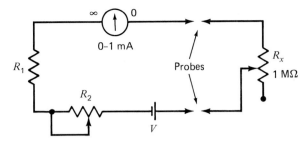

Fig. 8-3.5 Ohmmeter test circuit.

12. After calibrating the ohmmeter, measure the several resistances shown in Table 8-3.2 with an ohmmeter of known accuracy and your ohmmeter design.
13. Complete Table 8-3.3, and determine the percentage of accuracy. You will measure and record the resistance values in that table.

| NAME | DATE |

QUESTIONS FOR EXPERIMENT 8-3

Answer TRUE (T) or FALSE (F) for each question.

_____ 1. An ohmmeter is used to measure voltage and current.

_____ 2. An ohmmeter has an internal battery.

_____ 3. The infinity symbol (∞) on an ohmmeter indicates a short circuit.

_____ 4. The ohmmeter's leads are placed across the resistance to be measured.

_____ 5. When the ohmmeter leads are short-circuited, the needle will probably indicate zero.

_____ 6. Ohmmeters do not require internal current-limiting resistances or shunt paths.

_____ 7. The ohms or resistance scale that reads from left to right is called a *back-off scale*.

_____ 8. The zero-ohms adjustment should not be used when changing ranges.

_____ 9. For greater values of resistance, a less-sensitive meter is required to read lesser values of current.

_____ 10. An ohmmeter can be destroyed or have its fuse blown if it is used to measure resistance in a circuit where power is applied.

TABLES FOR EXPERIMENT 8-3

TABLE 8–3.1

| Steps 5, 6, and 7 |

$V_m = $ _____

$r_m = $ _____

$I_m = $ _____

| Step 8 |

$r_m = $ _____

TABLE 8–3.2

External R_x, Ω	Amount of Deflection	Scale Reading
0	_____	_____
750	_____	_____
1,500	_____	_____
3,000	_____	_____
150,000	_____	_____
500,000	_____	_____

TABLE 8–3.3

R	Known Meter	Design Meter	% Accuracy
100 Ω	_____	_____	_____
1 kΩ	_____	_____	_____
4.7 kΩ	_____	_____	_____
22 kΩ	_____	_____	_____
100 kΩ	_____	_____	_____
1 MΩ	_____	_____	_____

EXPERIMENT RESULTS REPORT FORM

Experiment No: _____ Name: _____
                                 Date: _____
Experiment Title: _____ Class: _____
_____   Instr: _____

Explain the purpose of the experiment:

List the first Learning Objective:
OBJECTIVE 1:

After reviewing the results, describe how the objective was validated by this experiment?

List the second Learning Objective:
OBJECTIVE 2:

After reviewing the results, describe how the objective was validated by this experiment?

List the third Learning Objective:
OBJECTIVE 3:

After reviewing the results, describe how the objective was validated by this experiment?

Conclusion:

If required, attach to this form: ☐ Answers to Questions, ☐ Tables, and ☐ Graphs.

CHAPTER 9

EXPERIMENT 9-1

KIRCHHOFF'S LAWS

OBJECTIVES

At the completion of this experiment, you will be able to:
- Validate Kirchhoff's current and voltage laws.
- Gain proficiency with lab equipment and technique.
- Predict the path of current flow using Kirchhoff's current law.

SUGGESTED READING

Chapter 9, *Basic Electronics*, Grob/Schultz, Tenth Edition

INTRODUCTION

In 1847, Gustav R. Kirchhoff formulated two laws that have become fundamental to the study of electronics. They are as follows:

1. The algebraic sum of the currents into and out of any point must be equal to zero.
2. The algebraic sum of the applied source voltages and the *IR* voltage drops on any closed path must be equal to zero.

At first, Kirchhoff's laws seem obvious. That is, it seems obvious to state that whatever goes into a circuit must also equal what comes out of it. However, Kirchhoff's laws are used to analyze circuits that are not simple series or parallel or series-parallel circuits. For example, circuits that contain more than one voltage source and that contain hundreds of transistors cannot be understood without using Kirchhoff's laws to analyze those circuits. In its simplest form, Kirchhoff's current law could be used to analyze the circuit in Fig. 9-1.1.

Simplified version:
$$I_1 = I_2 + I_6 \qquad I_2 = I_3 + I_4 \qquad I_5 = I_2$$
$$I_6 = I_1 - I_2 \qquad I_3 = I_2 - I_4 \qquad I_7 = I_6 + I_5$$
$$I_2 = I_1 - I_6 \qquad I_4 = I_5 - I_3 \qquad I_7 = I_1$$

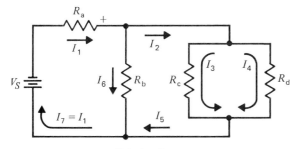

Fig. 9-1.1 Series-parallel circuit.

Algebraic version:
$$+I_1 - I_6 - I_2 = 0 \qquad +I_5 + I_6 - I_7 = 0$$
$$+I_2 - I_3 - I_4 = 0 \qquad I_7 = I_1 = I_7$$

Note: For the circuit of Fig. 9-1.1, if $V_S = 20$ V, the sum of the *IR* voltage drops across the series-parallel combination of R_a, R_b, R_c, and R_d will also equal 20 V.

Be sure that you understand the algebraic version of the circuit. It has become standard practice to assign positive and negative values to the currents as follows: The current into any point is positive (+), and the current out of any point is negative (−). Also, this would be true if there were more than one path for the current to enter or leave. Do not confuse this with electron flow.

Although the circuit of Fig. 9-1.1 is a series-parallel circuit, it was used to demonstrate how Kirchhoff's laws operate. Now consider the circuit of Fig. 9-1.2. This circuit could not be solved for its currents and *IR* voltage drops without using Kirchhoff's laws because of the two source voltages.

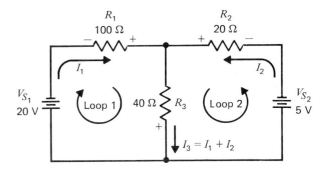

Fig. 9-1.2 Circuit with two voltage sources.

By using the loop method, Kirchhoff's laws can be used to determine the currents and *IR* voltages in the circuit by calculation. For example,

$$V_{S_1} = 20\text{ V} - V_{R_1} - V_{R_3} = 0 \qquad \text{(loop 1)}$$
$$V_{S_2} = 5\text{ V} - V_{R_2} - V_{R_3} = 0 \qquad \text{(loop 2)}$$

By following the loop around its path, Kirchhoff's law provides a method. Take the source as positive, and take the individual path resistance as negative. The sum is equal to zero. This means that circuits will be divided into separate loops as if the other voltage source and its corresponding path did not exist. The *IR* voltage drops would be calculated as follows.

$$V_{R_1} = I_1 R_1 = I_1 \times 100 \, \Omega$$
$$V_{R_2} = I_2 R_2 = I_2 \times 20 \, \Omega$$
$$V_{R_3} = I_3 R_3$$
$$= (I_1 + I_2) R_3$$
$$= (I_1 + I_2) \times 40 \, \Omega$$

Refer to the equations for loops 1 and 2. The IR voltages would now be replaced in the formulas as follows:

For loop 1,
$$20 \, V - 100(I_1) - 40(I_1 + I_2) = 0$$
$$-140 I_1 - 40 I_2 = -20 \, V$$

and for loop 2,
$$5 \, V - 20(I_2) - 40(I_1 + I_2) = 0$$
$$-60 I_2 - 40 I_1 = -5 \, V$$

Note that the final equations are transposed versions that can be simplified further by division:

$$\frac{-140 I_1 - 40 I_2}{-20 \, V} = \frac{-20 \, V}{-20 \, V} \text{ or } 7 I_1 + 2 I_2 = 1 \quad \text{(loop 1)}$$

$$\frac{-60 I_2 - 40 I_1}{-5 \, V} = \frac{-5 \, V}{-5 \, V} \text{ or } 12 I_2 + 8 I_1 = 1 \quad \text{(loop 2)}$$

To solve for the currents, isolate the I_2 currents by making them the same value. Multiply the loop 1 equation by 6:

$$42 I_1 + 12 I_2 = 6 \quad \text{(loop 1)}$$

Note:

$$6 = \frac{12 I_2}{2 I_2}$$

Note that either loop could be changed. In the case above, only loop 1 was changed (multiplied by 6) so that I_2 would have the same value for both loop equations:

$$42 I_1 + 12 I_2 = 6 \quad \text{(loop 1)}$$
$$8 I_1 + 12 I_2 = 1 \quad \text{(loop 2)}$$

Subtracting the two equations, term by term, will eliminate I_2 because $12 I_2 - 12 I_2 = 0$. Therefore,

$$34 I_1 = 5$$
$$I_1 = 147 \, mA$$

By using the loop equation method, Kirchhoff's law, the current through the resistance of R_1 is determined. Also, the direction of current flow is correct as assumed because the answer was a positive value.

To calculate I_2, substitute 147 mA for I_1 in either loop equation. Substituting in loop 2,

$$8(147 \, mA) + 12 I_2 = 1$$
$$1.18 \, A + 12 I_2 = 1$$
$$12 I_2 = 1 - 1.18 \, A$$
$$= -0.18 \, A$$
$$I_2 = -0.015 \, A$$
$$= -15 \, mA$$

Because the negative sign appears in the solution for the current I_2, it follows that the current through I_2 is opposite in direction from the polarity shown in the circuit of Fig. 9-1.2, the assumed direction using the loop method.

Now that Kirchhoff's law has been used to determine the currents through the circuit of Fig. 9-1.2, the circuit can be redrawn to reflect the correct values, as shown in Fig. 9-1.3.

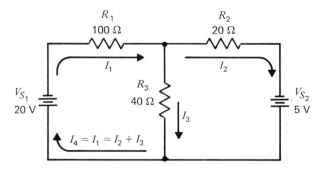

Fig. 9-1.3 Redrawn version of the circuit in Fig. 9-1.2.

In the case of the circuit in Fig. 9-1.3, the 20 V dc source actually overrides the 5 V dc source. This series opposition results in approximately 15 V dc.

Because R_1 is 100 Ω, it is easy to verify the Kirchhoff's law loop equation solution as follows:

$$\frac{15 \, V}{100 \, \Omega + (20 \, \Omega \parallel 40 \, \Omega)} = \frac{15 \, V}{113.3 \, \Omega} = 132 \, mA$$

$$\text{Loop 1 } I_T = 147 \, mA$$
$$\text{Loop 2 } I_T = -15 \, mA \quad \text{(Subtract negative value)}$$

$$\text{Total } I_T = 132 \, mA$$

Finally, note that the circuit in Fig. 9-1.4 is a bridge type of circuit. Kirchhoff's law could be used to calculate the unknown values in this circuit as well as circuits that contain many voltage sources and many current paths.

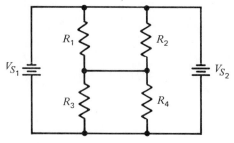

Fig. 9-1.4 Bridge-type circuit.

EQUIPMENT

DC power supply, 0–10 V
Ammeter (milliamp capability)
Voltmeter or VTVM

Springboard or protoboard
Connecting leads

COMPONENTS

Resistors (all 0.25 W):

    (1) 100 Ω   (1) 390 Ω
    (1) 220 Ω   (1) 470 Ω
    (1) 330 Ω

(2) Dry cells (flashlight batteries, 1.5 V)

PROCEDURE

1. Measure and record the values of voltage for each dry cell you are using. It may be necessary to solder connecting wires onto the ends of the batteries. Record the results in Table 9-1.1. Together, series-aiding, the two batteries should be approximately 3.0 V.

2. For each of the three circuits in Figs. 9-1.5 to 9-1.7, use batteries for voltages of 1.5 V and use the power supply for the other source voltage. If 3.0 V and 1.5 V are both used, connect the two 1.5 V batteries in series for 3.0 V. Measure and record the values of current in Table 9-1.2. Also, measure and record the value of the *IR* voltage drop across each resistance. Finally, use arrows to indicate the direction of current flow.

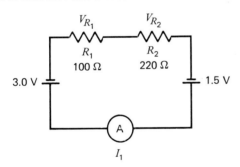

Fig. 9-1.5 Series-opposing voltage circuit.

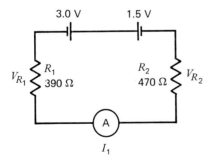

Fig. 9-1.6 Series-opposing voltage circuit.

3. For the circuit of Fig. 9-1.8, refer to the procedure outlined in the introduction to this experiment. Preferably, use the loop method of Kirchhoff's law to determine the values and polarities of both current and voltage for the circuit. Then use the lab to verify the values (within 10 to 15 percent). Record all the

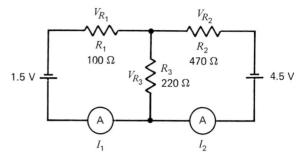

Fig. 9-1.7 Kirchhoff's circuit for analysis.

calculated values, and then measure and record in Table 9-1.3 all the values determined in the lab for the circuit. Be sure to include arrows alongside or directly above the meters marked A. Include Fig. 9-1.8 in your report. Note that Fig. 9-1.8 will be used to verify Kirchhoff's law. That is, you will compare your calculations to your measurements. Do not forget to include the values of *IR* drops across each resistance.

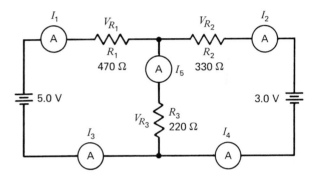

Fig. 9-1.8 Kirchhoff's circuit. Use loop method to determine the values and polarities of current and voltage.

4. Connect the circuit of Fig. 9-1.9. Measure and record in Table 9-1.4 all the values of *IR* voltage drops

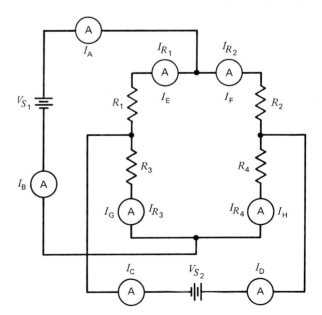

Fig. 9-1.9 Kirchhoff's law circuit (values to be given by instructor).

EXPERIMENT 9-1 **195**

and currents. This final circuit is more complex than the previous circuits. Check with your instructor about the calculations for this experiment. It is recommended for extra credit only. The important thing here is to be able to properly connect the circuit and properly measure the unknown values.

Note: Your instructor will assign values for V_{S_1}, V_{S_2}, R_1, R_2, R_3, and R_4. This ensures that the experiment cannot be precalculated.

NAME _____ DATE _____

QUESTIONS FOR EXPERIMENT 9-1

1. State Kirchhoff's current law, KCL.

2. State Kirchhoff's voltage law, KVL.

TABLES FOR EXPERIMENT 9-1

TABLE 9–1.1 Dry Cell Voltages

Battery	Measured Voltage
1	_____
2	_____

TABLE 9–1.2

Circuit	V_{R_1}	V_{R_2}	V_{R_3}	I_1	I_2
Fig. 9-1.5	_____	_____		_____	
Fig. 9-1.6	_____	_____		_____	
Fig. 9-1.7	_____	_____	_____	_____	_____

EXPERIMENT 9-1

TABLE 9–1.3

	Calculated	Measured
I_1		
I_2		
I_3		
I_4		
I_5		
V_{R_1}		
V_{R_2}		
V_{R_3}		

TABLE 9–1.4

	Current	Voltage
R_1		
R_2		
R_3		
R_4		
I_A		
I_B		
I_C		
I_D		
I_E		
I_F		
I_G		
I_H		

EXPERIMENT RESULTS REPORT FORM

Experiment No: _____   Name: _____
                                        Date: _____
Experiment Title: _____   Class: _____
_____   Instr: _____

Explain the purpose of the experiment:

List the first Learning Objective:
OBJECTIVE 1:

After reviewing the results, describe how the objective was validated by this experiment?

List the second Learning Objective:
OBJECTIVE 2:

After reviewing the results, describe how the objective was validated by this experiment?

List the third Learning Objective:
OBJECTIVE 3:

After reviewing the results, describe how the objective was validated by this experiment?

Conclusion:

If required, attach to this form: ☐ Answers to Questions, ☐ Tables, and ☐ Graphs.

CHAPTER 10

EXPERIMENT 10-1

NETWORK THEOREMS

OBJECTIVES

At the completion of this experiment, you will be able to:
- Thevenize a circuit.
- Nortonize a circuit.
- Identity the difference between a Thevenin and Norton circuit.

SUGGESTED READING

Chapter 10, *Basic Electronics*, Grob/Schultz, Tenth Edition

INTRODUCTION

The analysis of Ohm's law and Kirchhoff's laws have been of primary use in the solution of relatively simple solutions of dc circuits. In the analysis of relatively complex circuits, a more powerful method is required. In the case of simplifying complex circuits, Thevenin and Norton theorems are used. This technique involves reducing a complex network to a simple circuit, which acts like the original circuit. In general, any circuit with many voltage sources and components, with no regard made to interconnection, can be represented by an equivalent circuit with respect to a pair of terminals in the equivalent circuit.

Thevenin's theorem states that a circuit can be replaced by a single voltage source V_{Th} in series with a single resistance R_{Th} connected to two terminals. This is shown in Fig. 10-1.1.

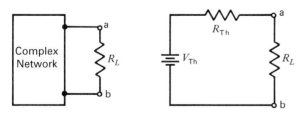

Fig. 10-1.1 Thevenin's circuit.

Norton's analysis is used to simplify a circuit in terms of currents rather than voltage, as is done in Thevenin circuits. Norton's circuit can be used to reduce a complex network into a simple parallel circuit that consists of a current source I_N and a parallel resistance to R_N. An example of this is shown in Fig. 10-1.2.

In the procedure that follows, the techniques of Thevenizing and Nortonizing simple voltage source–resistor networks will be developed as the procedure is completed.

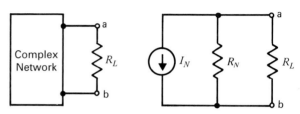

Fig. 10-1.2 Norton's circuit.

EQUIPMENT

DC power supply
Ammeter
Voltmeter
Protoboard or springboard
Test leads

COMPONENTS

Resistors:

(1) 100 Ω 1 W   (1) 270 Ω 0.25 W
(1) 220 Ω 0.25 W

PROCEDURE

Thevenizing

1. Construct the circuit shown in Fig. 10-1.3, where $R_1 = 100\ \Omega$, $R_2 = 270\ \Omega$, $R_L = 220\ \Omega$, and V is adjusted to 10 V.

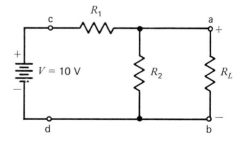

Fig. 10-1.3 Thevenizing a circuit.

2. Open the circuit at points a and b by disconnecting R_L from the circuit. The remainder of the circuit

connected to a and b will be Thevenized. Calculate, measure, and record in Table 10-1.1 the voltage across points ab. Note that $V_{ab} = V_{Th}$.

3. Turn off the power supply and completely remove it from the circuit.

4. With the power supply removed, connect points c and d.

5. Calculate, measure, and record in Table 10-1.1 the value of R_{ab}. Note that $R_{ab} = R_{Th}$.

V_{Th} has now been determined and is found to be in series with R_{Th}. Since R_L was disconnected, this Thevenin equivalent can be applied to any value of R_L.

6. Reconnect the circuit shown in Fig. 10-1.3. Calculate (using the voltage divider formula), measure, and record V_L and I_L in Table 10-1.1 by reconnecting R_L. V_L is defined by the voltage divider formula as

$$V_L = \frac{R_L}{R_L + R_{Th}} \times V_{applied}$$

and I_L can be determined as

$$I_L = \frac{V_L}{R_L}$$

The same answers could be determined by using Ohm's law. The advantage of Thevenizing the circuit is that the effect of R_L can be calculated easily for different values.

7. Complete Table 10-1.1 for the percentage of accuracy.

Nortonizing

8. Construct the circuit shown in Fig. 10-1.4, where $R_1 = 100 \, \Omega$, $R_2 = 270 \, \Omega$, $R_L = 220 \, \Omega$, and V is adjusted to 10 V.

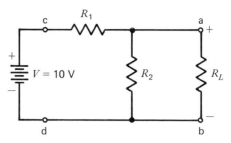

Fig. 10-1.4 Nortonizing a circuit.

9. Short-circuit points a and b together. This will also short-circuit R_2, and this will create a circuit condition in which resistor R_1 is in series with the power supply. Calculate, measure, and record in Table 10-1.2 the current flowing through R_1.

Note: $I_{R_1} = I_N$ (Norton)

10. Determine R_N by removing the short circuit, and remove R_L. This will leave points a and b unconnected to R_L.

11. Turn off the power supply and completely remove it from the circuit.

12. With the power supply removed, connect point c to d.

13. Calculate, measure, and record in Table 10-1.2 the value R_{ab}. Note that $R_{ab} = R_N$.

14. Reconnect the circuit shown in Fig. 10-1.4. Calculate, measure, and record I_L and V_L in Table 10-1.2 by reconnecting R_L.

15. Complete Table 10-1.2 for the percentage of accuracy.

NAME _____ DATE _____

QUESTIONS FOR EXPERIMENT 10-1

1. What are the primary use and importance of Thevenizing a circuit?

2. What are the primary use and importance of Nortonizing a circuit?

3. Draw a Thevenin equivalent of the circuit shown in Fig. 10-1.3.

4. Draw a Norton equivalent of Fig. 10-1.4.

5. Is the statement made at the end of procedure Step 6, which reads, "the advantage of Thevenizing the circuit is that the effect of R_L can be calculated easily for different values," valid? Explain and prove by example.

TABLES FOR EXPERIMENT 10-1

TABLE 10–1.1

	Calculated	Measured	% Accuracy
V_{Th}	_____	_____	_____
R_{Th}	_____	_____	_____
V_L	_____	_____	_____
I_L	_____	_____	_____

TABLE 10–1.2

	Calculated	Measured	% Accuracy
I_N	_____	_____	_____
R_N	_____	_____	_____
I_L	_____	_____	_____
V_L	_____	_____	_____

EXPERIMENT RESULTS REPORT FORM

Experiment No: _____   Name: _____
                                  Date: _____
Experiment Title: _____ Class: _____
_____   Instr: _____

Explain the purpose of the experiment:

List the first Learning Objective:
OBJECTIVE 1:

After reviewing the results, describe how the objective was validated by this experiment?

List the second Learning Objective:
OBJECTIVE 2:

After reviewing the results, describe how the objective was validated by this experiment?

List the third Learning Objective:
OBJECTIVE 3:

After reviewing the results, describe how the objective was validated by this experiment?

Conclusion:

If required, attach to this form: ☐ Answers to Questions, ☐ Tables, and ☐ Graphs.

CHAPTER 11

EXPERIMENT 11-1

CONDUCTORS AND INSULATORS

OBJECTIVES

At the completion of this experiment you will be able to:
- Identify the characteristics of conductors and insulators.
- Identify and validate the pole and throw of a switch.
- Describe the operating characteristics of a fuse.

SUGGESTED READING

Chapter 11, *Basic Electronics,* Grob/Schultz, Tenth Edition

INTRODUCTION

This experiment investigates the electrical characteristics of components and materials commonly found within the electronics industry and used by electronic technicians. Understanding these electrical characteristics promotes a general understanding of electronics while at the same time, protecting the safety of the electronic technician within the work place. This experiment will investigate four areas of interest: conductors and insulators, body resistance and safety, electrical switches, and electrical fuses.

Conductors, Insulators, and Semiconductors Electrical materials are generally classified into one of three areas: conductors, insulators, or semiconductors. When electrons can easily move from one atom to another in a material, it is considered a conductor. In general metals are a good example of conductors. In the electronics industry typical conductors are copper, aluminum, silver, and gold.

A material with atoms in which the electrons tend to stay in their own orbits, and cannot easily move from one atom to another is considered an insulator because insulators cannot easily conduct electricity. Insulators are a category of material that does not permit current to flow when voltage is applied, because of its high electrical resistance. However, insulators are able to hold or store electricity better than conductors are. Examples of typical insulators are rubber, certain plastics, paper, glass, and air.

The category called *semiconductors* are those materials which will conduct less than metal conductors and more than insulators. Examples of materials that are classified semiconductors are carbon, germanium, and silicon.

Body Resistance and Safety The human body contains various levels of electrical resistance. However, consider that a current of 100 mA through the heart will probably kill most people. The concept of maintaining electrical safety while working with electrical circuits is important. It is crucial to respect the potential harm of electrical circuits and test equipment. Body resistance can be easily and safely measured by the use of an ohmmeter. Skin resistance is typically 500 kΩ, while the resistance of blood may be less than 100 Ω.

One of the "rules of thumb" that is often used is the 1-10-100 rule of current. This rule states that 1 mA of current through the human body can be felt, 10 mA of current is sufficient to make muscles contract to the point where you cannot let go of a power source, and 100 mA is sufficient to stop the heart. According to Ohm's law,

$$V = IR$$

where V is voltage in volts, I is current in amps, and R is resistance in ohms.

Using a 9 V battery and assuming that 500 kΩ is a typical level of body resistance in the equation, we determine that 18 μA of current will flow. The level of 18 μA is below what is determined to be our "feel" threshold of 1 mA. However, if you were to have a cut finger, your resistance would be lower because blood electrolyte is a good conductor. At around 100 Ω, in fact, a current of 90 mA results—sufficient to stop the human heart, resulting in death. Therefore, it is for safety purposes that students are encouraged to increase bodily resistance while working on electrical circuitry. This can be accomplished by the use of insulated hand tools, gloves, and rubber-soled shoes.

Electrical Switches

A switch is a component that allows for the control of the flow of electrical current. The switch operates within two states, closed or open. When a switch is closed, it is considered in the ON position. When a switch is opened, it is considered in the OFF position. A closed switch exhibits practically zero resistance, and an open switch is considered to be practically an infinite amount of resistance.

Toggle type switches are usually described as having a certain number of poles and throws. The number of poles and throws that they contain generally classifies switches.

The most basic type of switch is a single-pole, single-throw switch (SPST). Other popular switch types include the single-pole, double-throw (SPDT); double-pole, single-throw (DPST); and the double-pole, double-throw (DPDT). The schematics for each type of switch is shown in Fig. 11-1.1.

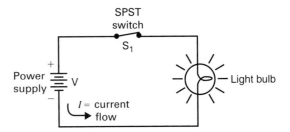

Fig. 11-1.2 Closed circuit.

The tungsten filament in the electric lamp is a conductor, but it is not as good a conductor as the connecting copper wires. Tungsten is used for the filament because its properties cause it to glow and give off light when an electric current heats it.

It is apparent that not all materials are equally good conductors of electricity. Specific materials are used for electrical components because of their unique characteristics as conductors and nonconductors.

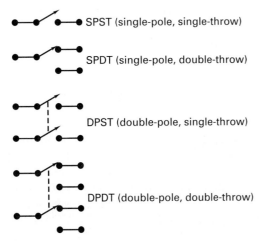

Fig. 11-1.1 Various toggle switch schematics.

EQUIPMENT

Digital (DMM) or analog (VOM, VTVM) multimeter
Glass cartridge fuse holder
DC power supply

COMPONENTS

SPST toggle switch
SPDT toggle switch
DPST toggle switch
DPDT toggle switch
Operating glass cartridge fuse (0.25 ampere)
Inoperative glass cartridge fuse (0.25 ampere)
47 Ω-2 W
56 Ω-2 W

SPECIAL MATERIALS

A beaker of water
A beaker of salt water
10 kΩ resistor
Capacitor (disc)
Inductor
Diode (1N4004)
Nichrome wire
Copper wire
Wood dowel
Plastic rod
Glass rod
Paper
Fuse (operational)
Fuse (nonoperational)

Electrical Fuses

Typically, electrical circuits use a fuse in series with a circuit as a protection against an electrical overload resulting from a short circuit. This is accomplished when excessive current melts the fuse element, blowing the fuse and opening the series circuit. The purpose is to let the fuse blow before the components are damaged.

A complete (closed) circuit or path for direct current consists of a voltage source such as a battery or power supply, closed switch, a load such as an electric lamp, and connecting conductors, such as copper wire (see Fig. 11-1.2). In such a closed circuit there is electric current, and if sufficient current flows, the lamp will light. If rubber cords instead of copper wires were used to tie the battery to the lamp, the lamp would not light, because rubber is an insulator and does not permit current to flow through it in the circuit.

PROCEDURE

Part A: Investigating the Resistive Characteristics of Insulators and Conductors

1. For the purposes of measuring resistance, obtain a digital (DMM) or analog multimeter (VOM, VTVM).

2. **Note for Analog Multimeters:** If you are using an analog multimeter, such as a VOM (Volt, Ohm, Milliampmeter), or VTVM (Vacuum Tube Volt-Meter), it will be necessary to calibrate the meter for correct resistance measurement. If during the course of the experiment you find it necessary to change the resistance range of the meter it will *also* be necessary to check the calibration of the meter at this new range. If found to be uncalibrated at this new range, then it will be necessary to recalibrate the meter for accurate resistance measurements. If you are unsure how to calibrate your meter, review Experiment 2-1, Resistance Measurement, or see your instructor.

3. **Note for either Digital or Analog Multimeters:** When using either digital or analog multimeters to measure resistance it is important to hold the meter probes by their insulation, not their metal tips. If you come into contact with the metal probes, the meter will attempt to measure your body resistance, which can interfere with accurate resistance measurements of the object you are measuring.

4. Measure and record the resistance of each of the 15 items listed in Table 11-1.1.

Note: It may be necessary to adjust your meter's range switch (RX scale) to make accurate measurements.

5. Reverse the meter's polarity by interchanging the test leads of the meter. Again, measure and record the resistance of each of the 14 items listed in Table 11-1.1.

Part B: Investigating the Body Resistance

6. Set your ohmmeter to the highest resistance measurement range of the meter. If you are using an analog meter, such as a VOM, or VTVM, check the calibration of the meter before it is used.

7. Under the supervision of your instructor, measure your body resistance by loosely holding the red (positive) and black (negative) test probe tip between the thumb and forefinger of each hand. Measure the ohmmeter's value and record your body resistance in Table 11-1.2.

8. Reverse the ohmmeter leads and read the ohmmeter's value and record your body resistance in Table 11-1.2.

9. Increase you grip by squeezing the probes (but be careful not to break the skin), and measure the ohmmeter's value and record your body resistance in Table 11-1.2.

10. Wet your thumb and finger and repeat the measurement. Read the ohmmeter's value and record your body resistance in Table 11-1.2.

Part C: The Switch

11. Through the use of an ohmmeter, identify each of the switches. Classify each of the switches as to SPST, SPDT, DPST, and DPDT. Draw their respective schematic diagram in Table 11-1.3 and identify their poles and throws.

Part D: The Fuse

12. Measure and record in Table 11-1.4, the resistance values of R_1 (47 Ω-2 W) and R_2 (56 Ω-2 W).

13. Using an *operating* glass cartridge fuse and holder, construct the circuit as shown in Fig. 11-1.3.

Note: Do not turn on the power supply yet.

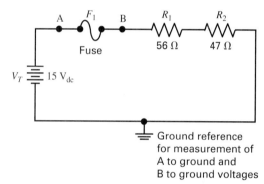

Fig. 11-1.3 Fused series circuit.

14. If the internal resistance of the fuse is assumed to be zero, calculate (from the measured values) the total series resistance R_T of the circuit as shown in Fig. 11-1.4, and record this information in Table 11-1.4.

15. If it would be unacceptable to exceed the current carrying capability of the circuit fuse, calculate the maximum voltage, V_{max}, that could be applied to this circuit without damaging the fuse. Record this information in Table 11-1.4.

16. Turn on the power supply and adjust the voltage to 10 V dc. Using Ohm's law, calculate the total expected series circuit current I_T and record in Table 11-1.4.

17. Calculate the expected voltage from point A to ground and record in Table 11-1.4.

18. Measure the voltage from point A to ground and record in Table 11-1.4.

19. Calculate the expected voltage from point B to ground and record in Table 11-1.4.

20. Measure the voltage from point B to ground and record in Table 11-1.4.
21. Calculate the expected voltage across R_1, and record in Table 11-1.4.
22. Measure the voltage across R_1, and record in Table 11-1.4.
23. Calculate the expected voltage across R_2, and record this information in Table 11-1.4.
24. Measure the voltage across R_2, and record in Table 11-1.4.
25. Calculate the expected voltage across points A and B and record in Table 11-1.4.
26. Measure the voltage across the fuse, or from points A to B and record in Table 11-1.4.
27. Turn off the power supply, and replace the *operable* fuse with an *inoperable* glass cartridge fuse. Repeat Procedure steps 17 through 26.

NAME	DATE

QUESTIONS FOR EXPERIMENT 11-1

1. Considering that a current of 100 mA through your heart will almost certainly kill you, how much voltage across your hands would be lethal if you have a body resistance of 250 kΩ?

2. How many connecting terminals does the SPST switch have? The SPDT? The DPST? The DPDT?

3. What is the voltage drop across a closed switch? What is the voltage drop across an open switch?

4. The double-pole, double-throw switch can be used to reverse the polarity of voltage across the terminals of a dc motor. From reading Chapter 11, *Basic Electronics*, Grob/Schultz, Tenth Edition, draw the schematic diagram that would accomplish this task.

5. When an electrical open appears with a component found within a series circuit, such as a fuse or resistor, what can the assumed voltage be across the circuit's open? Is this always true? Explain what the implication is for electrical safety.

6. Explain your results from Table 11-1.1 concerning the two fuses used in this experiment.

7. Explain your results from Table 11-1.2. Using this information, how could you improve your safety while working with electrical equipment?

8. Explain your results from procedural Steps 25 and 26 in Table 11-1.4. Using this information, how could you improve your safety while working with electrical equipment?

TABLES FOR EXPERIMENT 11-1

TABLE 11–1.1 Conductive Materials

Materials to Be Measured	Step 4 Resistance in Ω	Step 5 Resistance in Ω
Water		
Salt water		
10-kΩ resistor		
Disc capacitor		
Inductor		
1N4004 diode		
Nichrome wire		
Copper wire		
Wood dowel		
Plastic rod		
Glass rod		
Paper		
Operational fuse		
Nonoperational fuse		

TABLE 11–1.2 Body Resistance

Procedural Step and Measurement		Resistance Value
7	Loosely holding	_____ Ω
8	Reverse ohmmeter measurement	_____ Ω
9	Increased grip	_____ Ω
10	Wet measurement	_____ Ω

TABLE 11-1.3 Switches

Schematic Diagram Including Pole and Throw Identifications	Switch Classification
	SPST
	SPDT
	DPST
	DPDT

TABLE 11–1.4 Fused Series Circuits

Analysis of Fused (Operable) and Un-Fused (Inoperable) Series Circuits		
Procedural Step	**Resistor**	**Measured Value**
12	R_1	_____ Ω
12	R_2	_____ Ω
Establishing Maximum Circuit Parameters		
Procedural Step	**Maximum Parameters**	**Calculated/Measured Value with Appropriate Unit**
14	R_T	_____
15	V_{max}	_____
16	I_T	_____
Analysis with the Operable Fuse Circuit		
Procedural Step	**Location**	**Calculated/Measured Value with Appropriate Unit**
17	Calculated voltage A to ground	_____
18	Measured voltage A to ground	_____
19	Calculated voltage B to ground	_____
20	Measured voltage B to ground	_____
21	Calculated voltage R_1 to ground	_____
22	Measured voltage R_1 to ground	_____
23	Calculated voltage R_2 to ground	_____
24	Measured voltage R_2 to ground	_____
25	Calculated voltage A to B	_____
26	Measured voltage A to B	_____
Analysis with the Inoperable Fuse Circuit		
17-Repeated	Calculated voltage A to ground	_____
18-Repeated	Measured voltage A to ground	_____
19-Repeated	Calculated voltage B to ground	_____
20-Repeated	Measured voltage B to ground	_____
21-Repeated	Calculated voltage R_1 to ground	_____
22-Repeated	Measured voltage R_1 to ground	_____
23-Repeated	Calculated voltage R_2 to ground	_____
24-Repeated	Measured voltage R_2 to ground	_____
25-Repeated	Calculated voltage A to B	_____
26-Repeated	Measured voltage A to B	_____

EXPERIMENT RESULTS REPORT FORM

Experiment No: _____   Name: _____
                                  Date: _____
Experiment Title: _____ Class: _____
_____   Instr: _____

Explain the purpose of the experiment:

List the first Learning Objective:
OBJECTIVE 1:

After reviewing the results, describe how the objective was validated by this experiment?

List the second Learning Objective:
OBJECTIVE 2:

After reviewing the results, describe how the objective was validated by this experiment?

List the third Learning Objective:
OBJECTIVE 3:

After reviewing the results, describe how the objective was validated by this experiment?

Conclusion:

If required, attach to this form: ☐ Answers to Questions, ☐ Tables, and ☐ Graphs.

CHAPTER 12

EXPERIMENT 12-1

BATTERY INTERNAL RESISTANCE

OBJECTIVES

At the completion of this experiment, you will be able to:
- Validate the concept of internal resistance in a power source.
- Determine the internal resistance of a dry cell battery and a dc power supply (generator).
- Graph or plot decreasing terminal voltage versus load current.

SUGGESTED READING

Chapter 12, *Basic Electronics*, Grob/Schultz, Tenth Edition

INTRODUCTION

Any source of electric power that produces a continuous output voltage can be called a *generator*. All generators have some internal resistance, labeled r_i. This internal resistance has its own *IR* voltage drop, because it is in series with any load connected to the generator. In other words, the internal resistance of a source subtracts from the generated voltage, resulting in a decreased voltage across the output terminals. In a battery, r_i is due to the chemical makeup inside; in a power supply, r_i is due to the internal circuitry of the supply.

For example, Fig. 12-1.1 is a schematic representation of a 9-V battery with 100 Ω of internal resistance.

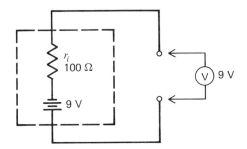

Fig. 12-1.1 A 9 V battery with $r_i = 100$ Ω.

Notice that the dotted line indicates that r_i is actually inside the battery. This battery has 9 V across its output terminals when it is measured with a voltmeter. If r_i were equal to 100 kΩ, the voltmeter would still measure 9 V across the output terminals. Thus, the value of r_i does not affect the output voltage. However, if a load is connected across the output terminals,

then the value of r_i becomes significant. In any case, Fig. 12-1.2 shows that the battery's internal resistance now becomes a resistance in series with the load.

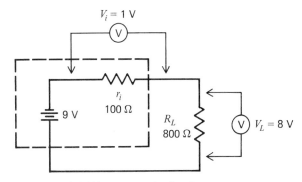

Fig. 12-1.2 A 9-V battery under load.

With a load resistance of 800 Ω connected across the output terminals, the voltmeter will now measure 8 V instead of 9 V. The other 1 V is now across the internal resistance of the battery. If r_i were equal to 100 kΩ, for example, the voltage across the output terminals would be almost 0 V due to the excessive value of r_i. In that case, the battery would be worn out or depleted.

Most bench power supplies have a fixed value of internal resistance that does not vary, regardless of the load value. Remember, without the load connected, the circuit is an open load. Therefore, the voltage drop across r_i equals zero. In this case, the total voltage is still available across the output, and it is called *open-circuit voltage*, or *no-load voltage*.

In the example of Fig. 12-1.2, the total circuit current I_T is equal to

$$I_T = \frac{V_L}{R_L} = \frac{8 \text{ V}}{800 \text{ Ω}} = 0.01 \text{ A}$$

As the load resistance decreases, more circuit current will flow. If R_L decreases to 350 Ω, the current will increase and the load will require more current. Also, the voltage drop across the load will decrease and the voltage drop across V_i will increase.

Notice that as the load resistance decreased, the circuit current increased. Thus, the terminal voltage (the same thing as the load *IR* voltage) decreased. Therefore, the terminal voltage drops with more load current.

There is a method for determining the internal resistance of a source (generator) based on the examples given. Simply put, it is as follows:

EXPERIMENT 12-1 **215**

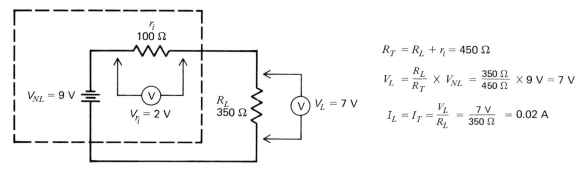

Fig. 12-1.3 Determining r_i.

1. Measure the no-load voltage.
2. Connect a load, and measure the voltage across the load and the circuit current.
3. Use the following formula to determine r_i:

$$r_i = \frac{V_{\text{no load}} - V_{\text{load}}}{I_{\text{load}}}$$

In Fig. 12-1.3, this would be

$$r_i = \frac{V_{NL} - V_L}{I_L}$$
$$= \frac{9\text{ V} - 7\text{ V}}{0.02\text{ A}}$$
$$= \frac{2\text{ V}}{0.02\text{ A}}$$
$$= 100\ \Omega$$

Finally, in general, if a generator has a very low internal resistance in relation to load resistance, it is considered a constant voltage source, because the voltage across r_i will subtract very little from the load voltage. If the value of r_i is very great in relation to load resistance, the generator is considered a constant current source, because the load resistance will have little effect upon the total resistance $(r_i + R_L)$ and the total circuit current.

EQUIPMENT

DC power supply
Voltmeter
Ammeter
VOM
1.5 V battery
Protoboard or springboard
Test leads

COMPONENTS

Resistors (all 0.25 W unless indicated otherwise):

(1) 220 Ω   (1) 2.2 kΩ
(2) 560 Ω   (1) 5.6 kΩ
(2) 1 kΩ   (1) 10 kΩ

PROCEDURE

1. Connect the circuit of Fig. 12-1.4. Do not connect the load yet. Measure and record in Table 12-1.1 the no-load voltage (V_{NL}) across the output terminals.

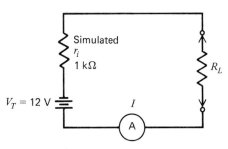

Fig. 12-1.4 Power supply with simulated r_i. Note that r_i is a 1 kΩ resistor in series.

2. Connect the following loads to the circuit of Fig. 12-1.4. Measure and record in Table 12-1.1 the load voltage and current for each load resistance.

$$R_L = 10\text{ k}\Omega$$
$$= 5.6\text{ k}\Omega$$
$$= 2.2\text{ k}\Omega$$
$$= 1\text{ k}\Omega$$
$$= 560\ \Omega$$
$$= 220\ \Omega$$

3. Calculate the value of r_i by using the measured values of load voltage and current for each load resistance. Show your calculations on a separate sheet of paper, and record the results in Table 12-1.1.
4. Change the value of r_i to 560 Ω and repeat steps 1 to 3 above. Record the results in Table 12-1.2.
5. Change the value of V_T to 6 V (using $r_i = 560\ \Omega$), and repeat steps 1 to 3. Record the results in Table 12-1.3.
6. Measure the voltage across a 1.5 V battery and record the value in Table 12-1.4.
7. Measure the short-circuit current of a 1.5 V battery by placing an ammeter across the output terminals of the battery for no longer than approximately 5 s, as shown in Fig. 12-1.5. Record the value in Table 12-1.4.

CAUTION: Use a VOM on its 10 or 12 A scale range. Some VOMs have special input jacks for this purpose.

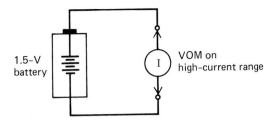

Fig. 12-1.5 Short-circuit method.

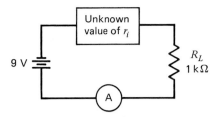

Fig. 12-1.7 Circuit for determining the internal resistance.

8. Calculate r_i by using Ohm's law. If available, repeat with a 22 V or any other size battery that will not damage the meter.

Note: You can do this for any value of battery, provided a meter of large enough current capacity is used.

9. Using the method for determining r_i (steps 1 to 3), use an unknown value of r_i (three times) and use a 1 kΩ load resistor. This can be done by disguising the value of a resistance with black electrical tape or by placing the resistance inside a chassis, as illustrated in Fig. 12-1.6. Use the circuit in Fig. 12-1.7 for this step. Record the results in Table 12-1.5.

10. *OPTIONAL:* Plot the results of steps 1 to 3 (Table 12-1.1). For example, V_L versus I_L. Use regular graph paper, *not* semilog graph paper. See Fig. 12-1.8. See Appendix G for suggestions on how to make graphs.

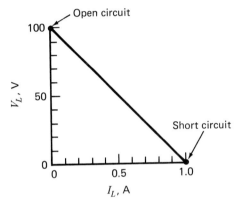

Fig. 12-1.8 How terminal voltage V_L drops with more load current I_L.

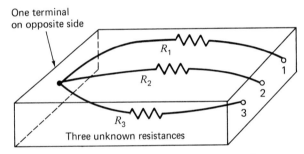

Fig. 12-1.6 Resistance box containing three unknown resistors.

QUESTIONS FOR EXPERIMENT 12-1

Answer TRUE (T) or FALSE (F) to the following.

_____ 1. Batteries have internal resistance, but dc power supplies do not.

_____ 2. As load resistance increases, the terminal voltage decreases.

_____ 3. As load current increases, terminal voltage increases.

_____ 4. Connecting four batteries in parallel, each with $r_i = 100\ \Omega$, would increase the total r_i four times.

_____ 5. Connecting four batteries in series, each with $r_i = 100\ \Omega$, would increase the total r_i four times.

_____ 6. The internal resistance of a generator is always in parallel with a load.

_____ 7. Subtracting the load voltage from the no-load voltage gives a remainder that is equal to the IR voltage drop in the internal resistance of the source.

_____ 8. Internal resistance is in series with the load resistance.

_____ 9. Short-circuiting a battery will not drain the battery.

_____ 10. A 1.5 V dry cell battery with $1\ \Omega$ of internal resistance is probably a depleted battery.

TABLES FOR EXPERIMENT 12-1

TABLE 12–1.1 $r_i = 1\ k\Omega$ (Steps 1–3)

R_L	Measured V_L, V	Measured I_L, A	Calculated r_i, Ω
10 kΩ	_____	_____	_____
5.6 kΩ	_____	_____	_____
2.2 kΩ	_____	_____	_____
1 kΩ	_____	_____	_____
560 Ω	_____	_____	_____
200 Ω	_____	_____	_____

V_{NL} = _____

TABLE 12–1.2 $r_i = 560\ \Omega$ (Step 4)

R_L	Measured V_L, V	Measured I_L, A	Calculated r_i, Ω
10 kΩ	_____	_____	_____
5.6 kΩ	_____	_____	_____
2.2 kΩ	_____	_____	_____
1 kΩ	_____	_____	_____
560 Ω	_____	_____	_____
220 Ω	_____	_____	_____

TABLE 12–1.3 $V_T = 6$ V; $r_i = 560\ \Omega$ (Step 5)

R_L	Measured V_L, V	Measured I_L, A	Calculated r_i, Ω
10 kΩ	_____	_____	_____
5.6 kΩ	_____	_____	_____
2.2 kΩ	_____	_____	_____
1 kΩ	_____	_____	_____
560 Ω	_____	_____	_____
220 Ω	_____	_____	_____

TABLE 12–1.4   Steps 6–8

	V_{NL}, V	Short-circuit I, A	Calculated r_i, Ω
1.5-V battery	_____	_____	_____
Additional ____-V battery	_____	_____	_____

Note: Only the instructor will know the value of the three unknown values of r_i. In this way, your lab techniques and your ability to follow procedures will be tested.

TABLE 12–1.5 r_i Unknown; $r_L = 1$ kΩ (Step 9)

	Measured V_L, V	Measured I_L, A	Calculated r_i, Ω
Measured V_{NL} = _____			
r_i No. 1	_____	_____	_____
r_i No. 2	_____	_____	_____
r_i No. 3	_____	_____	_____

Chassis or box number (if applicable): _____

EXPERIMENT RESULTS REPORT FORM

Experiment No: _____   Name: _____
                                  Date: _____
Experiment Title: _____   Class: _____
_____   Instr: _____

Explain the purpose of the experiment:

List the first Learning Objective:
OBJECTIVE 1:

After reviewing the results, describe how the objective was validated by this experiment?

List the second Learning Objective:
OBJECTIVE 2:

After reviewing the results, describe how the objective was validated by this experiment?

List the third Learning Objective:
OBJECTIVE 3:

After reviewing the results, describe how the objective was validated by this experiment?

Conclusion:

If required, attach to this form: ☐ Answers to Questions, ☐ Tables, and ☐ Graphs.

EXPERIMENT 12-2

LOAD MATCH AND MAXIMUM POWER

OBJECTIVES

At the completion of this experiment, you will be able to:
- Validate that maximum source power is transferred to a load when the value of source $r_i = R_L$.
- Plot a graph of load power for differing values of load resistance.
- Understand the concept of maximum efficiency versus maximum power.

SUGGESTED READING

Chapter 12, *Basic Electronics*, Grob/Schultz, Tenth Edition

INTRODUCTION

When the internal resistance of a generator is equal to the load resistance, the load is considered matched to the source. The matching of load to source resistance is significant because the source can then transfer maximum power to the load.

Whenever $R_L = r_i$, maximum power is transferred to the load. When load resistance is more than r_i, the output voltage is more but the circuit current is less. When the load resistance is less than r_i, the output voltage is less but the circuit current is more. This experiment will provide data that you can analyze and thus prove that these concepts are valid.

The circuit of Fig. 12-2.1 and the accompanying graph of Fig. 12-2.2 illustrate the concept of matching a load to an internal source resistance to obtain maximum power transfer.

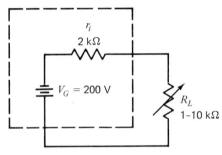

Fig. 12-2.1 Maximum power transfer circuit for analysis.

Because of the voltage divider formed by r_i and R_L, there is an equal voltage division: half of V_G is across r_i and half of V_G is across R_L. Under these circumstances, the load develops the maximum power that is possible using the particular source.

Referring to Fig. 12-2.1, as R_L increases, current decreases, resulting in less power dissipated in r_i. This results in more circuit efficiency because less power is lost across r_i. However, when $r_i = R_L$, the circuit efficiency is 50 percent.

$$\frac{P_L}{P_T} \times 100 = \text{circuit efficiency}$$

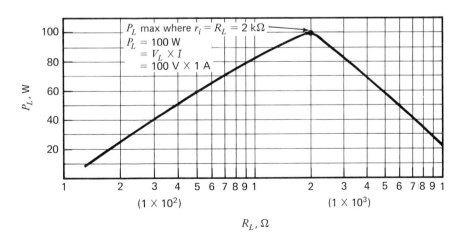

Fig. 12-2.2 Semilog graph of P_L versus R_L.

where P_T is the total power dissipated by the circuit, or

$$P_T = P_L + P_{r_i}$$

By this definition, 100 percent circuit efficiency means that absolutely no power is being dissipated.

EQUIPMENT

Ammeter
Voltmeter
DC power supply
Decade box
Protoboard or springboard
Leads

COMPONENTS

(1) 820 Ω, 0.25 W resistor

PROCEDURE

1. Connect the circuit of Fig. 12-2.3.

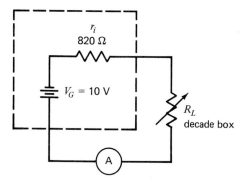

Fig. 12-2.3 Maximum power transfer circuit.

2. Increase the load resistance in 100 Ω steps from 100 Ω to 1 kΩ. Measure the voltage across R_L and the current at each step. Record the results in Table 12-2.1 for each step.

3. Increase the load resistance from 1 kΩ to 10 kΩ in 1 kΩ steps. Measure the voltage across R_L and the current at each step. Record the results in Table 12-2.1 for each step.

4. Calculate the *IR* voltage drop across r_i at each step, as $V_{r_i} = V_G - V_{R_1}$ and record the results in Table 12-2.1.

5. Calculate the load power dissipated at each step of R_L, as

$$P_L = V_L \times I$$

Record in Table 12-2.1.

6. Calculate the power dissipated across r_i at each step of R_L as

$$P_{r_i} = V_{r_i} \times I$$

Record in Table 12-2.1.

7. Calculate the total power dissipated in the circuit for each step of R_L as

$$P_T = V_G \times I$$

Note that

$$V_G = V_L + V_{r_i}$$

Record in Table 12-2.1.

8. Calculate circuit efficiency for each step of R_L as

$$\frac{P_L}{P_T} \times 100$$

expressed as a percentage.

9. Plot a graph of load resistance versus load power, using your data. Use two-cycle semilog paper. Prepare this graph as if it were to be used in a professional situation. It should be neat and well-organized, it should include a title and all possible values, and critical parameters should be labeled.

NAME	DATE

QUESTIONS FOR EXPERIMENT 12-2

1. Explain the difference between circuit efficiencies of 1, 50, and 100 percent. In other words, explain what is meant by *circuit efficiency* as it relates to transfer of maximum power.

2. Explain what would happen if the circuit of Fig. 12-2.3 had an internal resistance of 100 kΩ.

3. Explain what would happen if the circuit of Fig. 12-2.3 had an internal resistance of 0.001 Ω.

4. Explain why semilog paper is used to graph the data.

5. Explain how you could get maximum power transferred to a 15 kΩ load if the internal resistance of your source were 10 kΩ.

TABLE FOR EXPERIMENT 12-2

TABLE 12–2.1 Data for Circuit Fig. 12-2.3

R_L	Measured V_L, V	Measured I, A	Calculated V_{r_i}, V	Calculated P_L, W	Calculated P_{r_i}, W	Calculated P_T, W	% Efficiency
100 Ω							
200 Ω							
300 Ω							
400 Ω							
500 Ω							
600 Ω							
700 Ω							
800 Ω							
900 Ω							
1 kΩ							
2 kΩ							
3 kΩ							
4 kΩ							
5 kΩ							
6 kΩ							
7 kΩ							
8 kΩ							
9 kΩ							
10 kΩ							

EXPERIMENT RESULTS REPORT FORM

Experiment No: _____   Name: _____
                                          Date: _____
Experiment Title: _____   Class: _____
_____   Instr: _____

Explain the purpose of the experiment:

List the first Learning Objective:
OBJECTIVE 1:

After reviewing the results, describe how the objective was validated by this experiment?

List the second Learning Objective:
OBJECTIVE 2:

After reviewing the results, describe how the objective was validated by this experiment?

List the third Learning Objective:
OBJECTIVE 3:

After reviewing the results, describe how the objective was validated by this experiment?

Conclusion:

If required, attach to this form: ☐ Answers to Questions, ☐ Tables, and ☐ Graphs.

CHAPTER 13

EXPERIMENT 13-1

MAGNETISM

OBJECTIVES

At the completion of this experiment, you will be able to:
- Validate that current in a conductor has an associated magnetic field.
- Understand the concept of shielding.
- Examine the left-hand rule to determine magnetic polarity.

SUGGESTED READING

Chapters 13, *Basic Electronics*, Grob/Schultz, Tenth Edition

INTRODUCTION

Any electric current has an associated magnetic field that can do the work of attraction or repulsion. Not only is the magnetic field useful for doing work, it is also the cause of unwanted attraction and repulsion. Thus, it is often necessary to shield particular circuits to prevent one component from affecting another.

The most common example of magnetic force is that produced by a magnet. The magnet, with its north and south poles, acts as a generator that produces an external magnetic field provided by the opposite magnetic poles of the magnet. The idea is like the two opposite terminals of a battery that have opposite charges. Also, the earth itself is a huge natural magnet, having both north and south poles. Thus, the needle of a compass (also a magnet) is attracted to the north pole, because the atoms that make up the needle have been aligned in such a way that their magnetic field is attracted to the magnetic field of the earth's north pole.

It is these magnetic fields that are the subject of electromagnetism. These fields are thought of as lines of force, called *magnetic flux*, as shown in Fig. 13-1.1.

If current is flowing in a conductor, there is a similar magnetic field that can be used in conjunction with the fields of a magnet. For example, PM (permanent magnet) loudspeakers found in most radios, televisions, and public address systems all use the principles of magnetism to produce the audible sound we listen to.

Finally, the opposite effect of current moving through a conductor is a magnetic field in motion, forcing electrons to move. This action is called *induction*. Inductance is produced by the motion of magnetic lines of flux cutting across a conductor, thus forcing free electrons in the conductor to move, as shown in Fig. 13-1.2.

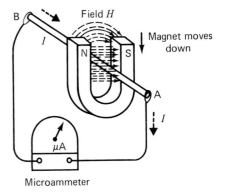

Fig. 13-1.2 Magnetically induced current in a conductor. Current *I* is electron flow.

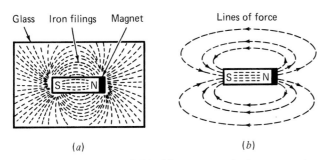

Fig. 13-1.1 Magnetic field of force around a bar magnet. (*a*) Field outlined by iron filings. (*b*) Field indicated by lines of force.

EQUIPMENT

DC power supply
Galvanometer or microammeter
Heavy-duty horseshoe magnet (> 20 lb pull)
Magnetic compass
Shield (6 × 6 in.) conductance sheet metal

COMPONENTS

2 to 3 ft of thin insulated wire
(1) No. 18 iron nail
Iron filings

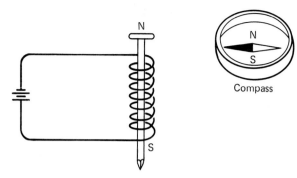

Fig. 13-1.3 Electromagnetic circuit for step 1.

PROCEDURE

1. Connect the circuit of Fig. 13-1.3. Have the compass and the iron filings nearby. Wrap the insulated wire evenly around the nail (about 10 to 15 turns).
2. The left-hand rule states that if a coil is grasped with the fingers of the left hand curled around the coil in the direction of electron flow, the thumb (extended) points to the north pole of the coil. Imagine your left hand grasped around the coil of wire wound around the nail. Determine which end is the north pole.
3. Turn the power on, and slowly move the compass close to both ends of the needle. Determine which end of the nail is north and which is south. Compare the results to step 2 above.
4. Place the shield between the compass and the nail in the circuit and repeat step 3. Note the results.
5. Turn the power off. Remove the nail and pass it through the iron filings. Note the results. Disconnect the circuit.
6. Connect the circuit of Fig. 13-1.4. Using the same wire as in steps 1 to 4 above, loop the wire around the magnet many times, making sure that the galvanometer is on the lowest range.
7. Move the horseshoe magnet up and down, as necessary, and note the amount of current produced. Try moving the magnet more rapidly, then slowly.
8. Disconnect the circuit.

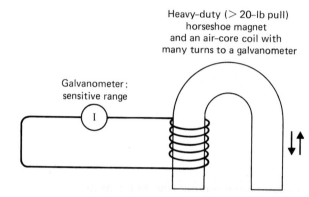

Fig. 13-1.4 Electromagnetic generator circuit for step 6.

QUESTIONS FOR EXPERIMENT 13-1

1. Explain what is meant by shielding.

2. Discuss the results of moving the magnet (in Step 7) faster or slower.

3. Explain which end of the nail attracted the iron filings, and why.

4. Discuss any differences between the results of Steps 1 to 4 and Steps 6 to 8.

5. Explain the left-hand rule as it was applied in this experiment.

EXPERIMENT RESULTS REPORT FORM

Experiment No: _____ Name: _____
                                 Date: _____
Experiment Title: _____ Class: _____
_____   Instr: _____

Explain the purpose of the experiment:

List the first Learning Objective:
OBJECTIVE 1:

After reviewing the results, describe how the objective was validated by this experiment?

List the second Learning Objective:
OBJECTIVE 2:

After reviewing the results, describe how the objective was validated by this experiment?

List the third Learning Objective:
OBJECTIVE 3:

After reviewing the results, describe how the objective was validated by this experiment?

Conclusion:

If required, attach to this form: ☐ Answers to Questions, ☐ Tables, and ☐ Graphs.

EXPERIMENT 13-2

ELECTROMAGNETISM AND COILS

OBJECTIVES

At the completion of this experiment you will be able to:
- Identify the physical characteristics of a solenoid coil.
- Understand what circuit characteristics influence magnetomotive force.
- Compare the dc characteristics of a solenoid coil.

SUGGESTED READING

Chapter 13, *Basic Electronics,* Grob/Schultz, Tenth Edition

INTRODUCTION

Electromagnetic fields are always associated with the flow of current in an electric circuit. From this current flow, magnetic units are defined. A coil of wire is also known as a *solenoid* and as current flows through the coil, magnetic relationships are also defined.

The movement of current through a coil provides a magnetizing force known as *magnetomotive force,* or *mmf.* This magnetomotive force is directly dependent upon the amount of current flow. Also, magnetomotive force creates a magnetic field density (H) that will decrease with the length of the solenoid. In addition, the field density (H) creates flux density (B) that is directly related to the permeability (μ) of the core of the solenoid.

An ideal solenoid is one that enhances the ability to create magnetic flux density (H). This is accomplished by creating a solenoid that has a length much greater that its diameter. Like the magnetic field that is created by passing a current through a single wire loop, the solenoid concentrates these magnetic field inside the coil and provides opposing magnetic poles at the respective ends. These effects are enhanced by the number of wire turns around the coils, the length of the coil, and the type of core material used within the coil.

In addition, the strength of the magnetic field depends upon the level of current that flows in the coil. This can be summarized as:
- The *larger* current, the *stronger* the magnetic field.
- The *greater* the number of wire-turns in a specific length of solenoid, the *greater* the concentration of the field.

This experiment analyzes the importance of the size (number of turns) and the type of core material used within a coil.

EQUIPMENT

Breadboard
24 V dc power supply
Ammeter
Voltmeter

COMPONENTS

Resistor: 33 Ω 2 W
Normally Open Push Button Switch

SPECIAL MATERIALS

12-inch wooden or plastic ruler
Solenoid coil (150 turns)
Solenoid coil (300 turns)
5-inch length of 1/4-inch diameter plastic rod
5-inch length of 1/4-inch diameter iron rod
5-inch length of 1/4-inch diameter wood rod
5-inch length of 1/4-inch diameter brass rod
5-inch length of 1/4-inch diameter aluminum rod
5-inch length of 1/4-inch diameter glass rod
Compass

PROCEDURE

1. Construct the circuit from the pictorial diagram as shown in Fig. 13-2.1.

2. Position the magnetic compass approximately 2 inches from the 150-turn coil. Turn the power supply on and adjust the supply's voltage until the ammeter reads 200 mA. Move the magnetic compass in such a way that there is just a slight needle deflection noted.

3. Insert the plastic rod into the center of the solenoid coil and note any changes in the position of the magnetic compass. Is there any change in the circuit's series current? Record your observation in Table 13-2.1.

4. Insert the iron rod into the center of the solenoid coil and note any changes in the position of the magnetic compass. Is there any change in the circuit's series current? Record your observation in Table 13-2.1.

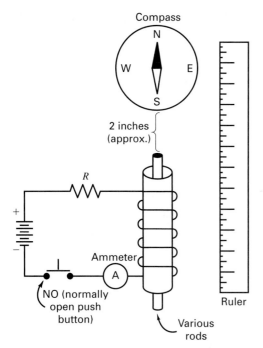

Fig. 13-2.1 Solenoid coil apparatus.

5. Insert the wood dowel into the center of the solenoid coil and note any changes in the position of the magnetic compass. Is there any change in the circuit's series current? Record your observation in Table 13-2.1.

6. Insert the brass rod into the center of the solenoid coil and note any changes in the position of the magnetic compass. Is there any change in the circuit's series current? Record your observation in Table 13-2.1.

7. Insert the aluminum rod into the center of the solenoid coil and note any changes in the position of the magnetic compass. Is there any change in the circuit's series current? Record your observation in Table 13-2.1.

8. Insert the glass rod into the center of the solenoid coil and note any changes in the position of the magnetic compass. Is there any change in the circuit's series current? Record your observation in Table 13-2.1.

9. Reverse the polarity of the power supply and repeat Procedure Steps 2 through 8. Record your observations in Table 13-2.2.

10. Reverse the polarity of the power supply, and substitute the 300-turn coil for the 150-turn coil and repeat Procedure Steps 2 through 8. Record your observations in Table 13-2.3.

11. Reverse the polarity of the power supply and repeat Procedure Steps 2 through 8. Record your observations in Table 13-2.4.

NAME	DATE

QUESTIONS FOR EXPERIMENT 13-2

1. Which of the materials inserted into the solenoid coil had the greatest effect on the compass? Which of the materials had the least effect?

2. Explain why some of the materials inserted into the solenoid had a greater effect than other materials.

3. What were the observable changes, if any, when the power supply's polarity was reversed?

4. Explain any observable changes when a 300-turn coil replaced the 150-turn coil.

5. What circuit characteristics influence the level of magnetomotive force?

TABLES FOR EXPERIMENT 13-2

TABLE 13–2.1 Circuit Characteristics of the 150-Turn Coil

Procedural Step	Record the Observed Change in Compass Deflection	Record the Observed Change in Measured Circuit Current
3	_____	_____
4	_____	_____
5	_____	_____
6	_____	_____
7	_____	_____
8	_____	_____

TABLE 13–2.2   150-Turn Coil with a Reversed Supply Polarity

Procedural Step	Record the Observed Change in Compass Deflection	Record the Observed Change in Measured Circuit Current
3	_____	_____
4	_____	_____
5	_____	_____
6	_____	_____
7	_____	_____
8	_____	_____

TABLE 13–2.3   Circuit Characteristics of the 300-Turn Coil

Procedural Step	Record the Observed Change in Compass Deflection	Record the Observed Change in Measured Circuit Current
3	_____	_____
4	_____	_____
5	_____	_____
6	_____	_____
7	_____	_____
8	_____	_____

TABLE 13–2.4   300-Turn Coil with a Reversed Supply Polarity

Procedural Step	Record the Observed Change in Compass Deflection	Record the Observed Change in Measured Circuit Current
3	_____	_____
4	_____	_____
5	_____	_____
6	_____	_____
7	_____	_____
8	_____	_____

EXPERIMENT RESULTS REPORT FORM

Experiment No: _____   Name: _____
                                   Date: _____
Experiment Title: _____ Class: _____
_____   Instr: _____

Explain the purpose of the experiment:

List the first Learning Objective:
OBJECTIVE 1:

After reviewing the results, describe how the objective was validated by this experiment?

List the second Learning Objective:
OBJECTIVE 2:

After reviewing the results, describe how the objective was validated by this experiment?

List the third Learning Objective:
OBJECTIVE 3:

After reviewing the results, describe how the objective was validated by this experiment?

Conclusion:

If required, attach to this form: ☐ Answers to Questions, ☐ Tables, and ☐ Graphs.

CHAPTER 14

EXPERIMENT 14-1

RELAYS

OBJECTIVES

At the completion of this experiment you will be able to:
- Identify a relay.
- Describe the function of a magnetic field associated with the flow of electric current.
- List and explain significant relay ratings.

SUGGESTED READING

Chapter 14, *Basic Electronics*, Grob/Schultz, Tenth Edition

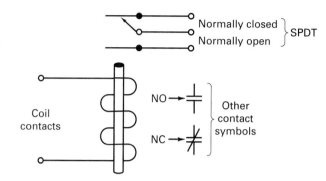

Fig. 14-1.1 Basic SPDT relay and related schematic symbols.

INTRODUCTION

The link between electricity and magnetism was discovered in 1820 by Hans Christian Oersted, who found that current moving through a conductor could move a compass's magnetic needle. Later the opposite effect was discovered; that a magnetic field in motion forces electrons (current) to move through a conductor.

Electromagnetism combines the characteristics of an electric current and magnetism. Electrons in motion always have an associated magnetic field. These electromagnetic effects have many practical applications that form the basis of motors, generators, solenoids, and relays.

A relay is an electromechanical device which uses either an ac or dc actuated electromagnet to open or close one or more sets of contacts. Relay contacts which are open when the relay is not energized are called Normally Open (NO) contacts, and relay contacts which are closed when the relay is not energized are called Normally Closed (NC) contacts. Relay contacts are held in their resting position either by a spring or some other mechanism.

Figure 14-1.1 details the basic relay and schematic symbols that are used to represent relay contacts. Similar to mechanical switches, the switching contacts of a relay can have any number of poles and throws. A relay will always have two electrical connections for the coil or electromagnet. Depending upon the type, or configuration, the relay can have any number of poles and throws. For example Fig. 14-1.1 depicts a SPDT (single-pole, double-throw) relay.

When terminals 1 and 2 are connected to a source voltage, current will flow through the relay coil and an electromagnet is formed. If there is sufficient current in the relay coil, the normally open (NO) contacts 3 and 4 close, and the normally closed (NC) contacts 4 and 5 open.

Relay Specifications

Manufacturers of electromechanical relays always supply a specification sheet for each of their relays. The specification sheet contains voltage and current ratings for both the relay coil and its switching contacts. The specification sheet also includes information regarding the location of the relay coil and switching contact terminals.

The following is an explanation of a relay's most important ratings.

Pickup current: The minimum amount of relay coil current necessary to energize or operate the relay.

Holding current: The minimum amount of current required keeping a relay energized or operating. (The holding current is less than the pickup current.)

Dropout voltage: The maximum relay coil voltage at which the relay is no longer energized.

Contact voltage rating: The maximum voltage the relay contacts are capable of switching safely.

Contact current rating: The maximum current the relay contacts are capable of switching safely.

EXPERIMENT 14-1 **239**

Contact voltage drop: The voltage drop across the closed contacts of a relay when operating.

Insulation resistance: The resistance measured across the relay contacts in the open position.

EQUIPMENT

24 V dc power supply
Ohmmeter
Ammeter
Voltmeter

COMPONENTS

24 V dc 3PDT relay (DAYTON #1A488-M, or equivalent)
Square relay socket, DIN/Screw Mounting 11 mounting pins
Normally open (NO) push-button switch
Normally closed (NC) push-button switch

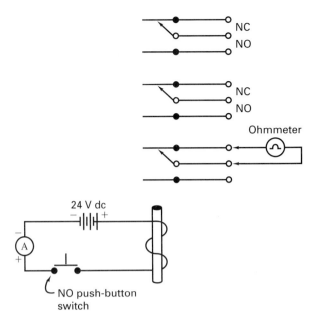

Fig. 14-1.2 Relay schematic diagram.

PROCEDURE

1. Obtain and inspect a 24 V dc 3PDT relay.
2. Obtain and inspect a compatible relay socket.
3. Align and plug the relay into the relay socket. Observe, inspect, and determine the terminals associated with the relay's coil (both normally open and closed contacts). On the schematic diagram shown in Table 14-1.1, identify and record the pin numbers for the coil, poles, and NO/NC contacts for the type of relay you are using for this experiment. Also in this table, associate the coil, poles, NO, and NC contacts for this relay to the relay's pin numbers.

OPTIONAL:

4. Obtain the manufacturer's specification sheet for the type of relay that you will be using for this experiment. Determine from the specification sheet the operating voltage, current and other electrical characteristics of the relay. Record this information in Table 14-1.2.
5. Connect the schematic diagram shown in Figure 14-1.2.
6. Turn on the power supply and increase the voltage level until the relay's coil just becomes energized. You will be able to determine this condition by hearing the relay "click" or by observing the relay's internal electrical contacts moving. When a relay is energized it is referred to as being "tripped." Note the power supply's voltage and circuit current level at this juncture, and record your results in Table 14-1.3. These measurements are known as the *pickup voltage* and *pickup current*.

7. Reduce the power supply voltage until the relay's coil is de-energized. Note the power supply's voltage at this point, and record this value in Table 14-1.3. This measurement is known as the *dropout voltage*. Through experimentation, by increasing and decreasing the power supply's voltage, determine the minimum amount of current necessary to keep the relay energized. Note this current level and record its value in Table 14-1.3. This measurement is known as the *holding current*.
8. Construct the circuit as shown in Figure 14-1.3.
9. Turn on and adjust power supply to 24 V dc. Press the normally open (NO) push button switch A and note the circuit action.
10. Press the normally closed (NC) push-button switch B and note the circuit action.
11. Repeat Procedural steps 9 and 10 several times.
12. Describe the observed circuit operation in Table 14-1.4.

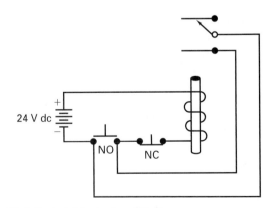

Fig. 14-1.3 Latching relay circuit.

NAME	DATE

QUESTIONS FOR EXPERIMENT 14-1

1. Describe how an ohmmeter could be used to determine if the relay's coil had opened.

2. Understanding that the relay is an electromechanical device, what electrical problems could you see developing when using the device over a long period of time?

3. If a set of normally closed (NC) relay contacts are closed when the relay's coil is energized, is the relay operating correctly?

4. It has been stated that one of the main advantages of using a relay is its ability to control high-power loads with a low amount of input power. Could you describe why this is an advantage?

5. What is the advantage of using a relay over a mechanical switch in remote control applications?

6. What is the purpose (and/or probable use of) the schematic circuit shown in Figure 14-1.3?

TABLES FOR EXPERIMENT 14-1

TABLE 14–1.1 Identifying the Pins and Functions of a Relay

Using the pin numbers on the schematic diagram shown below, identify the function of each pin. If you are using a different type of relay, other than a 3PDT, create your own table.	Pin Numbers and Relay Function. Use the terms, coil, pole, NO (normally open), and NC (normally closed) in completing this section of the Table.	
	Pin Number	Relay Term*
	1	_____
	2	_____
	3	_____
	4	_____
	5	_____
	6	_____
	7	_____
	8	_____
	9	_____
	10	_____
	11	_____

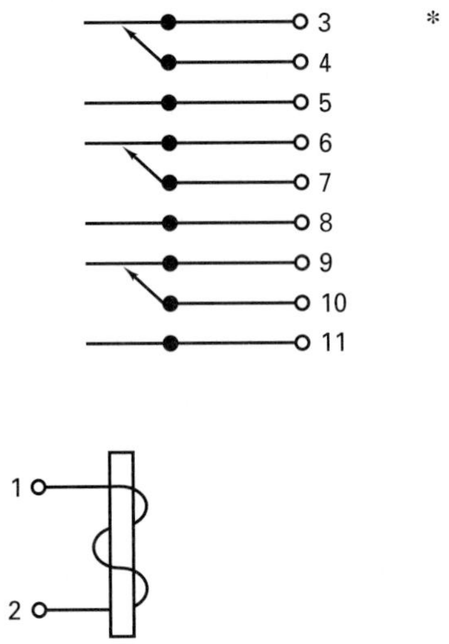

*Answers can vary

TABLE 14–1.2 Instructors Option-Operational Characteristics

Electrical Specification	Electrical Rating	Short Written Decription that Describes the Electrical Specification
Pickup voltage	_____ Volts	
Pickup current	_____ Amps	
Holding current	_____ Amps	
Dropout voltage	_____ Volts	
Contact voltage rating	_____ Volts	
Contact current rating	_____ Amps	
Contact voltage drop	_____ Volts	
Insulation resistance	_____ Ohms	

TABLE 14-1.3 Measured Operational Characteristics

Procedural Step	Operational Characteristic	Measured Value
6	_____ Pickup voltage	_____ Volts
6	_____ Pickup current	_____ Amps
7	_____ Dropout voltage	_____ Volts
7	_____ Holding current	_____ Amps

TABLE 14-1.4 Measured Operational Characteristics

Procedural Step 12	For the circuit constructed from the latching relay schematic shown in Figure 14-1.3, describe below the observed circuit action.

I observed... _____

_____

_____

_____

_____

EXPERIMENT RESULTS REPORT FORM

Experiment No: _____   Name: _____
                                  Date: _____
Experiment Title: _____   Class: _____
_____   Instr: _____

Explain the purpose of the experiment:

List the first Learning Objective:
OBJECTIVE 1:

After reviewing the results, describe how the objective was validated by this experiment?

List the second Learning Objective:
OBJECTIVE 2:

After reviewing the results, describe how the objective was validated by this experiment?

List the third Learning Objective:
OBJECTIVE 3:

After reviewing the results, describe how the objective was validated by this experiment?

Conclusion:

If required, attach to this form: ☐ Answers to Questions, ☐ Tables, and ☐ Graphs.

CHAPTER 15

EXPERIMENT 15-1

AC VOLTAGE AND OHM'S LAW

OBJECTIVES

At the completion of this experiment, you will be able to:

- Validate the Ohm's law expression for alternating current, where:

$$V = I \times R$$
$$I = \frac{V}{R}$$
$$R = \frac{V}{I}$$

- Determine the power dissipation in an alternating current circuit.
- Operate an ac oscillator or signal generator.

SUGGESTED READING

Chapter 15, *Basic Electronics*, Grob/Schultz, Tenth Edition

INTRODUCTION

This experiment is designed to introduce you to ac voltages and to validate that Ohm's law can still be used to determine current, voltage, or resistance if two of the three terms of an ac circuit are known. Also the amount of electric power, measured in watts, can be determined for ac circuits by using Ohm's law in the same way as for dc circuits.

Alternating current is electron flow in two directions: positive and negative. As shown in Fig. 15-1.1,

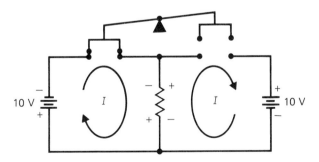

Fig. 15-1.1 Rocker switch alternates current flow.

two batteries can be switched on and off alternately to achieve the effect of ac voltage. Notice that, depending on which battery is switched on, the direction of current flow through the resistor will alternate. Now, imagine that you could move the rocker switch back and forth 100 times in 1 s. If you could do it, you would create an alternating current of 100 cycles/s, expressed as 100 hertz (Hz). Note that 1 Hz = 1 cycle/s, named after the 19th-century scientist Hertz.

The concept of alternating current is described not only by the frequency (in hertz) at which it alternates, but also by the amount of voltage that is being alternated. In the circuit of Fig. 15-1.1, the two batteries supply a voltage to the resistor at 10 V in each polarity or direction. Therefore, it is valid to say that the voltage across the resistor has +10 or −10 V across it at a particular time. In fact, this is a total potential of 20 V from one peak value (+10 or −10 V) to another.

Because it takes some amount of time for the voltage to reach the resistor, depending on the frequency of the alternating current, each potential voltage rises to its peak value (+10 or −10 V) and returns to zero before rising to the alternate peak value. And because this alternation of current flow occurs in a back-and-forth or oscillating manner, it is represented by a sine-wave symbol. Notice that Fig. 15-1.2 shows the number line of Fig. 15-1.3 combined with a sine wave to illustrate how ac voltage is represented.

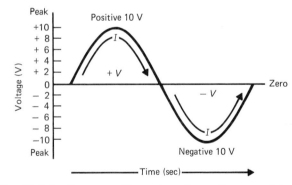

Fig. 15-1.2 Sine-wave illustration of ac voltage: peak to peak.

Fig. 15-1.3 Peak values on a number line.

The most common form of applying ac voltage to a circuit is by using an oscillator or signal generator. This instrument can be simple or complex, depending on

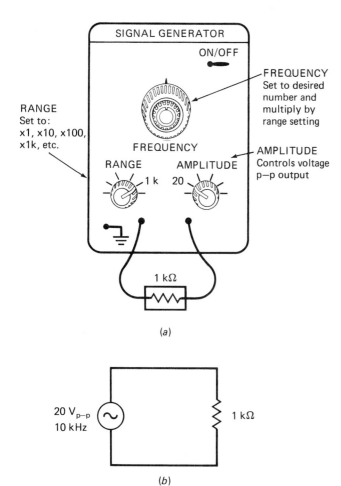

Fig. 15-1.4 (a) Simplified signal generator or oscillator connected to a 1 kΩ resistor. (b) Schematic.

its style and manufacture. However, Fig. 15-1.4a shows a simplified version of such an instrument.

Study the signal generator in your lab. If necessary, refer to its operating manual for details or instructions.

Notice that Fig. 15-1.4b shows the schematic version of the drawing where the sine-wave symbol represents the signal generator. Also notice that the signal generator is set to 10 kHz (10,000 Hz) using the range switch (×100) and the tuning control set to 10. The amplitude adjustment control is a voltage adjust knob that is used to set the output to the desired voltage (p-p) level.

In the ac circuit of Fig. 15-1.4, the 1 kΩ resistor has the same resistance as it would in a dc circuit. Thus, Ohm's law still dictates that

$$I = \frac{V}{R} = \frac{20 \text{ V}_{\text{p-p}}}{1 \text{ k}\Omega} = 20 \text{ mA}$$

The only difference is that the current is labeled as I (p-p) to show that it is the result of an ac voltage (p-p).

In ac circuits, an ammeter is not used to measure current because it is only built to measure current flow in one direction (direct current). Do not put a dc ammeter in an ac circuit.

Finally, power in an ac circuit is calculated slightly differently from that in a dc circuit. To calculate the power dissipation, the p-p voltage must be converted to its dc equivalent value. This is also called the *rms* (root-mean-square) *value* and is equal to 70.7 percent of the peak voltage. For example, a 20 V p-p voltage will produce the same heating or lighting power (in watts) as a 7 V dc battery. Thus, ac power is calculated as

$$\frac{V_{\text{p-p}}}{2 \times 0.707} = V_{\text{dc}} \text{ or } V_{\text{rms}} \text{ for use in the}$$

power equation

$$\text{Power (watts)} = \frac{V^2}{R}$$

$$\text{where } V = \frac{V_{\text{p-p}}}{2} \times 0.707$$

In summary, ac voltage can be used in any Ohm's law equation where I or V is labeled as a p-p value and converted to its dc equivalent (rms) for power calculations. Although ac voltages are usually measured on an oscilloscope, the following procedure will use a voltmeter (VOM or DVM) to validate Ohm's law.

EQUIPMENT

AC voltmeter (DVM or VOM) using the ac scale
Signal generator or oscillator
DC power supply
Springboard
Leads

COMPONENTS

Resistors (all 0.25 W):

(1) 100 Ω
(1) 1 kΩ

PROCEDURE

1. Connect the circuit of Fig. 15-1.5.

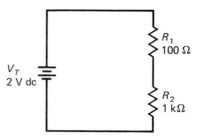

Fig. 15-1.5 Series circuit with dc voltage.

2. Measure and record in Table 15-1.1 the total voltage V_{R_1} and V_{R_2}. Compute the total current I_T and I_{R_1} and I_{R_2} by using Ohm's law.

3. Connect the circuit of Fig. 15-1.6.

248 EXPERIMENT 15-1

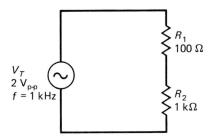

Fig. 15-1.6 Series circuit with ac voltage.

Note: Connect the entire circuit and then measure V_T to be sure it is 2 V p-p.

4. Repeat step 2 for the circuit of Fig. 15-1.6. Current should be calculated as:

$$I_{\text{p-p}} = \frac{V_{\text{p-p}}}{R}$$

5. Change the frequency to 5 kHz and repeat step 2 for Fig. 15-1.6.

6. Calculate the power dissipated by each circuit as shown in Table 15-1.1.

EXPERIMENT 15-1

QUESTIONS FOR EXPERIMENT 15-1

Answer TRUE (T) or FALSE (F) to the following:

_____ 1. Resistors do not function the same in ac circuits as in dc circuits.

_____ 2. A 10 V battery and a 7 V (p-p) signal will produce the same amount of power across a 2 kΩ resistor.

_____ 3. Ohm's law can be used only to find current in dc circuits.

_____ 4. Current in an ac circuit can be measured with an ammeter just as a dc circuit can.

_____ 5. A signal generator does not produce any current in an ac circuit.

TABLES FOR EXPERIMENT 15-1

TABLE 15-1.1

Resistance	Nominal Value	Volts Measured	Current $I = V/R$	Power, Watts, V^2/R
	**DC CIRCUIT—Fig. 15-1.5**
R_1	100 Ω	_____	_____	_____
R_2	1 kΩ	_____	_____	_____
R_T	1.1 kΩ	_____	_____	_____
	**AC CIRCUIT—Fig. 15-1.6* f = 1 kHz**
R_1	100 Ω	_____	_____	_____
R_2	1 kΩ	_____	_____	_____
R_T	1.1 kΩ	_____	_____	_____
	**AC CIRCUIT—Fig. 15-1.6* f = 5 kHz**
R_1	100 Ω	_____	_____	_____
R_2	1 kΩ	_____	_____	_____
R_T	1.1 kΩ	_____	_____	_____

*Note: Use p-p values. For example, 10 V p-p ÷ 100 Ω = 100 mA, 1 p-p.

Also, power = $\left(\dfrac{V\text{ p-p}}{2} \times 0.707\right)^2 \div R$ for ac circuits.

EXPERIMENT RESULTS REPORT FORM

Experiment No: _____ Name: _____
                                        Date: _____
Experiment Title: _____ Class: _____
_____   Instr: _____

Explain the purpose of the experiment:

List the first Learning Objective:
OBJECTIVE 1:

After reviewing the results, describe how the objective was validated by this experiment?

List the second Learning Objective:
OBJECTIVE 2:

After reviewing the results, describe how the objective was validated by this experiment?

List the third Learning Objective:
OBJECTIVE 3:

After reviewing the results, describe how the objective was validated by this experiment?

Conclusion:

If required, attach to this form: ☐ Answers to Questions, ☐ Tables, and ☐ Graphs.

CHAPTER 15

EXPERIMENT 15-2

BASIC OSCILLOSCOPE MEASUREMENTS FOR AC CIRCUITS

OBJECTIVES

At the completion of this experiment, you will be able to:
- Operate an oscilloscope.
- Measure dc voltages.
- Measure ac voltages.

SUGGESTED READING

Chapter 15, *Basic Electronics*, Grob/Schultz, Tenth Edition; the operating instructions for your oscilloscope; and Appendix F of this lab manual.

INTRODUCTION

This experiment continues the ac portion of your laboratory studies. The same rules for series and parallel circuits and Ohm's law still apply to ac circuits as they did for dc circuits. However, there is a difference. As the name *alternating current* implies, current is alternating its direction. That is, electron flow reverses its direction, and therefore positive and negative polarities will alternate every time electron flow changes direction.

Using the Oscilloscope

To study ac voltages in the lab, it is necessary to learn how to operate an oscilloscope. Although the oscilloscope is a complex instrument, it is basically a voltmeter. It measures and displays ac voltages as well as dc voltages.

There are some aspects of ac voltage that must be understood before attempting to measure them. AC voltages have frequency; that is, one alternation occurs over some period of time. For example, the ac wall outlet has a frequency of 60 hertz (Hz). This means that 60 complete cycles of electron flow, from negative to positive and back again, occur during 1 s. A frequency of 20 kHz means that 20,000 cycles occur during 1 s.

DC voltage does not have frequency. It is a steady, or direct, current flow that does not change over time. This difference in frequency is a major factor in understanding ac voltages.

The other important difference between ac and dc voltage is the way that magnitude is measured. DC voltage is usually steady and constant. AC voltage,

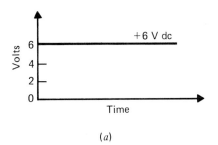

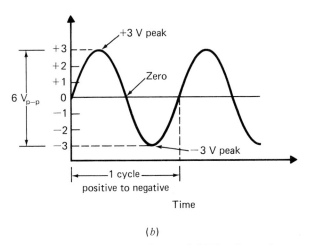

Fig. 15-2.1 DC versus ac voltages. (*a*) DC voltage plot. (*b*) AC voltage plot.

because of its alternating character, takes some time to reach a peak value, then it reverses direction and becomes a zero value before it reaches a peak value in the other direction. Figure 15-2.1 illustrates how this works. Notice how the dc and ac voltages differ. They differ in magnitude and frequency. The +6 V dc is referenced from zero. The 6 V peak-to-peak (p-p) ac voltage is referenced from zero but is described by its peak-to-peak relationship. Zero volts still exists, but the ac voltage goes above it and below it. The other difference is that the dc voltage has no frequency, whereas the ac voltage occurs during some period of time. If, for example, the cycle of Fig. 15-2.1*b* occurred during 1 ms, the frequency would be

$$f = \frac{1}{t} = \frac{1}{0.001 \text{ s}} = 1000 \text{ Hz or } 1 \text{ kHz}$$

where f = frequency
t = time it takes for a complete alternation

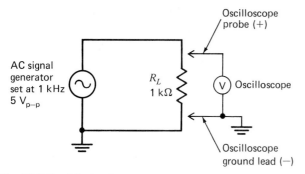

Fig. 15-2.2 AC circuit.

The circuit of Fig. 15-2.2 shows the ac signal generator with its sine-wave symbol. Notice that the ground symbol indicates that one side of the signal generator is grounded. This means that there is a true earth ground on the signal generator. This is where the ground lead of the oscilloscope is also connected. Always remember that the grounds must be connected together. If, for example, the oscilloscope leads in Fig. 15-2.2 were reversed, the load resistor would be between two earth grounds. Therefore, it would be effectively out of the circuit, and no voltage would be dropped across it. Properly connected, the oscilloscope will display all the applied voltage dropped across R_L, as shown in Fig. 15-2.3.

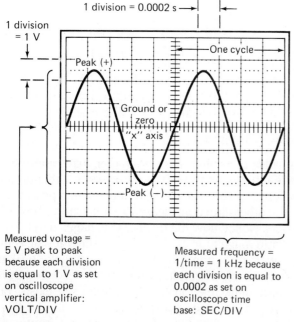

Fig. 15-2.3 Typical oscilloscope display for circuit shown in Fig. 15-2.2: 5 V_{p-p} sine wave at 1 kHz. Note that $f = 1/t$. Here, five divisions = 5×0.0002 s = 0.001. Thus $f = 1/0.001 = 1$ kHz.

If a dc voltage were measured the oscilloscope display would be different. Figure 15-2.4b shows how the oscilloscope would display the measured voltage. Notice that the dc measurement shows a trace that is a

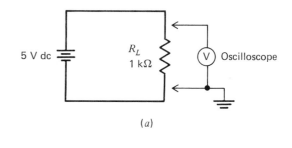

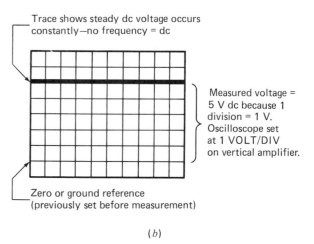

Fig. 15-2.4 (a) Oscilloscope measurement of dc voltage. (b) Oscilloscope display of dc measurement.

straight line. No frequency and no peak-to-peak value are displayed because there is no ac voltage, only a steady dc voltage.

For both ac and dc measurements, the oscilloscope must be adjusted prior to reading the trace from the display. In the same way that it may have taken you several attempts before you could adjust an ohmmeter and properly measure a resistor, it may take several attempts before you will be able to measure voltages by using the oscilloscope. The following procedures are meant to instruct you on how to use a scope. However, because there are so many different brands of oscilloscopes, you may receive some preliminary instructions from your laboratory instructor or you can read the operator's instructions for your oscilloscope. Also, you can read Appendix H in this manual. Most oscilloscope manuals have a section that is intended to get you started. It is a good idea to spend some time examining your oscilloscope before you start.

Keep in mind the following main points when you start the following procedures:

1. An oscilloscope is a voltmeter.
2. It measures magnitude and frequency of a signal.
3. The display is like a television display: you adjust it for the best picture. If a sine wave appears larger or smaller, it is due to the operator's adjustment.
4. The magnitude and frequency of the input signal are adjusted by using the signal generator controls.

5. Avoid ground loops by keeping all ground connections at the same point when measuring ac voltages.

Signal Generators Supply the Current and Voltage

A signal generator is like a power supply because it is a source of power—voltage and current—but it supplies ac voltages instead of dc voltages. The voltages are in the lower RF (radio frequency) range, usually less than 1 GHz, with very safe or low values of voltage and current. Some signal generators supply only sine waves and are often called oscillators, but many are capable of supplying many other signals, including nonsinusoidal waveforms such as square waves and sawtooth waves. These instruments are called *function generators*. Regardless of the type you have, you will need to learn how to use its front panel to supply ac voltages. If it is extremely complex, you may need to ask for help or refer to the operating manual.

EQUIPMENT

Oscilloscope (preferably, a late-model, solid-state, auto-triggering type, including an operator's manual)
Signal generator or function generator
VTVM/VOM/DMM
Voltmeter
DC power supply
Springboard
Leads

COMPONENTS

Resistors (all 0.25 W):

(1) 2.2 kΩ   (1) 10 kΩ
(1) 4.7 kΩ

PROCEDURE

Measuring DC Voltages

1. Set up the oscilloscope; locate the oscilloscope probe. It should be connected to the channel 1 (or A) input for dual channel scopes. The probe should have a positive and a negative lead. Be sure the scope is plugged into the ac power line. Do not turn on the power yet. Set the front panel controls as follows:

Set the INTENSITY to midrange.
For dual trace scopes, set the VERTICAL MODE (trigger) to channel 1 or channel A.
Set the VOLTS/DIV (vertical amplifier, channel 1) to 1 V per division.
Set the AC-DC-GROUND switch (input coupling) to the ground position.
Set the SEC/DIV (time base) to 0.2 s.
Set the TRIGGERING for auto, or for normal (manual) if no auto trigger exists.
Also, be sure the trigger control is set to the INTERNAL mode, not to external or line.
Turn the FOCUS control to midrange.
Be sure the vertical amplifier and time base controls are in the CAL position (typically, fully clock-wise).

Now, turn the power on.

Note: The remaining front panel controls will vary, depending upon the model. Therefore, be sure to consult your scope manual or your instructor if you have any concerns.

2. The trace should appear as a straight line across the display graticule. Note that the graticule is eight divisions vertically and ten divisions horizontally. If the trace does not appear, push the button marked BEAM, or BEAM FINDER, etc. If some type of trace appears when you push this, it means the scope is probably working. In any case, now use the following controls to adjust the trace so that a sharply defined trace is displayed at the middle of the XY axis, as in Fig. 15-2.5.

VERTICAL POSITION control ① Adjusts trace up and down.
FOCUS control ② sharpens the trace.
HORIZONTAL POSITION control ③ Adjusts trace right to left.

Here, in Fig. 15-2.5, the trace is a sharply defined straight line. Because the input coupling switch is set to GROUND, this trace means that ground, or zero volts, is at the division line. Thus, each division line above the ground line equals +1 V, and each division line below equals −1 V. This is similar to zeroing a VTVM.

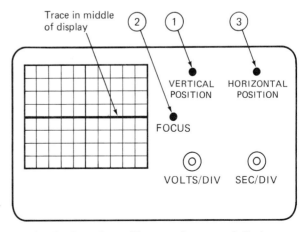

Fig. 15-2.5 Generic oscilloscope front panel display.

EXPERIMENT 15-2 **255**

If your trace looks like the one in Fig. 15-2.5, then you are doing fine. If not, repeat steps 1 and 2 or ask for assistance.

3. Leave the scope as it is. Connect the circuit of Fig. 15-2.6.

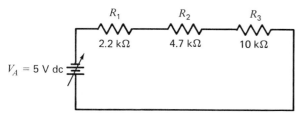

Fig. 15-2.6 Simple dc series circuit.

Note: Use the dc power supply and a voltmeter.

4. Using the voltmeter, measure and record all the voltages for the circuit of Fig. 15-2.6. Refer to Table 15-2.1. These voltmeter-measured dc voltages will be compared to the same voltages measured with the scope.

5. Now, use the oscilloscope to measure the same dc voltages. Connect the ×1 (times 1) scope probe across R_1 (2.2 kΩ). Move the resistors around to avoid group loops if necessary. Usually, for this type of experiment it is not necessary.

With the scope leads across R_1, you should still see no change in the scope display because the input coupling is set to ground. Therefore, now switch this control to the dc position. The trace should now move slightly higher on the graticule display. Here is how you would read it.

Because the channel 1 VOLTS/DIV control is set at 1.0 V per division, each major vertical division on the graticule is equal to 1.0 V. Therefore, if the trace appears at 0.7 divisions above ground, the dc measured voltage is equal to 0.7 divisions multiplied by 1.0 V (0.7 × 1.0 = 0.7 V). Multiply the VOLTS/DIV value by the number of divisions. Using 0.1 V per division is more accurate.

Record the voltage in Table 15-2.1.

6. Go ahead and measure the remaining dc voltages for R_2 and R_3 of Fig. 15-2.6, using the oscilloscope. These values should be the same as those measured with the voltmeter, plus or minus 15 percent. Do not measure V_T yet.

7. Reverse the power supply leads on the circuit of Fig. 15-2.6, and remeasure each resistor voltage. Notice that the trace appears below the center zero/ground reference in the same proportions as it appeared in steps 5 and 6 above. These are the same voltages except that, with respect to ground, they are opposite in polarity, and thus, they are read as negative voltages. Do not record these measurements.

8. To measure the total voltage, connect the scope across all three resistors with the ground lead at the negative side. Notice that the trace has disappeared from the graticule display. Now, readjust the vertical amplifier VOLTS/DIV control to 2 V per division. The trace should now appear. This is like changing ranges on the VTVM/VOM. Read the trace and record the value. Because the total voltage is +5 V dc, and the VOLTS/DIV = 2, you should be reading 2.5 divisions positive (above ground reference).

9. With the oscilloscope probe still across V_T, slowly increase and then decrease the applied voltage from the dc power supply. Notice how the trace rises and falls on the display in proportion to the voltage adjustment.

After completing this portion (above) of the experiment, you should be able to understand that the oscilloscope is basically a voltmeter.

Measuring AC Voltages

The oscilloscope will display ac voltages so that their waveshapes can be observed. In this way, the amplitude and the frequency can be measured, referenced to an XY axis: the amplitude is read by analyzing the vertical Y axis; the frequency is read by analyzing the horizontal X axis.

Notice in Fig. 15-2.3 in the introduction that the sine-wave amplitude is five vertical divisions from peak to peak. Because the VOLTS/DIV control is set at 1 V per division, each peak is about 2.5 divisions with respect to ground, which equals 5 V peak to peak. From the peak-to-peak value, the following values, typical for describing ac voltages, can be determined:

$$V_{peak} \text{ or } V_{max} = \frac{V_{p\text{-}p}}{2}$$

$$V_{av} = \frac{V_{p\text{-}p}}{2} \times 0.636$$

$$V_{rms} = \frac{V_{p\text{-}p}}{2} \times 0.707$$

Frequency is read as the number of cycles that occur during 1 s (cycles per second = hertz). Notice in Fig. 15-2.3 that one cycle occurs over five horizontal divisions. This is measured from zero to zero on the X axis as long as ground (zero reference) is set at a desired place and noted. Here, time base control (sometimes called the *horizontal amplifier*) is set for 0.2 ms per division. Thus, one cycle occurs over five divisions, or 5 × 0.2 ms = 0.001 s (1 ms). Because $f = 1/t$, the frequency equals 1 kHz, or 1000 cycles per second.

1. Readjust the oscilloscope. Set the VOLTS/DIV to 1. Set the input coupling switch back to the ground

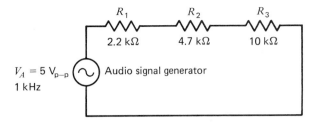

Fig. 15-2.7 Simple ac series circuit.

position, and set the trace so that ground is, once again, at midscale XY, with a sharply defined trace.
2. Connect the circuit of Fig. 15-2.7.

Note: The 5 V_{p-p} 1-kHz sine wave is applied to the circuit by using a signal generator (oscillator). Do not be concerned if you cannot adjust it yet. Simply connect the output of the signal generator as you did the dc power supply. However, there are no positive or negative terminals; they are both the same because it is alternating current. If there is a grounded terminal, avoid using it if possible.

3. Connect the scope leads across the entire circuit. Then, set the input coupling for alternating current. The display should be a sine wave. Adjust the amplitude and frequency controls of the signal generator so that the trace is approximately 5 V_{p-p} at 1 kHz. The display should be very similar to the one shown in Fig. 15-2.3.
4. Go ahead and measure all the voltages in the circuit—R_1, R_2, R_3—and of course V_T that is now being displayed. Record the values in Table 15-2.2. Change the VOLTS/DIV, or any other setting, so that you get the biggest display without going off the screen. Record all the values in Table 15-2.2, and calculate the remaining values.
5. Reconnect the scope leads across the entire circuit and adjust the signal generator so that 4 V_{p-p} at 20 kHz is the applied voltage. Then, adjust the scope so that the biggest amplitude possible without going off the CRT is displayed. Also, adjust it so that only two complete cycles appear on the screen. Draw a picture of the trace on the graticule of Fig. 15-2.8, and record the VOLTS/DIV and SEC/DIV setting you used

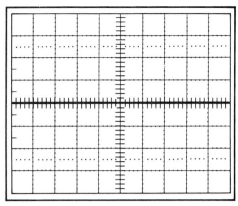

VOLTS/DIV = _____ SEC/DIV = _____

Fig. 15-2.8 Graticule. Displayed waveform = 4 V_{p-p} at 20 kHz.

to get this display. If your lab instructor requires that you hand in Fig. 15-2.8, you can use one of the graticule sheets in Appendix F, or you can use engineering paper and draw your own.
6. Spend a few moments adjusting the signal generation to other dc and ac values and measure them with the scope. Use this time to become more familiar with measuring ac voltages of differing frequencies and waveshapes.

FURTHER EXPLORATION

1. Disconnect the above circuit. Connect a permanent magnet speaker to the input leads of the oscilloscope as shown in Fig. 15-2.9.
2. Set the input coupling switch to alternating current.
3. While whistling into the speaker, adjust the VOLTS/DIV, or any other oscilloscope setting, so that you get the largest display without going off the screen. As you review the circuit action, determine the operating nature of the speaker. For example, is the speaker producing and direct of alternating current? How can the frequency of the signal be changed? Predict how the display will change if the oscilloscope has its input coupling switch changed from ac to dc coupling.

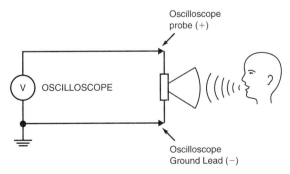

Fig. 15-2.9 Oscilloscope connected to a permanent magnet speaker.

EXPERIMENT 15-2

QUESTIONS FOR EXPERIMENT 15-2

Answer TRUE (T) or FALSE (F) to the following:

_____ 1. An oscilloscope can measure both ac and dc voltages.

_____ 2. The horizontal amplifier section controls the amplitude of the display on the CRT.

_____ 3. The vertical amplifier section can be adjusted to increase or decrease the peak-to-peak voltage display.

_____ 4. The oscilloscope probe does not have a grounded side.

_____ 5. The oscilloscope input coupling switch is usually set to GND, AC, or DC, depending upon the desired usage.

TABLES FOR EXPERIMENT 15-2

TABLE 15–2.1 Measured DC Voltages for Fig. 15-2.6

	Voltmeter	Scope
V_{R_1}		
V_{R_2}		
V_{R_3}		
V_T		

TABLE 15–2.2 AC Voltages for Fig. 15-2.7

	Measured V_{p-p}	Calculated V_{peak}	Calculated V_{av}	Calculated V_{rms}
$R_1 = 2.2\ k\Omega$				
$R_2 = 4.7\ k\Omega$				
$R_3 = 10\ k\Omega$				
V_T				

EXPERIMENT RESULTS REPORT FORM

Experiment No: _____ Name: _____
                                         Date: _____
Experiment Title: _____   Class: _____
_____   Instr: _____

Explain the purpose of the experiment:

List the first Learning Objective:
OBJECTIVE 1:

After reviewing the results, describe how the objective was validated by this experiment?

List the second Learning Objective:
OBJECTIVE 2:

After reviewing the results, describe how the objective was validated by this experiment?

List the third Learning Objective:
OBJECTIVE 3:

After reviewing the results, describe how the objective was validated by this experiment?

Conclusion:

If required, attach to this form: ☐ Answers to Questions, ☐ Tables, and ☐ Graphs.

CHAPTER 15

EXPERIMENT 15-3

ADDITIONAL AC OSCILLOSCOPE MEASUREMENTS

OBJECTIVES

At the completion of this experiment, you will be able to:
- Improve your ability to operate a signal generator.
- Operate various oscilloscope front panel controls.
- Measure various waveshapes and calculate values.

SUGGESTED READING

Chapter 15, *Basic Electronics,* Grob/Schultz, Tenth Edition

INTRODUCTION

An oscilloscope, commonly called a *scope,* is a sophisticated voltmeter that operates in the time domain. It measures and displays both ac and dc signals as they occur over time. The previous experiment introduced you to basic dc and ac measurements in a series circuit. You measured amplitude (voltage) and time, computing frequency (Hz) from the time measurement and calculating peak, rms, and average values from the p-p voltage measurements. You learned the basic operation, including how to set the trace to ground (0 V) for both dc and ac measurements. This experiment will allow you to practice the skills introduced in the previous experiment, and it will introduce new ones that will prepare you for future labs.

Understanding the Oscilloscope System

Every scope has four basic systems: display, horizontal, vertical, and trigger. Each of the four systems is controlled or set up by front panel inputs, switches, and adjustments. Knowing what these systems do will help you understand how to use the front panel. Appendix H in this manual describes those systems in a manner similar to the descriptions in oscilloscope manuals. If you have not done so already, read Appendix H or read your oscilloscope manual to gain a general idea of how those systems work.

The Oscilloscope Probe

When you use an oscilloscope to measure higher frequencies, the probe prevents other signals such as radio signals and 60 Hz power line radiated signals from getting into the measurement. In addition, the oscilloscope does not have the high resistance (impedance) of a dc voltmeter, so it could allow also too much circuit current to flow through it in some cases. All of these effects would *load down* the circuit, which means that the oscilloscope would, in effect, become part of the circuit rather than an isolated measuring device. Your measurements would therefore be inaccurate. To avoid this, the probe is specially designed to block out unwanted signals and increase the resistance. Resistive probes are sometimes called *attenuator probes* because they decrease the signal amplitude by 10 times or more (×10 probe). Other probes are also available for specific measurements, usually of higher frequencies or lower circuit impedances (50 Ω).

Signal Generators

You cannot learn to use an oscilloscope without using a signal generator to supply signals to the circuit to be measured. A signal generator is like a power supply because it is a source of power—voltage and current—but it supplies ac voltages instead of dc voltages. The voltages are in the lower RF (radio frequency) range, usually less than 1 GHz, with very safe or low values of voltage and current. As mentioned previously, some signal generators supply only sine waves and are often called oscillators, but many are capable of supplying many other signals, including nonsinusoidal waveforms such as square waves and sawtooth waves. These instruments are called *function generators* and they are invaluable for testing digital circuits.

EQUIPMENT

Signal generator (function generator)
Oscilloscope: auto-triggering, properly grounded
Probe: ×1

COMPONENTS

Resistors: 2.2 kΩ, 4.7 kΩ, 10 kΩ

PROCEDURE

Front Panel Identification

1. Identify the front panel elements of your scope by writing down which section they belong to as follows. On a separate sheet of paper, make four columns and label them as: DISPLAY, VERTICAL, HORIZONTAL, and TRIGGER. In each column, list the switches, buttons, knobs, input connections, etc., that are in that system.

Setup for a Free-Running Display

A free-running display has no signal applied, and the trace is not triggered (it does not start at any particular time). Setting up the scope in this manner, every time you use it, will help you learn its operation, provide a consistent starting point or reference, and help you to avoid incorrect measurements.

2. With the scope power off, set the following controls:

INTENSITY: completely off (full counterclockwise)
VOLTS/DIV: largest number possible (least sensitivity)
INPUT COUPLING (AC-DC-GROUND): ground
SEC/DIV (time base): 1 ms
TRIGGERING: set to INTERNAL (source) and AUTO (mode)
CAL: fully clockwise for all knobs
SLOPE: (+)

3. Turn power on.
4. Press the BEAM finder and hold it down. At the same time, increase the INTENSITY slowly until the trace appears and then release the beam switch—use the POSITION knob to center the beam and adjust the FOCUS for a sharp display in the center of the screen. You now have a starting point to begin making measurements.
5. Turn off the scope and change all the settings randomly. Repeat steps 2, 3, and 4.

Measuring the AC Signals (Fields) in the Air

This simple measurement should allow you to see the 60 Hz signals that are present in the air because they are radiated everywhere around you. By using your body as a receiving antenna, the scope can measure the 60 Hz signals present in the air.

6. With the probe and lead connected to the vertical input, set the following:

INPUT COUPLING: AC
SEC/DIV: 5 ms

7. Hold the end of the probe tip and adjust the VOLTS/DIV until a signal appears. If the trace is not stable, switch the trigger mode from AUTO to NORMAL and set the SLOPE to (+).

Observations: Note the waveform. Is it 60 Hz? If so, what is the amplitude? Switch the INPUT COUPLING to the dc position and notice any change.

8. Return the scope to the free-running setup.

Using Front Panel Controls

9. Measure the resistors used in this experiment and replace any that are not within tolerance. Then connect the circuit of Fig. 15-3.1.

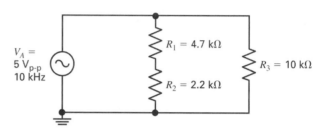

Fig. 15-3.1 Sine-wave series-parallel circuit.

10. Adjust the oscilloscope as necessary to measure the p-p voltages across each resistor. Record the values in Table 15-3.1.

11. Vertical CAL—This control is often used to determine when a signal reaches the point where its power is decreased by one-half. With the scope leads connected across R_1, turn the vertical VOLTS/DIV CAL knob back and forth to see what effect it has on the displayed trace. Adjust the knob so the display shows a measured valued across R_1 that is 70.7 percent (half-power) of the previously measured p-p value. Record the value. Leaving the knob in this position, measure the values across the other resistors and record the values in Table 15-3.1. They should all be 70 percent of their previous values. Afterward, return the CAL knob to the calibrated position.

12. Horizontal CAL—This control is useful for comparing sine waves or aligning them. With the scope leads across R_1, turn the horizontal (SEC/DIV) CAL knob back and forth to see what effect it has on the displayed trace. Adjust the CAL so that the frequency appears to be one-half the original frequency. Record the frequency and the V_{p-p} value. Repeat the measurements across each resistor and record the values in Table 15-3.1. Return the SEC/DIV CAL knob to the calibrated position.

Note: If your scope does not have this control, look for the magnification factor (MAG) switch and set it to ×5 or ×10. Note the factor in the table, measure across all resistors, and record the results.

13. SLOPE and LEVEL—This trigger control determines where the trace starts. With the scope leads across R_1, turn the slope LEVEL knob back and forth to see what effect it has on the display. Then set the SLOPE button to (−) or opposite of what it was. Measure $V_{p\text{-}p}$ across R_1, R_2, and R_3 and record the results.

Measuring Square Waves or Rectangular Waves

Square waves are displayed just like sine waves except that the measurements are different. Both have p-p, peak, and Hz, but the square wave does not have rms or average voltage values. Instead, square-wave measurements have the values shown in Fig. 15-3.2.

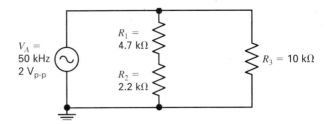

Fig. 15-3.3 Square-wave measurement circuit.

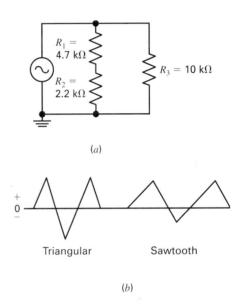

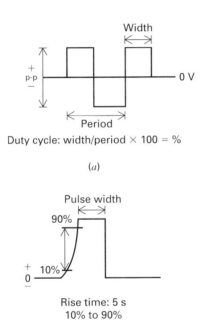

Fig. 15-3.2 Square- or rectangular-wave measurements. (*a*) Square or rectangular wave. (*b*) Pulse.

Fig. 15-3.4 Sawtooth- or triangular-wave circuit and wave-shape. (*a*) Circuit. (*b*) Waveshape.

Measuring Sawtooth Waves or Triangular Waves

16. Return the scope to the free-running setup and connect the circuit of Fig. 15-3.4, where the signal generator (function generator) now supplies a sawtooth wave or triangular wave.

17. Measure the $V_{p\text{-}p}$ value across R_1, R_2, and R_3. Record the results in Table 15-3.2.

18. PROBE ADJUST SIGNAL—This control is used to compensate attenuator probes to make accurate measurements of digital circuits. Locate this control on the front panel and try to measure the signal by connecting the tip of the scope probe (+) to it and the negative side to any ground connection. Observe the signal—it is used for adjusting ×10 probes. If you are using a ×1 probe, no adjustment is required. If you have a ×10 probe, refer to the scope manual for information on how to adjust a ×10 probe. In Table 15-3.2, record the values of the signal (×1 probe) as you did for the square-wave measurements.

14. Return the scope to the free-running setup and connect the circuit of Fig. 15-3.3, where the signal generator (function generator) now supplies a square wave or a rectangular wave. Note the type of wave in Table 15-3.2.

15. For the circuit of Fig. 15-3.3, measure the following values across R_1, R_2, and R_3: $V_{p\text{-}p}$, period (s), pulse width (s), and rise time. Rise time is the time it takes to go from 10 to 90 percent of the magnitude (see Fig. 15-3.2). Record the values in Table 15-3.2.

EXPERIMENT 15-3 **263**

NAME	DATE

QUESTIONS FOR EXPERIMENT 15-3

Provide a short answer each of these questions, referring to the measurements.

1. What is the purpose of starting with a free-running setup?

2. What does the vertical CAL knob do?

3. What does the horizontal CAL knob do?

4. What does the SLOPE button do?

5. What does the LEVEL adjust do?

6. What observations did you make in Step 7?

7. What observations did you make in Step 18?

8. What is rise time?

9. Does a square wave have an rms value? Why?

10. Explain why the dual-trace scope has become so popular.

TABLES FOR EXPERIMENT 15-3

TABLE 15–3.1 Sine-Wave Measurements

$V_A = 5\,V_{p\text{-}p}$ $f = 10\,kHz$	$V_{p\text{-}p}$	VERT CAL 70.7% $V_{p\text{-}p}$	HORIZ CAL ½ FREQ (Hz)	HORIZ CAL ½ FREQ $V_{p\text{-}p}$	SLOPE — $V_{p\text{-}p}$
R_1					
R_2					
R_3					

TABLE 15–3.2 Wave Type = _____ (Square or Rectangular)

$V_A = 2\,V_{p\text{-}p}$ $f = 50\,kHz$	$V_{p\text{-}p}$	Period (s)	Pulse Width (s)	Rise Time 10% to 90% (s)
R_1				
R_2				
R_3				

Wave Type = _____ (Sawtooth or Triangular)

R_1 _____

R_2 _____

R_3 _____

PROBE ADJ Signal

$V_{p\text{-}p}$	Period	Pulse Width	Rise Time

EXPERIMENT RESULTS REPORT FORM

Experiment No: _____　　Name: _____
　　　　　　　　　　　　　　　　　Date: _____
Experiment Title: _____　　Class: _____
_____　　Instr: _____

Explain the purpose of the experiment:

List the first Learning Objective:
OBJECTIVE 1:

After reviewing the results, describe how the objective was validated by this experiment?

List the second Learning Objective:
OBJECTIVE 2:

After reviewing the results, describe how the objective was validated by this experiment?

List the third Learning Objective:
OBJECTIVE 3:

After reviewing the results, describe how the objective was validated by this experiment?

Conclusion:

If required, attach to this form: ☐ Answers to Questions, ☐ Tables, and ☐ Graphs.

CHAPTER 15

EXPERIMENT 15-4

OSCILLOSCOPE LISSAJOUS PATTERNS

OBJECTIVES

At the completion of this experiment, you will:
- Be familiar with the operation of a frequency counter.
- Be familiar with the XY operation of an oscilloscope.
- Be able to measure frequency by using Lissajous patterns.

SUGGESTED READING

Chapter 15, *Basic Electronics,* Grob/Schultz, Tenth Edition

INTRODUCTION

This experiment demonstrates how a frequency measurement can be made by using different techniques. In previous experiments you learned how to measure the frequency of an ac signal by using the time base of the oscilloscope. That method allowed you to derive the frequency f by observing the time (seconds per division) it took for 1 cycle to occur: $f = 1/\text{time}$. In most cases, that technique is adequate for determining the approximate frequency of an ac signal.

However, the easiest and most accurate method for measuring frequency is to use a digital frequency counter. Digital frequency counters usually have LED (light-emitting diode) displays that are easy to read. In addition, the digital frequency counter is extremely accurate because it uses a crystal oscillator as a reference to count or compare the frequency you are measuring.

Another method for measuring the frequency of an ac signal is to create a Lissajous pattern on an oscilloscope. These patterns are named after Lissajous, a well-known French physicist. The pattern is created by simultaneously applying two ac signals to the horizontal and vertical deflection circuits of an oscilloscope.

To do this, the time base of the oscilloscope must be disabled. Most oscilloscopes have a setting on the time base marked XY. When the time base is set to this feature, it separates the two channels (using a dualtrace oscilloscope) so that one channel goes to the horizontal deflection circuit and the other channel goes to the vertical deflection circuit. During normal operation (not XY), both channel 1 and channel 2 send signals to the horizontal section, and the vertical section is controlled by the time base. However, when the XY mode is set, the oscilloscope will display a pattern that shows the frequency relationship of two ac signals. Figure 15-4.1 shows an example of four Lissajous patterns where the ratio of two sine-wave signals is determined by the number of horizontal and vertical peaks or lobes.

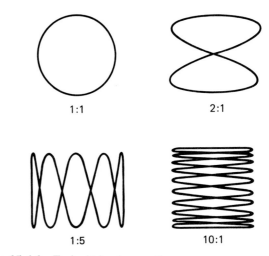

Fig. 15-4.1 Typical Lissajous patterns.

Although this experiment introduces you to Lissajous patterns and the XY operation of an oscilloscope, you should refer to the oscilloscope's operating manual for detailed information about XY operation. This experiment will simply introduce you to the concepts.

Figure 15-4.2 shows the three techniques discussed in this experiment to make frequency measurements. Notice that each technique shows a 1 kHz sine-wave signal except for the Lissajous pattern technique, where a 1 kHz signal is compared to another 1 kHz signal to create the 1:1 circle pattern as in Fig. 15-4.1.

Because this experiment concentrates on the Lissajous pattern method for measuring frequency on the oscilloscope, you must have a good idea of what you expect to see on the CRT. Figure 15-4.3 shows how to read the Lissajous pattern. In addition, remember that Lissajous patterns will not be stable or readable unless there is an integer (whole-number)

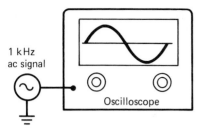

(a) Oscilloscope time-base method

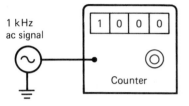

(b) Frequency counter method

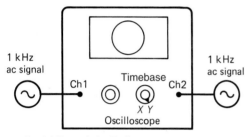

$f = 1$ kHz = circle display
When channel 1 and 2 both have signals of equal value in magnitude

(c) Lissajous pattern (X Y) method

Fig. 15-4.2 Three techniques for measuring frequency.

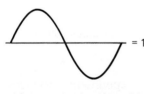

(a) Input signal to channel 1

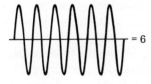

(b) Input signal to channel 2

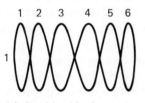

(c) Resulting Lissajous pattern

Fig. 15-4.3 Lissajous pattern 1:6.

relationship between the two frequencies. Also if either channel 1 or channel 2 frequencies are changed even slightly, the Lissajous pattern will rotate and change shape.

Notice that Fig. 15-4.3 shows the two signals input to channels 1 and 2. If channel 1 is 100 Hz, then channel 2 must be 600 Hz; or channel 1 could be 500 Hz, and channel 2 could be 3 kHz. In every case, the frequency measurement technique, using Lissajous patterns, requires that the frequencies be exact and have low integer ratios. If they do not, the XY relationship will not yield a pattern where the lobes are easy to read or count.

The following procedure does not require that you be an expert on the oscilloscope or frequency counter. It requires only that you have performed the previous ac measurements in this lab manual, using an oscilloscope. However, you can refer to your oscilloscope and frequency counter manuals. These manuals usually contain information about making frequency measurements. More information about XY measurements is given in the capacitive phase measurements experiment, Experiment 18-2.

EQUIPMENT

Oscilloscope with XY capability
Voltmeter
Digital frequency counter
(2) Audio signal generators

COMPONENTS

Leads for connecting equipment as required

PROCEDURE

1. Connect the equipment as shown in Fig. 15-4.4 and turn on the power.

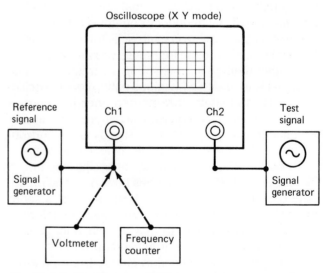

Fig. 15-4.4 Oscilloscope XY frequency measurement.

2. Set the oscilloscope time base to XY, and set channels 1 and 2 to 0.5 V per division. Make sure that all controls are in the CAL position.

3. Set the two audio generators to approximately 3 V p-p at 500 Hz. Use the voltmeter and frequency counter to verify the settings.

4. Adjust the oscilloscope position controls and the fine tuning (vernier) of the test audio signal generator (channel 2) until the most stable and most circular pattern is displayed on the CRT. Any rotation is due to phase differences between the two signals.

Note: In the following procedures, adjust for the most stable display possible to count and record the peaks of the "bow tie" patterns.

5. Adjust the test signal generator (channel 2) to 1 kHz. Then use the fine tuning to create a stable pattern. Draw the pattern in Table 15-4.1. Measure and record the two signal generator frequencies, using the frequency counter, in Table 15-4.1.

6. Adjust the test signal generator (channel 2) to 1.5 kHz. Draw the pattern and record both signal frequencies measured with the counter in Table 15-4.1.

7. Adjust the reference signal (channel 1) to 200 Hz and the test signal (channel 2) to 1200 Hz. Draw the pattern and record the signal frequencies in Table 15-4.1.

8. Adjust the test signal (channel 2) to 100 Hz, leaving the reference at 200 Hz. In Table 15-4.1 draw the pattern, record the measured frequencies, and determine the ratio.

9. Randomly change voltages, frequencies, and oscilloscope settings. Observe any results or consistent occurrences and note them for use in your report.

| NAME | DATE |

QUESTIONS FOR EXPERIMENT 15-4

1. Which method of measuring frequency is the easiest and most accurate? Why?

2. What happens to the oscilloscope when the XY setting is chosen?

3. Why do you think the Lissajous patterns are often unstable (moving)?

4. Does it matter if the X or Y axis is used as the reference or standard frequency? Why or why not?

5. If the test and reference signals had different amplitudes (voltages), could you still use Lissajous patterns to measure frequency? How?

TABLES FOR EXPERIMENT 15-4

TABLE 15-4.1

Procedure Step		
5	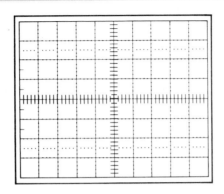	Reference = _____ Hz Test = _____ Hz Ratio = _____ to _____
6	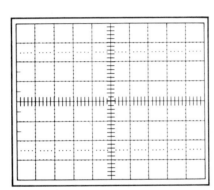	Reference = _____ Hz Test = _____ Hz Ratio = _____ to _____
7	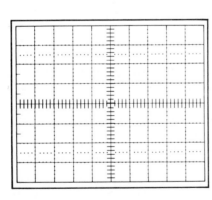	Reference = _____ Hz Test = _____ Hz Ratio = _____ to _____
8	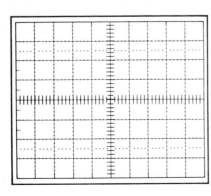	Reference = _____ Hz Test = _____ Hz Ratio = _____ to _____

EXPERIMENT RESULTS REPORT FORM

Experiment No: _____ Name: _____
                                        Date: _____
Experiment Title: _____ Class: _____
_____   Instr: _____

Explain the purpose of the experiment:

List the first Learning Objective:
OBJECTIVE 1:

After reviewing the results, describe how the objective was validated by this experiment?

List the second Learning Objective:
OBJECTIVE 2:

After reviewing the results, describe how the objective was validated by this experiment?

List the third Learning Objective:
OBJECTIVE 3:

After reviewing the results, describe how the objective was validated by this experiment?

Conclusion:

If required, attach to this form: ☐ Answers to Questions, ☐ Tables, and ☐ Graphs.

CHAPTER 15

EXPERIMENT 15-5

OSCILLOSCOPE MEASUREMENTS: SUPERPOSING AC ON DC

OBJECTIVES

At the completion of this experiment, you will be able to:

- Investigate the effective ac resistance of a power supply.
- Study the function of a bypass capacitor and a coupling capacitor.
- Develop a method for measuring both ac and dc components in the same circuit.

SUGGESTED READING

Chapter 15, *Basic Electronics*, Grob/Schultz, Tenth Edition

INTRODUCTION

Applied voltage will be dropped proportionally across individual resistances, depending upon their ohmic value. The voltage distribution can be solved as shown in Fig. 15-5.1a and b.

If a circuit has both an ac and a dc voltage source, as shown in Fig. 15-5.2, there will be both an ac and a dc voltage distribution. The voltage distribution is solved in the same manner as in Fig. 15-5.1.

If a bypass capacitor is connected across one of the series resistors, the capacitor can act as a short circuit to the ac component. Only a negligible amount of reactance will be created as a result of the chosen

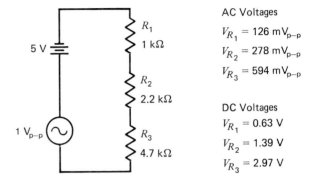

Fig. 15-5.2 AC and dc resistive circuit.

capacitor size and the frequency of the ac signal. The capacitor will become a parallel path for the ac component while blocking the dc component, as shown in Fig. 15-5.3. Note that the dc component is restricted to R_3 (4.7 kΩ) for a current path. The ac signal will make use of the capacitive path, because the charging and discharging action of the capacitor, in this case, creates only 3.39 Ω of reactance.

Another method of superposing an ac signal onto a dc voltage is through capacitive coupling. In Fig. 15-5.4, dc voltages are developed across R_1, R_2, and R_3 but are isolated from the signal generator by the coupling capacitor C_1. The dc resistance of the power supply can be considered approximately 0 Ω. The ac signal is applied to the parallel combination of $R_2 + R_3$ and R_1.

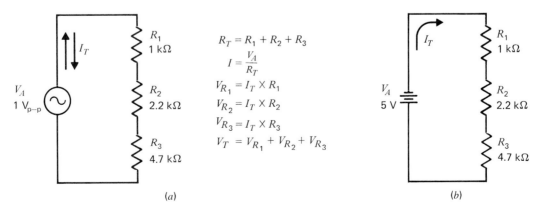

Fig. 15-5.1 (a) AC resistive circuit. (b) DC resistive circuit.

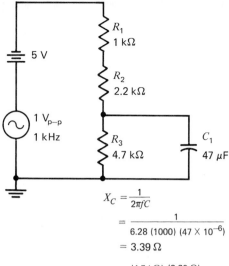

$$X_C = \frac{1}{2\pi f C}$$
$$= \frac{1}{6.28 \,(1000)\,(47 \times 10^{-6})}$$
$$= 3.39\ \Omega$$

AC equivalent resistance: $= \dfrac{(4.7\ \text{k}\Omega)\,(3.39\ \Omega)}{4.7\ \text{k}\Omega + 3.39\ \Omega}$

$= 3.39\ \Omega$

Fig. 15-5.3 Bypassing for alternating current.

Note: When measuring a dc component while an ac component is present, the oscilloscope will display the ac component at the level of the dc component. This means that a sine wave will appear above or below the zero reference when the input coupling is switched to measure direct current. It will be necessary to develop a technique to read the dc component. Therefore, the sine-wave zero (X axis) value must be determined in relation to the dc zero reference. This will require adjusting the oscilloscope for each reading. Be sure that you develop this technique. Refer to Fig. 15-5.5.

EQUIPMENT

Oscilloscope
Audio-frequency signal generator
VTVM
Protoboard or springboard
Test leads

COMPONENTS

Resistors (all 0.25 W):

(1) 1 kΩ (1) 4.7 kΩ
(1) 2.2 kΩ

(1) 47 μF electrolytic capacitor

PROCEDURE

1. Connect the circuit of Fig. 15-5.1a. Measure and record the voltages across each resistance as indicated in Table 15-5.1.

2. Connect the circuit of Fig. 15-5.1b. Measure and record the voltages across each resistance as indicated in Table 15-5.1.

3. Repeat procedures 1 and 2 for the circuit shown in Fig. 15-5.2, and complete the information requested in Table 15-5.2.

4. After recording the measured voltages, calculate the same voltages in accordance with the techniques reviewed in the introduction and record these in Tables 15-5.1 and 15-5.2. Compare the measured and calculated values.

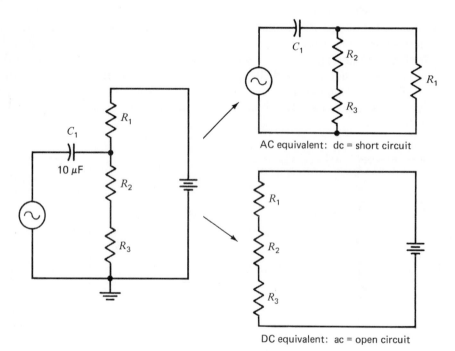

Fig. 15-5.4 Superposing an ac signal on a dc voltage by capacitive coupling with equivalent circuits shown.

REMINDER: Be sure to refer to the note (previous page) at this point in the procedure.

Optional Procedure Steps

Note: Be sure to avoid ground loops.

5. Connect the circuit of Fig. 15-5.3. Measure and record all ac and dc voltages. Prepare your own data table, similar to Table 15-5.2, and label it Table 15-5.3.
6. Connect the circuit of Fig. 15-5.4. Measure and record all ac and dc voltages. Prepare your own data table, and label it Table 15-5.4.

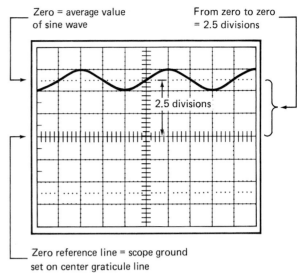

Fig. 15-5.5 Oscilloscope graticule display.

NAME	DATE

QUESTIONS FOR EXPERIMENT 15-5

1. What function does the coupling capacitor perform?

2. What function does the bypassing capacitor perform?

3. Was the effective ac resistance of the power supply equal to 0 Ω?

4. Draw the ac and dc equivalent circuit diagrams for the circuit shown in Fig. 15-5.2.

5. Is ac resistance the same as dc resistance? Explain.

TABLES FOR EXPERIMENT 15-5

TABLE 15–5.1   Measured and Calculated Voltages for Fig. 15-5.1a and b

Procedure Step		V_{R_1}	V_{R_2}	V_{R_3}	V_T
1	AC measured				
2	DC measured				
4	AC calculated				
	DC calculated				

TABLE 15–5.2 Measured and Calculated Voltages for Fig. 15-5.2

Procedure Step		V_{R_1}	V_{R_2}	V_{R_3}	V_T
3	AC measured				
	DC measured				
4	AC calculated				
	DC calculated				

EXPERIMENT RESULTS REPORT FORM

Experiment No: _____ Name: _____
                                       Date: _____
Experiment Title: _____ Class: _____
_____ Instr: _____

Explain the purpose of the experiment:

List the first Learning Objective:
OBJECTIVE 1:

After reviewing the results, describe how the objective was validated by this experiment?

List the second Learning Objective:
OBJECTIVE 2:

After reviewing the results, describe how the objective was validated by this experiment?

List the third Learning Objective:
OBJECTIVE 3:

After reviewing the results, describe how the objective was validated by this experiment?

Conclusion:

If required, attach to this form: ☐ Answers to Questions, ☐ Tables, and ☐ Graphs.

CHAPTER 16

EXPERIMENT 16-1

CAPACITORS

OBJECTIVES

At the completion of this experiment, you will be able to:
- Troubleshoot a capacitor to verify that it is functional.
- Verify the dc and ac response of a capacitor.
- Use a capacitor to route or remove high frequencies.

SUGGESTED READING

Chapter 16, *Basic Electronics*, Grob/Schultz, Tenth Edition

INTRODUCTION

A capacitor is made of two conductors separated by an insulator. Most capacitors are made of two metal plates with some type of insulating material such as air, ceramic, mica, paper, or an electrolyte. The insulating material is called a *dielectric*. Capacitors are like batteries because, at low frequencies (dc), they can store a charge. Therefore, it is always best to be cautious when working with larger capacitors because they can be a hazard and cause shock. To avoid this potential hazard, be sure to discharge capacitors safely, by shorting them out while you remain well insulated.

Like resistors, capacitors have a code which shows their value, but the code is different from resistor color codes. Capacitors are rated in units called *farads* (F), which is a measure of how much charge exists in the insulator's electrical field between the two plates. Some manufacturers label capacitors by printing the numerical value directly on the capacitor. Other capacitors are labeled with specific color codes, which may include temperature ratings. A special type of capacitor, the electrolytic capacitor, is polarity sensitive and has a minus sign (negative) printed on it to show which way to connect it with respect to the dc source. Be careful: **If it is connected backward, it can explode** if enough voltage is applied. Figure 16-1.1 shows some typical capacitors and their codes.

The two plates of a capacitor have opposite charges that create a potential difference (voltage) when a dc source is applied. If you remove the capacitor from the source of the charge, it will still hold the charge. However, it will have very little current unless it is an extremely large capacitor, in which case you should be careful not to discharge it with your hands or other parts of your body. Figure 16-1.2 is a circuit diagram of a capacitor storing a dc charge.

Capacitors are an open circuit (large resistance) to direct current. The dc voltage is stored on the plates, but no current passes through from one plate to another because of the insulation. Therefore, there is no current. When some capacitors fail, however, they can allow current to flow. Such capacitors are often called *leaky*.

You can use an ohmmeter to verify the operation of a capacitor. The resistance reading will be several million ohms or more (an open circuit). In the case of electrolytic capacitors, you may find one that is leaky. Other types of capacitors can also be damaged and will then give a measurement less than infinite ohms. Most good capacitors will measure at least 10 MΩ. Figure 16-1.3 shows a circuit for checking a capacitor with an ohmmeter.

If a capacitor is connected in series with an ac signal source, the capacitor will lose its insulating quality and may become a short circuit to the ac signal. The higher the frequency, the more a capacitor will act like a short circuit. Its plates will charge and discharge at the same rate as the ac signal source. The greater the value of the capacitor, the more easily it will charge and discharge; and the higher the frequency, the more the capacitor will act like a short circuit.

The reason the capacitor acts like a short is that the plates charge and discharge as the polarity of the source alternates between positive and negative. No current ever actually crosses the dielectric, but the effect is the same as if it did. This concept of the capacitor reacting to ac frequency will be examined more thoroughly in a future experiment.

Figure 16-1.4 shows a capacitor acting like a short in an ac circuit where the voltmeter reads 0 V, there is no resistance, and there is no *IR* voltage drop across the capacitor.

A capacitor can block dc but allow ac to pass when both ac and dc are present in a circuit. Therefore, a capacitor can be used to block dc from entering a circuit where only the ac signal is desired.

Type	Symbol	Uses
Fixed	○—│├—○ C	Stores charge in dielectric; passes ac voltage but blocks dc voltage
Electrolytic	○— −)│(+ —○ C	Fixed value with large C but has polarity
Variable	○—⧸│├—○ C	Variable capacitor; used for tuning

Color Codes for Tubular and Disc Ceramic Capacitors

BAND OR DOT COLOR	CAPACITANCE IN PICOFARADS		TOLERANCE		TEMPERATURE COEFFICIENT pp °C (5-DOT SYSTEM)	TEMPERATURE COEFFICIENT 6-DOT SYSTEM SIG. FIG.	TEMPERATURE COEFFICIENT MULTIPLIER
	SIGNIFICANT DIGITS	CAPACITANCE MULTIPLIER	≤10 pF	>10 pF			
	1st 2nd						
Black	0 0	1	±2.0 pF	±20%	0	0.0	−1
Brown	1 1	10	±0.1 pF	±1%	−33		−10
Red	2 2	100		±2%	−75	1.0	−100
Orange	3 3	1000		±3%	−150	1.5	−1000
Yellow	4 4				−230	2.0	−10000
Green	5 5		±0.5 pF	±5%	−330	3.3	+1
Blue	6 6				−470	4.7	+10
Violet	7 7				−750	7.5	+100
Gray	8 8	0.01	±0.25 pF		+150 to −1500		+1000
White	9 9	0.1	±1.0 pF	±10%	+100 to −75		+10000
Silver	— —						
Gold	— —						

Fig. 16-1.1 Capacitors and codes.

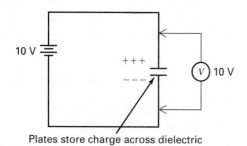

Fig. 16-1.2 Capacitor storing a dc charge.

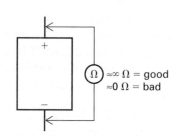

≈∞ Ω = good
≈0 Ω = bad

Fig. 16-1.3 Checking a capacitor with an ohmmeter.

286 EXPERIMENT 16-1

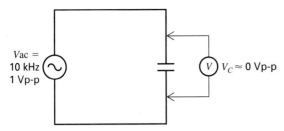

Fig. 16-1.4 Capacitor is a short to alternating current at high frequencies.

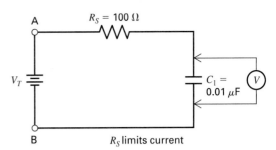

Fig. 16-1.6 A dc circuit with a capacitor in series.

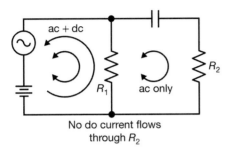

Fig. 16-1.5 Capacitor blocking direct current in a circuit.

Figure 16-1.5 shows the dc voltage blocked from the other part of the circuit where alternating current is present.

EQUIPMENT

Protoboard
DC power supply
AC signal generator
Ohmmeter
Oscilloscope

COMPONENTS

All resistors 0.25 W unless indicated otherwise:

(2) 100 Ω resistor
(1) 330 Ω resistor
(2) 1 kΩ resistor
(1) 0.01 μF capacitor
(1) 0.068 μF capacitor
(1) 0.1 μF capacitor
(1) 10 μF capacitor
(1) 100 μF capacitor

PROCEDURE

1. Verify the operation of each capacitor listed in Table 16-1.1 with an ohmmeter to be sure the capacitor is an open to direct current. Put a check mark next to each value in Table 16-1.1 to show that the capacitor is good. Be sure to replace any bad components.
2. Connect the circuit of Fig. 16-1.6, but do not connect the capacitor.

3. Apply 1 V V_T across A–B and then turn the power supply off. Insert the capacitor in the circuit and turn the power supply on.
4. Measure the voltage across the capacitor and record the voltage in Table 16-1.1.
5. Increase the power supply to 5 V (across A–B) and measure the voltage across the capacitor. Record the value in Table 16-1.1.
6. Go to the next value of capacitor in Table 16-1.1. Measure the voltage across the capacitor at 1 V and at 5 V as in the steps above and record the results in the table.
7. Repeat step 6 for all the remaining capacitors.
8. Connect the circuit of Fig. 16-1.7, but not connect the capacitor. Adjust the signal generator to 1 Vp-p at 1 kHz across A–B.

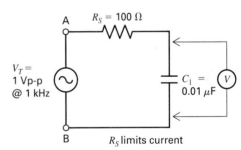

Fig. 16-1.7 An ac circuit with a capacitor in series.

9. Insert the first capacitor and measure the p-p voltage across it. Record the value in Table 16-1.1.
10. Increase the voltage to 5 Vp-p (at 1 kHz). Measure and record the voltage across the capacitor.
11. Increase the frequency to 100 kHz. Repeat the measurements at 1 Vp-p and 5 Vp-p at this new frequency and record the results in Table 16-1.1. Be sure to monitor the voltage as you adjust the frequency.
12. Carefully replace the capacitor with the next value of capacitor and repeat all the measurements at 1 Vp-p and at 5 Vp-p for frequencies of 1 kHz and 100 kHz, being sure to record the values in Table 16-1.1. Do this for all the remaining capacitors so that you have all the required data.

EXPERIMENT 16-1

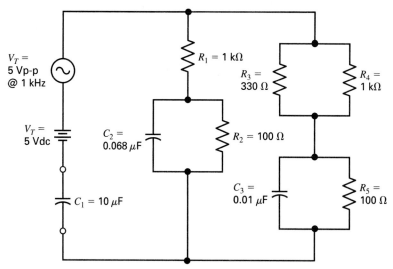

Fig. 16-1.8 A dc-ac circuit with a capacitor.

13. Connect the circuit of Fig. 16-1.8.
14. Measure the dc and ac voltages across the resistors. Record the results in Table 16-1.2. Measure the capacitor voltage and note it separately.

15. Carefully remove capacitor C_1 from the circuit and reconnect the circuit.
16. Measure all the dc and ac voltages across the resistors again. Record the results in Table 16-1.2.

QUESTIONS FOR EXPERIMENT 16-1

1. Why is a capacitor able to block direct current?

2. When is a capacitor like a 100 MΩ resistor?

3. When is a capacitor like a 0.001 Ω resistor?

4. Why is a capacitor able to pass high frequencies?

5. How can you verify the operation of a capacitor?

6. Is it possible for a capacitor to be leaky?

7. What type of capacitor needs to be connected with respect to polarity?

8. What precautions should you take to be sure a capacitor cannot be a hazard?

9. What differences, if any, did you notice between the capacitor voltage at 1 kHz and at 100 kHz in the circuit?

10. Describe the effects of removing the blocking capacitor C_1 from the last circuit, including any measurements that were significant.

TABLES FOR EXPERIMENT 16-1

TABLE 16–1.1

C (μF) (✓)	$V_T = 1$ V V_C	$V_T = 5$ V V_C	$f = 1$ kHz $V_{ac} = 1$ Vp-p	$V_{ac} = 5$ Vp-p	$f = 100$ kHz $V_{ac} = 1$ Vp-p	$V_{ac} = 5$ Vp-p
0.01						
0.068						
0.1						
10						
100						

TABLE 16–1.2 V_{C_1} = _____

R	With C_1 V_{dc}	V_{ac}	Without C_1 V_{dc}	V_{ac}
$R_1 = 1$ kΩ				
$R_2 = 100$ Ω				
$R_3 = 330$ Ω				
$R_4 = 1$ kΩ				
$R_5 = 100$ Ω				

EXPERIMENT RESULTS REPORT FORM

Experiment No: _____   Name: _____
                                   Date: _____
Experiment Title: _____ Class: _____
_____   Instr: _____

Explain the purpose of the experiment:

List the first Learning Objective:
OBJECTIVE 1:

After reviewing the results, describe how the objective was validated by this experiment?

List the second Learning Objective:
OBJECTIVE 2:

After reviewing the results, describe how the objective was validated by this experiment?

List the third Learning Objective:
OBJECTIVE 3:

After reviewing the results, describe how the objective was validated by this experiment?

Conclusion:

If required, attach to this form: ☐ Answers to Questions, ☐ Tables, and ☐ Graphs.

CHAPTER 17

EXPERIMENT 17-1

CAPACITIVE REACTANCE

OBJECTIVES

At the completion of this experiment, you will be able to:
- Understand how a capacitor reacts in a series RC circuit by calculation and measurement.
- Understand the effects of changing frequency upon an RC series circuit.
- Plot a graph of frequency versus V_C and V_R for an RC series circuit.

SUGGESTED READING

Chapters 16, 17, and 18, *Basic Electronics*, Grob/Schultz, Tenth Edition

INTRODUCTION

Capacitors have the ability to store an electric charge, much like a battery. The charge exists between the two plates made of conductive material. These plates are separated by an insulator, or dielectric. In a dc circuit, a capacitor will charge up to the potential difference across it and act as an open circuit, blocking any dc current flow. In an ac circuit, the capacitor will charge and discharge in proportion to the frequency of the alternating current, acting like a resistor or, at high frequencies, a short circuit to the alternating current. In fact, one definition of a capacitor is any two conductors separated by an insulator.

A charged capacitor is easily discharged by connecting a conducting path across the dielectric. Placing your fingers across a charged capacitor will discharge the capacitor through your fingers. With large capacitors, this will result in electric shock. *Never* assume that a capacitor is discharged.

Capacitors are manufactured and rated in units called *farads* (F), named after Michael Faraday. When 1 C (coulomb) is stored in the dielectric, with a potential difference of 1 V, the capacitance is equal to 1 F. Typically, capacitors are most often found in picofarad (pF) and microfarad (μF) values. The voltage rating of a capacitor is a maximum voltage that can be placed across the capacitor without damage. For example, a 10-μF capacitor, rated at 10 V, could be easily ruptured (exploded) if 20 V was across the capacitor.

When two capacitors of equal value are placed in series, the value of capacitance is reduced by one-half. This occurs because the dielectric thickness is increased and the plates are farther apart. Similar to resistors in parallel, the formula for capacitors in series is calculated by using the reciprocal method:

$$\frac{1}{C_T} = \frac{1}{C_1} + \frac{1}{C_2} + \cdots$$

When two capacitors of equal value are placed in parallel, the value of capacitance is increased by twice as much. Similar to resistors in series, the formula for capacitors in parallel is calculated by using simple addition:

$$C_T = C_1 + C_2 + \cdots$$

In an ac circuit, a capacitor will allow current to flow. The current does not actually flow through the capacitor; it flows because of the charging and discharging action of the capacitor. Therefore, the capacitive current varies with the frequency of applied ac voltage. As the frequency increases, the capacitor's opposition (in ohms) to current flow decreases. As frequency decreases, the opposition increases. That is, the capacitor reacts to changes in frequency. This is what is meant by capacitive reactance X_C. The formula for the capacitive reactance (in ohms) of any capacitor is

$$X_C = \frac{1}{2\pi f C}$$

where 2π is the sine-wave rotation, f is the frequency in hertz, and C is the value of capacitance in farads. Also, because $1/2\pi = 0.159$,

$$X_C = \frac{0.159}{fC}$$

$$f = \frac{0.159}{CX_C}$$

$$C = \frac{0.159}{fX_C}$$

Similar to an inductor, a capacitor will have a phase difference between the voltage across it and its current. Capacitive current i_C leads v_C by 90°. The term *ICE* is an easy way to remember that I (current) leads C (capacitive voltage E). Figure 17-1.1 illustrates this.

In an RC series circuit, the value of current is the same in all parts of the circuit. However, each has its own series voltage drop:

$$V_R = I_T \times R \quad \text{and} \quad V_C = I_T \times X_C$$

EXPERIMENT 17-1 **293**

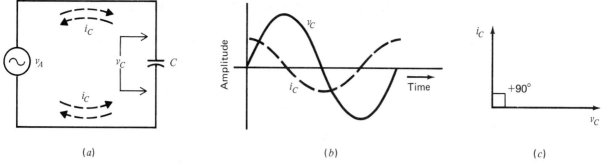

Fig. 17-1.1 *RC* circuit. (*a*) Purely capacitive circuit. (*b*) Waveshapes of i_C lead v_C by 90°. (*c*) Phasor diagram.

Again, similar to an inductor, the voltages must be added by phasor addition because of the 90° phase shift due to the lagging capacitive voltage. Figure 17-1.2 illustrates these points. In accordance with the triangles in Fig. 17-1.2, the total voltage in an *RC* series circuit is

$$V_T = (V_R^2 + V_C^2)^{1/2} \text{ or } V_T = \sqrt{V_R^2 + V_C^2}$$

and the total impedance (opposition to current flow) is

$$Z_T = (R^2 + X_C^2)^{1/2} \text{ or } Z_T = \sqrt{R^2 + X_C^2}$$

Remember that the circuit phase angle is between the series current and the generator ac voltage as follows:

$$\text{Inverse tan } \frac{X_C}{R} = \text{circuit phase angle}$$

This is because the phase angle is negative due to V_C lagging I_C.

The following procedures will allow you to validate the information given in this introduction.

EQUIPMENT

Audio signal generator
Oscilloscope

COMPONENTS

(1) 4.7 kΩ resistor
(1) 47 kΩ resistor
(1) 0.0068 µF capacitor
(1) 0.1 µF capacitor
(1) 0.01 µF capacitor

PROCEDURE

1. Connect the circuit of Fig. 17-1.3.

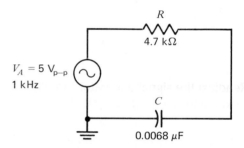

Fig. 17-1.3 *RC* circuit for measurement.

Note: Remember, the applied voltage (5 Vp-p) must be adjusted and maintained only after the entire circuit is connected. The signal generator has its own internal resistance and *IR* voltage drop that can subtract from the applied voltage. Therefore, constantly monitor the input voltage whenever you change frequency in the following steps.

2. Calculate X_C for this circuit in Table 17-1.1 as

$$X_C = \frac{1}{2\pi f C}$$

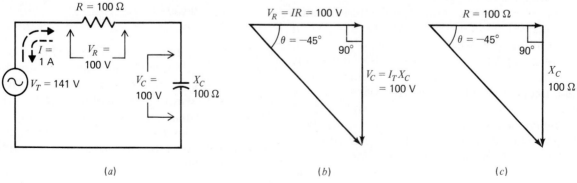

Fig. 17-1.2 (*a*) *RC* series circuit. (*b*) Phasor voltages. (*c*) Impedance triangle.

3. Measure and record in Table 17-1.1 the peak-to-peak voltage across the capacitor. Be sure the ground points are kept in the same place.

4. Measure and record in Table 17-1.1 the peak-to-peak voltage across the resistor. If necessary, exchange places with the capacitor to keep the ground points in a common place.

5. Calculate and record the circuit current I_T in Table 17-1.1, using the Ohm's law formula. Do this by using the measured voltage for both components.

(a) $I_T = \dfrac{V_R}{R}$

(b) $I_T = \dfrac{V_C}{X_C}$

6. Calculate and record in Table 17-1.1 the total circuit impedance, using the formula

$$Z_T = (R^2 + X_C^2)^{1/2}$$

7. Refer to the circuit of Fig. 17-1.3. Change the value of the capacitor to 0.1 µF and repeat steps 1 to 6.

8. Refer to the circuit of Fig. 17-1.3. Change the value of the capacitor to 0.01 µF and change the value of the resistor to 47 kΩ.

9. Readjust the signal generator to 200 Hz and increase the frequency in 200 Hz steps from 200 Hz to 1 kHz. Measure and record the voltage across the resistor V_R at each frequency step.

10. Increase the frequency from 1 to 15 kHz in 1 kHz steps and measure and record V_R at each frequency step.

11. Calculate and record X_C at each frequency in steps 9 and 10 above, as $X_C = 1/(2\pi f C)$.

12. Calculate the circuit current I_T as V_R/R for each frequency in step 11 above. Also, calculate V_C as $I_T X_C$ for each frequency.

13. Plot a graph of V_C and V_R versus frequency for the data in steps 10 to 12. Indicate where the V_C and V_R curves intersect at 45°.

OPTIONAL PROCEDURES

Series Capacitance

Connect two capacitors of equal value in series, and repeat steps 9 to 12. Make your own data table and record all values.

Parallel Capacitance

Connect two capacitors of equal value in parallel, and repeat steps 9 to 12. Make your own data table and record all values.

Note: Use $R = 10$ kΩ and $V_A = 5$ Vp-p.

QUESTIONS FOR EXPERIMENT 17-1

Answer the following questions on a separate sheet of paper. Show all work.

1. Using your graphed data, calculate the circuit phase angle at 8 Hz.

2. Using your graphed data, calculate the circuit phase angle at 600 Hz.

3. For a series RC circuit, at what frequency would a 10 μF capacitor have $X_C = 100\ \Omega$?

4. For a series RC circuit, what value of capacitor would have 31.8 Ω of X_C at 5 Hz?

5. For the circuit in Fig. 17-1.4, the voltage across $C = 25$ Vp-p. The circuit phase angle is 45°. What is the voltage across R and V_A?

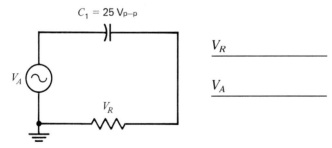

V_R _____

V_A _____

Fig. 17-1.4 RC circuit at 45°.

TABLE FOR EXPERIMENT 17-1

TABLE 17-1.1

Procedure Step	Measurement	Value	Calculations
2	X_C		_____ Ω
3	V_C	_____ Vp-p	
4	V_R	_____ Vp-p	
5a	I_T		_____ mA
5b	I_T		_____ mA
6	Z_T		_____ Ω
7	X_C		_____ Ω
	V_C	_____ Vp-p	
	V_R	_____ Vp-p	
	$I_T(a)$		_____ mA
	$I_T(b)$		_____ mA
	Z_T		_____ Ω

The students should prepare their own tables for $R = 47$ kΩ (Step 8) and Steps 9 to 13.

EXPERIMENT RESULTS REPORT FORM

Experiment No: _____ Name: _____
                                Date: _____
Experiment Title: _____ Class: _____
_____   Instr: _____

Explain the purpose of the experiment:

List the first Learning Objective:
OBJECTIVE 1:

After reviewing the results, describe how the objective was validated by this experiment?

List the second Learning Objective:
OBJECTIVE 2:

After reviewing the results, describe how the objective was validated by this experiment?

List the third Learning Objective:
OBJECTIVE 3:

After reviewing the results, describe how the objective was validated by this experiment?

Conclusion:

If required, attach to this form: ☐ Answers to Questions, ☐ Tables, and ☐ Graphs.

EXPERIMENT 18-1

CAPACITIVE COUPLING

OBJECTIVES

At the completion of this experiment, you will be able to:
- Describe capacitor coupling action.
- Measure the demonstrated effects of capacitive coupling.
- Determine the approximate frequency where a capacitor begins coupling.

SUGGESTED READING

Chapter 18, *Basic Electronics*, Grob/Schultz, Tenth Edition

INTRODUCTION

The most common type of coupling in amplifier circuits (Fig. 18-1.1) is *capacitive*. In this type of coupling, the output of one stage is connected to the input of another stage while the signals are sent through a capacitor. As in any coupling circuit which has sensitivity to frequency changes, the main goal of a good capacitive coupling circuit is to pass the signal with little attenuation of desired signals while attenuating the undesired frequencies. An advantage of the capacitive coupling circuit is its ability to block the dc voltage which may be present in the output signal of the first stage. This is important in that, as in transistor amplifiers, the dc operational characteristics of the second stage would be affected if its input signal were not free of possible dc components.

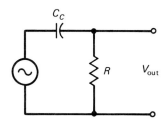

Fig. 18-1.1 AF coupling circuit.

EQUIPMENT

Protoboard
AF signal generator
Oscilloscope
Test leads

COMPONENTS

(1) 0.01 μF capacitor
(1) 22 kΩ 0.25 W resistor

PROCEDURE

1. Connect the circuit of Fig. 18-1.2. The capacitor in this circuit is used in a coupling application. The relatively low reactance of this capacitor allows practically all the generated ac voltage to be dropped across the resistor. Ideally very little of the generated ac voltage is dropped across the coupling capacitor.

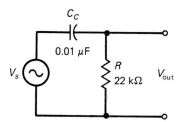

Fig. 18-1.2 AF coupling circuit.

2. Turn on the AF signal generator. Adjust the generator to 5 V p-p at 100 Hz. Measure and record in Table 18-1.1 the following ac voltages: source generator, coupling capacitor, and resistor.

3. Complete the measurements, and record the data in Table 18-1.1.

4. The dividing line for C_c to be a coupling capacitor at a specific frequency can be determined when X_c equals one-tenth (or less) of R. At this point the series RC circuit becomes primarily resistive, and all the voltage drop of the AF signal generator is across the series resistance, with very little voltage dropped across the coupling capacitor. Using the one-tenth rule described above and the information contained in Table 18-1.1, determine the frequency at which the capacitor becomes a *coupling* capacitor. Adjust the signal generator to this frequency, and complete the measurements required in Table 18-1.2.

NAME	DATE

QUESTIONS FOR EXPERIMENT 18-1

1. After reading the capacitor coupling section in your book, describe what a coupling capacitor is used for.

2. What importance does the one-tenth rule have, and where is it used?

TABLES FOR EXPERIMENT 18-1

TABLE 18–1.1

Frequency	V_s (p-p)	V_c (p-p)	V_R (p-p)
100 Hz			
200 Hz			
300 Hz			
400 Hz			
500 Hz			
600 Hz			
700 Hz			
800 Hz			
900 Hz			
1.0 kHz			
1.1 kHz			
1.2 kHz			
1.3 kHz			
1.4 kHz			
1.5 kHz			
1.6 kHz			
1.7 kHz			
1.8 kHz			
1.9 kHz			
2.0 kHz			
3.0 kHz			
4.0 kHz			
5.0 kHz			
6.0 kHz			
7.0 kHz			
8.0 kHz			
9.0 kHz			
10 kHz			
15 kHz			
20 kHz			

TABLE 18–1.2

Frequency Where C_c Becomes a Coupling Capacitor	V_s (p-p)	V_c (p-p)	V_R (p-p)
————	————	————	————

EXPERIMENT RESULTS REPORT FORM

Experiment No: _____   Name: _____
                                 Date: _____
Experiment Title: _____   Class: _____
_____   Instr: _____

Explain the purpose of the experiment:

List the first Learning Objective:
OBJECTIVE 1:

After reviewing the results, describe how the objective was validated by this experiment?

List the second Learning Objective:
OBJECTIVE 2:

After reviewing the results, describe how the objective was validated by this experiment?

List the third Learning Objective:
OBJECTIVE 3:

After reviewing the results, describe how the objective was validated by this experiment?

Conclusion:

If required, attach to this form: ☐ Answers to Questions, ☐ Tables, and ☐ Graphs.

CHAPTER 18

EXPERIMENT 18-2

CAPACITIVE PHASE MEASUREMENTS: USING AN OSCILLOSCOPE

OBJECTIVES

At the completion of this experiment, you will:

- Be able to make phase measurements on the oscilloscope, using the dual-trace method.
- Be able to make phase measurements on the oscilloscope, using Lissajous patterns.
- Be able to determine the phase angle by measuring voltages and calculating the phase shift.

SUGGESTED READING

Chapter 15 and 18, *Basic Electronics,* Grob/Schultz, Tenth Edition

INTRODUCTION

This experiment concentrates on the measurement of a phase angle. Unlike frequency measurements, phase measurements compare two sine-wave signals of similar frequency; the result is a phase angle between two waves.

Consider the drawing (Fig. 18-2.1) where two voltage waveforms are shown 90° out of phase as a dual-trace measurement. Figure 18-2.1a shows the two sine waves as they might be displayed on a dual-trace oscilloscope. Sine wave A starts at 0 V and 0°. Therefore, it is considered the reference signal.

Sine wave B actually begins 90° before sine wave A. In other words, the peak voltage of sine wave B occurs when sine wave A is at zero. The phasor diagram of Fig. 18-2.1b shows that both signals have about equal magnitude (voltage), indicated by the length of the arrow. The angle is shown with respect to the reference signal A on the horizontal axis. Therefore, the phase angle has magnitude and direction; here the direction is +90° compared to 0° of the reference signal. This corresponds to the standard practice of using a counterclockwise rotation as the positive direction of rotation. This also means that signal B leads signal A by +90°.

Although it is easier to describe phase differences when the signals have the same voltage, consider the drawing in Fig. 18-2.2. Figure 18-2.2a shows two sine waves with different magnitudes, but having the same frequency. Figure 18-2.2b shows two sine waves 180° out of phase, also with different magnitudes. Notice that both phase measurements can be represented by their corresponding phasor diagrams.

Although using the dual-trace capability of an oscilloscope is the easiest way to measure the phase of like frequencies, Lissajous patterns can also be used. The oscilloscope is used exactly as in Experiment 15-4, "Oscilloscope Lissajous Patterns." This means that the XY mode of the time base is selected. As Fig. 18-2.3 shows, two signals of equal frequency display any phase difference as a Lissajous pattern created by two signals of the exact same amplitude.

Notice that a pattern is created where signal A is the maximum vertical number of divisions. Signal B is the vertical measurement between the two points where the trace pattern crosses the centerline (vertical) on the graticule. Signal B is then divided by signal A to obtain the sine of the two signals. And the arcsine is calculated to obtain the phase angle in degrees.

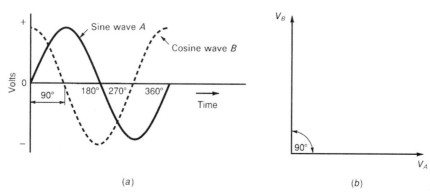

Fig. 18-2.1 Phase angle of two sine waves.

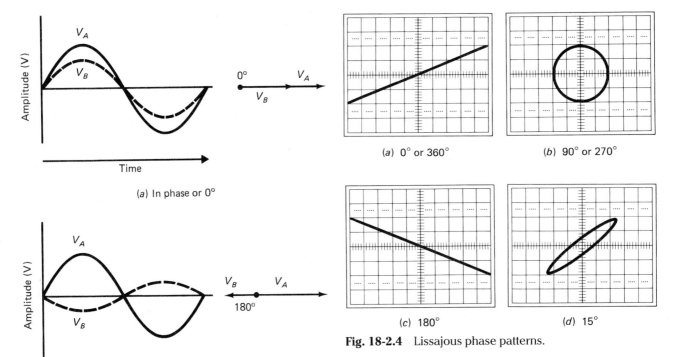

(a) In phase or 0°

(b) 180° out of phase

Fig. 18-2.2 Phase angles with different amplitudes.

(a) 0° or 360°

(b) 90° or 270°

(c) 180°

(d) 15°

Fig. 18-2.4 Lissajous phase patterns.

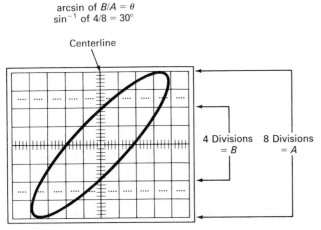

arcsin of $B/A = \theta$
$\sin^{-1}$ of $4/8 = 30°$

Fig. 18-2.3 XY phase measurement of two signals with equal frequency and magnitude equals 30°.

Figure 18-2.4 shows some typical Lissajous patterns for common phase angles. Remember that this type of Lissajous pattern is based on two signals of equal frequency and amplitude measured on an oscilloscope in the XY mode. Because the procedure in this experiment focuses on the use of an oscilloscope to measure phase, you can refer to the oscilloscope's operating manual for more information, if necessary. Also this experiment should be performed after you have completed Experiment 15-4, since the use of the equipment is similar.

Finally, to measure phase, a phase-shift network will have to be created. This will be a simple RC circuit where the difference between the two signals is measured across R and the signal generator.

EQUIPMENT

Oscilloscope with XY capability
Voltmeter
Digital frequency counter (optional)
Audio signal generator

COMPONENTS

Leads for connecting equipment as required
(1) 1.2 kΩ 0.25 W resistor
(1) 0.01 μF capacitor

PROCEDURE

1. Connect the circuit as shown in Fig. 18-2.5, and turn on the power.
2. Adjust the oscilloscope to XY (time base), and set channels 1 and 2 to 0.5 V per division each.
3. Set the signal generator to 23 kHz at 2 Vp-p. Verify this by using the frequency counter and voltmeter (or oscilloscope). Adjust the trace position.
4. The oscilloscope should now display the Lissajous pattern for the phase shift between R and the generator. Record the display in Table 18-2.1 by drawing it as close as you possibly can.
5. Determine the phase angle, using the arcsin (B/A) method described in the Introduction. Record the calculations and results in Table 18-2.1.
6. Measure the p-p voltage across the resistor, and record the results in Table 18-2.1. Because the measured phase shift is between R and V_G, calculate the

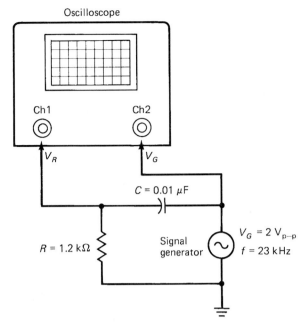

Fig. 18-2.5 Phase measurement of a phase-shift circuit.

phase shift as the arcsine or inverse cosine of V_R/V_G. Record the result in Table 18-2.1.

7. Adjust the oscilloscope back to time-base operation, and set it to display one or two complete cycles of each wave. Set both channels to the same ground at the center graticule. Then measure the phase difference between the two signals: channel 2 = V_G and channel 1 = V_R. If necessary, use the time-base CAL adjustment to display one cycle (V_G) over an equal number of divisions. Also you can use the position knobs and, if applicable, choose the CHOP mode instead of the ALTERNATE or ADD mode.

Draw the traces as best you can in Table 18-2.1. Label the two traces and show the phase difference, in degrees, where the V_G wave crosses the zero or middle graticule lines. Also show the degree per division.

8. Change the frequency of the signal generator to 8 kHz, and repeat steps 1 through 7. Use Table 18-2.2 to record your results.

NAME _____ DATE _____

QUESTIONS FOR EXPERIMENT 18-2

1. Which method of measuring phase is the easiest and fastest? Why?

2. If you had to measure both frequency and phase, could you use the Lissajous pattern method for two signals 500 Hz apart? Why or why not?

3. Why did changing the frequency of V_G cause a different phase angle?

4. If you didn't have any test equipment, could you calculate the phase angle between V_G and V_R? How?

5. Draw the phasor showing the results for a phase angle where $V_G = 12$ V and $V_R = 8.4$ V.

TABLES FOR EXPERIMENT 18-2

TABLE 18–2.1

Procedure Step	V_G = 23 kHz at 2 Vp-p
4	
5	$A =$ _____ DIV $B =$ _____ DIV arcsin $B/A =$ _____ °
6	$V_R =$ _____ Vp-p $V_G =$ _____ Vp-p arccos $V_R/V_G =$ _____ °
7	

Degrees per DIV = _____

TABLE 18–2.2

Procedure Step	V_G = 8 kHz at 2 Vp-p
4	
5	$A =$ _____ DIV $B =$ _____ DIV arcsin $B/A =$ _____ °
6	$V_R =$ _____ Vp-p $V_G =$ _____ Vp-p arccos $V_R/V_G =$ _____ °
7	

Degrees per DIV = _____

EXPERIMENT RESULTS REPORT FORM

Experiment No: _____   Name: _____
                                  Date: _____
Experiment Title: _____   Class: _____
_____   Instr: _____

Explain the purpose of the experiment:

List the first Learning Objective:
OBJECTIVE 1:

After reviewing the results, describe how the objective was validated by this experiment?

List the second Learning Objective:
OBJECTIVE 2:

After reviewing the results, describe how the objective was validated by this experiment?

List the third Learning Objective:
OBJECTIVE 3:

After reviewing the results, describe how the objective was validated by this experiment?

Conclusion:

If required, attach to this form: ☐ Answers to Questions, ☐ Tables, and ☐ Graphs.

CHAPTER 19

EXPERIMENT 19-1

INDUCTORS

OBJECTIVES

At the completion of this experiment, you will be able to:
- Troubleshoot an inductor to verify that it is functional for continuity.
- Verify the dc and ac response of an inductor.
- Induce a voltage with an inductor and light a neon bulb.

SUGGESTED READING

Chapter 19, *Basic Electronics,* Grob/Schultz, Tenth Edition

INTRODUCTION

Inductors are coils of insulated wire. Hence, they are often called *coils*. Coils have a specific number of coated and insulated wire windings around a center core, which is often made of air or iron. Inductors can produce (induce) voltage because of this winding construction, which can oppose or choke off higher-frequency currents. Inductors are rated in units of henrys (H), named after the scientist Joseph Henry. For example, a 1-H coil has the ability to induce 1 V when the current changes at the rate of 1 A/s. Whenever current is flowing in a wire or conductor which is next to another conductor, it is by definition an inductor. Figure 19-1.1 shows some typical inductors.

The current flowing in the windings of an inductor produces a magnetic field. With so many turns in close proximity, a large polarized magnetic field builds up, which has the ability to create a force field. A common example is a door bell which has a magnet in the center of an inductor. The free-moving magnet is forced out by the field's magnetic opposition and strikes a bell. When the current is stopped, the field collapses and the iron magnet falls back into the center of the inductor.

More windings or turns in a inductor result in a larger magnetic field when current flows. When a dc voltage is applied to an inductor, only the resistance of the wire exists and the current flows easily. In this case, the inductor is a short circuit to dc, with only the resistance of the coil. Because of this, you can verify that a coil is working by using an ohmmeter and checking for continuity. Figure 19-1.2 shows the circuit for verifying an inductor with an ohmmeter.

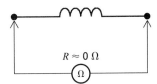

Fig. 19-1.2 Verifying an inductor with an ohmmeter.

When an ac signal is applied to the inductor, there is a reaction to the change in frequency. This reaction is greater when the frequency increases because the inductor opposes the flow of current or *chokes* it off. That is why an inductor is often called a *choke*. Another common use of an inductor, however, is in a power supply, where the buildup of the magnetic field in a primary winding is used to induce a voltage in a secondary winding, often called a *tap*. This is one method of stepping down a higher voltage to a lower voltage; the ratio of the number of turns in the primary to the number of turns in the secondary determines the value of induced voltage. Figure 19-1.3 shows a transformer, used to step down a voltage.

Another form of inductor is very small and is often used in microwave applications. It is called a *toroid* and is used to control or filter high frequencies. A toroid, as shown in Figure 19-1.4, has only a few turns wound around a magnetic ring.

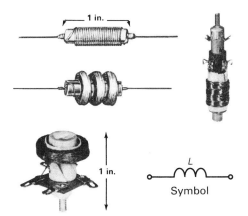

Fig. 19-1.1 Types of inductors and symbol.

EXPERIMENT 19-1 **315**

Fig. 19-1.3 Transformer.

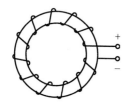

Fig. 19-1.4 Toroid.

EQUIPMENT

DC power supply
AC signal generator
Ohmmeter
Voltmeter
Oscilloscope

COMPONENTS

Resistors (0.25 W): 100 Ω
Inductors:
    (4) coils of varying values from 10 mH to 0.5 H
    (1) 0.5 H choke
Transformers: Standard 60 Hz power transformer
  with two or three taps or a center tap
Neon bulb: Ne2 bulb lighting at approximately 60 V
  or a similar bulb

PROCEDURE

1. Obtain a step-down transformer and record its values in Table 19-1.1. Then verify the resistance of the primary and secondary windings by checking the resistance (continuity). Record the value of dc resistance for each winding in Table 19-1.1. You should record the color of the wires (taps) in the table to identify them.

2. Check the continuity of each inductor and record the resistance value and the value of the inductor in Table 19-1.2. If any inductor is bad, replace it.

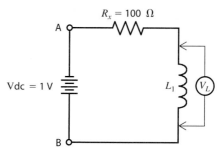

Fig. 19-1.5 Inductor in a series circuit (with current-limiting resistor).

3. Connect the circuit of Fig. 19-1.5, but do not insert the inductor.

4. Apply 1 Vdc across points A–B without the inductor in the circuit. Turn the power supply off and insert the inductor. Turn the power supply on and measure the voltage across the inductor. Record the value in Table 19-1.2.

5. Increase the applied voltage to 5 V. Measure and record the value in Table 19-1.2.

6. Replace the inductor with the next value. Measure the voltage across the inductor at 1 V and at 5 V and record the results in Table 19-1.2.

7. Repeat step 6 for the remaining inductors.

8. Connect the circuit of Fig. 19-1.6, but do not connect the inductor. Adjust the power supply so that 1 Vp-p at 100 Hz appears across A–B.

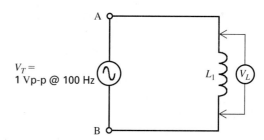

Fig. 19-1.6 Inductor in a series with ac source.

9. Measure the voltage across the inductor at 1 Vp-p and 100 Hz. Record the value in Table 19-1.2.

10. Increase the voltage to 5 Vp-p. Measure and record the value in Table 19-1.2.

11. Increase the frequency to 15 kHz. Measure the inductor voltage at 1 Vp-p and at 5 Vp-p applied. Record the results in Table 19-1.2.

12. Replace the inductor with the next value of inductor and repeat the measurements in steps 9, 10, and 11. Then repeat this procedure for each value of inductor you have.

13. Obtain the neon bulb and connect the circuit of Fig. 19-1.7, but do not apply power yet.

14. Apply exactly 2 Vdc to the circuit.

15. Toggle (close and open quickly) switch S_1 in an instantaneous manner to create a current pulse. This

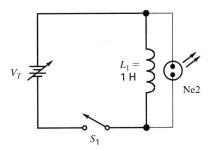

Fig. 19-1.7 Inductor with neon bulb circuit.

will induce a voltage in the inductor which should be greater than the source.

16. Did the bulb light or not? Record the results in Table 19-1.3.

17. Increase the voltage in 0.5 V steps and repeat steps 15 and 16 until the bulb lights. Note the applied voltage that lights the bulb in Table 19-1.3. *Do not exceed the allowable voltage rating (about 10 V is average max) for the neon bulb.*

18. Refer to Table 19-1.1. With the step-down transformer connected to an ac source, verify that the taps are lowering the applied voltage.

Note: Perform this step only with your instructor's approval. Use caution when measuring the outputs of the taps. Do not do this step until you have checked with your instructor about how to safely step down the voltage with a safe input of alternating current.

NAME	DATE

QUESTIONS FOR EXPERIMENT 19-1

1. How is an inductor constructed?

2. How can you check if an inductor is working?

3. How does a transformer tap work to produce various voltages?

4. Is an inductor sensitive to changes in frequency?

5. If an inductor measured 75 Ω, would it still be good?

6. What difference was there between the inductor voltage at 100 Hz and at 15 kHz?

7. Why is an inductor called a choke?

8. How is voltage induced in a coil?

9. How could you keep the neon bulb lit longer?

10. Is an inductor capable of causing a shock if you touch it?

TABLES FOR EXPERIMENT 19-1

TABLE 19–1.1 Transformer

Inductor	Wire Colors	Resistance
Primary	_____	_____
Secondary	_____	_____
_____	_____	_____
_____	_____	_____
_____	_____	_____

TABLE 19–1.2 Inductors

Value (H) and Resistance	$V_T = 1\,V$ V_L	$V_T = 5\,V$ V_L	$f = 100$ Hz Vac = 1 Vp-p	$f = 100$ Hz Vac = 5 Vp-p	$f = 15$ kHz Vac = 1 Vp-p	$f = 15$ kHz Vac = 5 Vp-p
_____	_____	_____	_____	_____	_____	_____
_____	_____	_____	_____	_____	_____	_____
_____	_____	_____	_____	_____	_____	_____
_____	_____	_____	_____	_____	_____	_____

TABLE 19–1.3 Induced Voltage Circuit (Neon Bulb)

Applied DC Voltage
Volts 2 2.5 3 3.5 4 4.5 5 5.5 6 6.5 7 7.5
Volts 8 8.5 9 9.5 10 ___ ___ ___ ___ ___ ___ ___

Circle the approximate voltage that lights the bulb.

EXPERIMENT RESULTS REPORT FORM

Experiment No: _____   Name: _____
                                   Date: _____
Experiment Title: _____ Class: _____
_____   Instr: _____

Explain the purpose of the experiment:

List the first Learning Objective:
OBJECTIVE 1:

After reviewing the results, describe how the objective was validated by this experiment?

List the second Learning Objective:
OBJECTIVE 2:

After reviewing the results, describe how the objective was validated by this experiment?

List the third Learning Objective:
OBJECTIVE 3:

After reviewing the results, describe how the objective was validated by this experiment?

Conclusion:

If required, attach to this form: ☐ Answers to Questions, ☐ Tables, and ☐ Graphs.

CHAPTER 20

EXPERIMENT 20-1

INDUCTIVE REACTANCE

OBJECTIVES

At the completion of this experiment, you will be able to:

- Understand the concept of inductive reactance and validate the formula $X_L = 2\pi fL$.
- Understand phase relationships.
- Plot the frequency response of a series RL circuit.

SUGGESTED READING

Chapters 19, 20, and 21, *Basic Electronics*, Grob/Schultz, Tenth Edition

INTRODUCTION

To study inductance, cosine waves, inductive reactance, inductive coupling, and many other aspects of an inductor (coil) in an ac circuit would require weeks of laboratory time. Therefore, this experiment will focus on the basic properties of an inductor that are fundamental to a technician.

As its name implies, an inductor produces its own (self-induced) voltage. In fact, any conductor that is close to another conductor (like two wires in close proximity) can be considered an inductor if current is flowing. Thus, as current flows in any inductor, a voltage is induced. This is accomplished because the inductor (coil) produces a magnetic field, and the magnetic field is greater with more turns of wire and/or more changes of current (frequency).

Remember that a coil's inductance uses the symbol L (from linkages of magnetic flux) and is measured in henrys (H). An 8 H coil, for example, is a rather large inductor and is used for 60-Hz ac power lines or other low-frequency (audio range) applications. A 1 μH coil, however, is a rather small inductor and is used at higher frequencies. Also, remember that as the frequency increases in an ac circuit, the inductive reactance X_L increases even though the value of the inductor (in henrys) remains the same. In other words, the coil reacts to, or opposes, a change in current. As frequency increases, X_L increases. Therefore, the formula for inductive reactance is

$$X_L = 2\pi fL$$

where f is the determining factor. Note that 2π is the constant circular motion from which a sine wave is derived (360°).

As X_L increases, the amount of circuit current decreases because the inductive reactance acts like a variable resistance. That is, for any given frequency, the inductive reactance will have a different resistance in ohms. For this reason, two series inductors of the same value will have twice as much reactance (in ohms), behaving like two series resistors. Similarly, two parallel inductors of the same value will have half as much reactance as one. And, according to Ohm's law, the value of circuit current is

$$I_T = \frac{V_T}{X_L}$$

where X_L is in ohms.

When a voltage is induced in a coil, the current lags the induced voltage by 90°. This lag exists in time, as is best represented by Fig. 20-1.1.

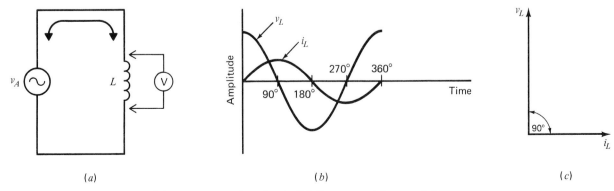

Fig. 20-1.1 Inductive circuit. (*a*) Purely inductive circuit. (*b*) Sine wave i_L lags v_L. (*c*) Phasor diagram.

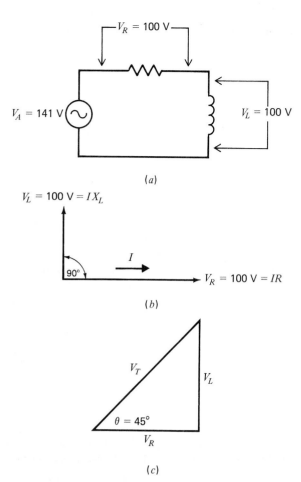

Fig 20-1.2 RL series circuit. (a) Series RL circuit with voltage drops shown. (b) Phasor diagram. (c) Resultant of two phasors = $V_T = \sqrt{V_R^2 + V_L^2} = 141$ V.

Although the current in the inductor lags by 90°, the value is still the same in all parts of a series circuit. The time lag only exists between the induced voltage and current, and is more theoretical than practical.

When a resistor and an inductor are in series with an ac source, the current is the same in all parts of the circuit, and the inductor and resistor each have their own voltage drop (IR) calculated by Ohm's law. However, the voltage across the inductor leads the voltage across the resistor by 90° due to the lagging current. This is often referred to as *ELI*, where *E* (voltage across *L*) leads *I*. Thus, an inductor may have a 100 V ($I \times X_L$) drop and the resistor may also have a 100 V *IR* drop, but the combined total voltage (*V* applied from the generator) will be the applied 141 V, as shown in Fig. 20-1.2.

Notice that in Fig. 20-1.2 the pythagorean theorem can easily be used to combine the two *IR* series voltages. The resultant of the phase addition of *R* and X_L (in ohms) is calculated the same way and is called the *impedance*, or total opposition to current, with the symbol *Z*:

$$Z = (R^2 + X_L^2)^{1/2} \text{ or } Z = \sqrt{R^2 + X_L^2}$$

Therefore, if $R = 1$ kΩ and $X_L = 1$ kΩ,

$$Z = (1000^2 + 1000^2)^{1/2} \text{ or } Z = \sqrt{1000^2 + 1000^2}$$
$$= 1.414 \text{ k}\Omega$$

The phase angle θ (theta) of the complete circuit is calculated as the inverse tangent of X_L/R. In the case of Fig. 20-1.2,

$$\frac{X_L}{R_L} = \frac{1000}{1000} = 1 \text{ and } \tan^{-1} = 45°$$

At 45°, notice that X_L and R are equal in resistance and also in *IR* voltages.

The procedures that follow will allow you to validate all the concepts discussed above. However, remember that the effects of inductive reactance are sometimes difficult to comprehend in the same way that magnetism produces somewhat mysterious effects.

EQUIPMENT

Oscilloscope
Audio signal generator
Protoboard
Leads
DC power supply

COMPONENTS

(1) 560 Ω resistor
(1) 33 mH inductor

PROCEDURE

Note: Show all calculations on a separate sheet and label which procedure step they apply to.

1. Using an ohmmeter, measure the dc resistance of the inductor and record the value in Table 20-1.1. Remember, this is the resistance at 0 Hz.

2. Connect the circuit of Fig. 20-1.3.

Note: Adjust the signal generator to 5 V p-p by placing the oscilloscope across points A and B only after the entire circuit is connected. This is because the signal generator has its own internal resistance. Therefore, V_A can never be adjusted correctly unless

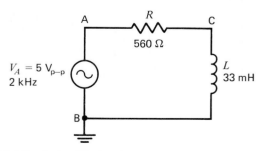

Fig 20-1.3 Series *RL* circuit.

it is connected in the circuit. The frequency does not have to be exact.

3. Calculate X_L for this circuit as

$$X_L = 2\pi fL$$

Record the value in Table 20-1.1.

4. Measure and record the peak-to-peak voltage across the inductor. Be sure the oscilloscope probe (+) is at point C.

5. Measure and record in Table 20-1.1 the peak-to-peak voltage across the resistor. Be sure to exchange the resistor's place with the coil so that the oscilloscope ground and the signal generator ground are at the same point.

6. Calculate the circuit current and record it in Table 20-1.1 by using the Ohm's law formula. Do this by using the measured voltages for both components:

(a) $I_T = \dfrac{V_R}{R}$ (b) $I_T = \dfrac{V_L}{X_L}$

7. Calculate and record in Table 20-1.1 the total circuit impedance, using two formulas:

(a) $Z_T = (R^2 + X_L^2)^{1/2}$ (b) $Z_T = \dfrac{V_G}{I_T}$

8. Calculate and record in Table 20-1.1 the frequency as

$$f = \dfrac{X_L}{2\pi L}$$

9. Calculate and record in Table 20-1.1 the inductances as

$$L = \dfrac{X_L}{2\pi f}$$

10. Refer to Fig. 20-1.3. Readjust the frequency to 100 Hz. Increase the frequency in 100 Hz steps from 100 Hz to 1 kHz. Measure and record V_L at each frequency step in Table 20-1.2.

11. Beginning at 1 kHz, increase the frequency in 1 kHz steps from 1 kHz to 10 kHz and measure and record V_L at each step in Table 20-1.2.

12. Calculate X_L for each frequency point in steps 10 and 11 above and record the values in Table 20-1.2.

13. Plot a graph of frequency versus V_L and X_L for the data of steps 11 and 12, using semilog paper.

OPTIONAL PROCEDURES

Series Inductance

Connect two inductors of equal value in series, and repeat steps 1 to 9. Make your own data table (similar to Table 20-1.1).

Parallel Inductance

Connect two inductors of equal value in parallel, and repeat steps 1 to 9. Make your own data table (similar to Table 20-1.1).

QUESTIONS FOR EXPERIMENT 20-1

Answer TRUE (T) or FALSE (F) to the following:

_____ 1. As frequency increases, X_L increases.

_____ 2. The dc resistance of a coil is always 0 Ω.

_____ 3. As frequency decreases, X_L increases.

_____ 4. Two parallel coils of equal value will have twice as much inductance as one coil.

_____ 5. Two series coils of equal value will have half as much inductance.

_____ 6. If, in a series RL circuit, $V_R = 2$ Vp-p and $V_L = 2$ Vp-p, the applied voltage would equal 4 Vp-p.

_____ 7. In a series RL circuit, the voltage across the resistor always lags the voltage across the inductor by 90°.

_____ 8. The current in an RL series circuit is the same in all parts of the circuit.

_____ 9. There is a phase relationship in a RL circuit because the resistor opposes a change in current.

_____ 10. The phase angle of an RL series circuit, where R and X_L are equal, will always be 45°.

TABLES FOR EXPERIMENT 20-1

TABLE 20-1.1

Procedure Step	Circuit Component	Value
1	R_L Measured	_____ Ω
3	X_L Calculated	_____ Ω
4	V_L Measured	_____ Vp-p
5	V_R Measured	_____ Vp-p
6a	I_T Calculated	_____ mA
6b	I_T Calculated	_____ mA
7a	Z_T Calculated	_____ Ω
7b	Z_T Calculated	_____ Ω
8	f Calculated	_____ kHz
9	L Calculated	_____ H

TABLE 20–1.2

F	V_L	X_L
100 Hz	_____	_____
200 Hz	_____	_____
300 Hz	_____	_____
400 Hz	_____	_____
500 Hz	_____	_____
600 Hz	_____	_____
700 Hz	_____	_____
800 Hz	_____	_____
900 Hz	_____	_____
1 kHz	_____	_____
2 kHz	_____	_____
3 kHz	_____	_____
4 kHz	_____	_____
5 kHz	_____	_____
6 kHz	_____	_____
7 kHz	_____	_____
8 kHz	_____	_____
9 kHz	_____	_____
10 kHz	_____	_____

EXPERIMENT RESULTS REPORT FORM

Experiment No: _____   Name: _____
                                          Date: _____
Experiment Title: _____   Class: _____
_____   Instr: _____

Explain the purpose of the experiment:

List the first Learning Objective:
OBJECTIVE 1:

After reviewing the results, describe how the objective was validated by this experiment?

List the second Learning Objective:
OBJECTIVE 2:

After reviewing the results, describe how the objective was validated by this experiment?

List the third Learning Objective:
OBJECTIVE 3:

After reviewing the results, describe how the objective was validated by this experiment?

Conclusion:

If required, attach to this form: ☐ Answers to Questions, ☐ Tables, and ☐ Graphs.

CHAPTER 21

EXPERIMENT 21-1

INDUCTIVE CIRCUITS

OBJECTIVES

At the completion of this experiment you will be able to:
- Measure time and voltage in an inductive circuit.
- Calculate phase shift between signals.
- Determine the Q of an inductor.

SUGGESTED READING

Chapter 21, *Basic Electronics,* Grob/Schultz, Tenth Edition

INTRODUCTION

Most inductors are combined with capacitors to create filters. However, inductors can also be used alone as chokes to keep an ac or RF signal out of some part of a circuit, usually the dc bias of an amplifier or mixer. In this lab, you verify the use of an inductor for feeding dc to one part of a circuit while keeping the RF or ac signal out.

As you previously learned, an ideal inductor is considered a short to dc. Therefore, when an inductor is in a circuit, it will pass dc but will oppose ac. This opposition to ac is a variable impedance, depending upon the ac frequency and the amount of inductance L. Inductors also have some internal resistance which is usually very small compared to the reactance of the inductor. In most cases, as frequency increases, the reactance of the inductor is usually much greater than its resistance.

Because the inductor is a reactive component, the concept of phase angle becomes important, especially with high-frequency or microwave circuits. Therefore, keep in mind the principle that the ac voltage always leads current in the inductor (which is the same as the current always lagging the voltage). And in both cases, the difference is always 90 degrees. This 90 degree phase angle is used to describe the constant delay in time between the peak voltage and the peak current in the inductor. This effect occurs because the inductor has many closely spaced turns or windings. However, regardless of the number of turns of the inductor or the frequency, the 90 degree phase difference stays constant between V and I in the inductor itself. For practical purposes current is not measured directly in an inductive circuit but can be calculated from the voltage and the reactance.

Another quality of an inductor is its Q. For any coil, this quality of merit of Q is calculated as X_L/R, where R is the internal dc resistance of the coil. Of course, the reactance X_L is dependent upon frequency. Therefore, the Q is only valid for a given frequency. In general, the higher the Q, the greater the quality of the coil. Because Q is a ratio, less R has the most effect.

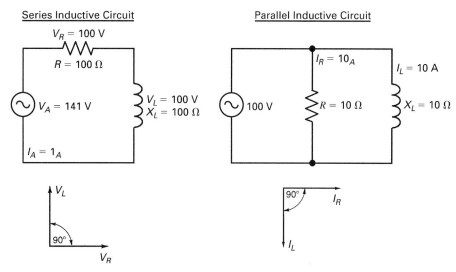

Fig. 21-1.1 Phasor diagram. Left, *R-L* series circuit impedance. Right, parallel *R-L* circuit response.

EXPERIMENT 21-1 **331**

In review, keep in mind that when R and L are in series, the amount of current is the same in all parts of the circuit but the voltages across L and R are 90 degrees out of phase. This means that at the same point in time their voltages are different: when the resistor voltage is at its peak, the inductor voltage is at zero. To describe this, a phasor diagram is used (Fig. 21-1.1) where the two voltages V_L and V_R are shown on a right triangle. This is also used to describe the series circuit impedance Z of an R-L series circuit. When L and R are in parallel, the 90 degree phase angle (shift in time) applies to the branch currents but the voltages are the same. Again, the right triangle and phasor diagram (Fig. 21-1.1) can be used to describe this parallel R-L circuit response.

Note: The ideal phase shift of 90 degrees between R and L is not a practical measurement because the scope is grounded and cannot easily be made to measure across the individual components R and L at the same time. For this reason, measurements in this lab will have phase shifts less than 90 degrees due to the combined R-L effects.

EQUIPMENT

Voltmeter
Oscilloscope: dual trace
Signal generator
Protoboard
Leads

COMPONENTS

(1) 100 mH inductor
(1) 33 mH inductor
(2) 100 Ω resistor
(2) 1 kΩ resistor

PROCEDURE

1. With an ohmmeter, measure and verify that the resistors used in this experiment are accurate. If any resistor is more than 10 percent from nominal, replace it.
2. With an ohmmeter, measure the dc resistance of the inductors (33 mH and 100 mH) and record the values in Table 21-1.1. This will also verify that the inductors (coils) are not open. Your coils may have 40 or 50 Ω of R_i at dc. Calculate the Q of each coil, assuming a reactance of 400 Ω at 1 kHz and record the results in Table 21-1.1.
3. Connect the circuit of Fig. 21-1.2.
4. Using a dc voltmeter, measure and record, in Table 21-1.2, the dc voltages to ground across bias resistor R_1 at point A, inductor L_1 at point B, and the 1 kΩ load resistor at point C.
5. Adjust the signal generator to each frequency step shown in Table 21-1.3 and measure the peak

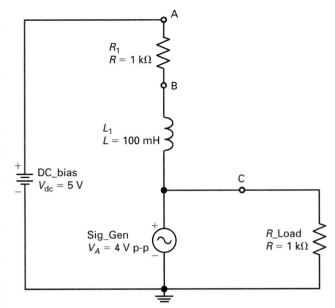

Fig. 21-1.2 Inductive circuit. Step 4.

voltages (not p-p) to ground across bias resistor R_1 at point A, inductor L_1 at point B, and the load resistor at point C. Record the results in Table 21-1.3.

6. Exchange the position of the resistor R_1 and the inductor L_1 so that the circuit looks like the one shown here in Fig. 21-1.3. Repeat the previous steps for this new configuration, measuring and recording the peak voltage at point C, across the 1 kΩ load resistor as you adjust the signal generator to 4 V p-p for each step. Refer to Tables 21-1.4 and 21-1.5 for recording results and varying the frequency.
7. Slowly turn down the dc power supply voltage from 5 to 0 volts and look at the traces on the scope.

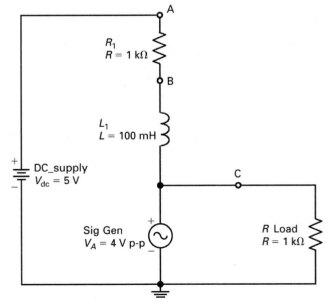

Fig. 21-1.3 Inductive circuit. Step 6.

332 EXPERIMENT 21-1

Note any change that occurs to the ac signal and be prepared to include this information in your report.

8. Connect the circuit of Fig. 21-1.4 shown here. The signal generator should be set beginning at 200 Hz with 4 volts p-p (peak-to-peak) across the circuit (ground to point A). You will need to connect the oscilloscope (Channel 1) to the circuit to measure the applied signal to get the correct results. Remember that the signal generator has its own impedance that is in parallel with the circuit—so connect it first and then measure 4 V p-p across the circuit.

Note: The scope time-base will be at about 1 ms per/div on the x axis for 200 Hz and you may have to adjust the triggering. The y axis for magnitude (p-p value) can be 0.5 volts/div for Channel 1 (4 V p-p) and 0.5 volts/div for Channel 2.

9. With the dual trace oscilloscope displaying about 2 periods each of the p-p signals from ground to point A (Channel 1) and from ground to point C (Channel 2), draw the two signals as best you can on the oscilloscope graphs in Table 21-1.6. This will show their phase relationship or time on the x axis. On the graph, label the two signals: V_A is 4 V p-p and the other is V_C. Also, indicate the difference on the x axis. For example, if the difference is ½ division (between peaks), at 1 ms per division, then the difference in time is about 0.5 ms. Using that value, you can calculate the phase shift by using the example formula shown in Table 21-1.6.

10. Decrease the signal generator frequency to 100 Hz and lower and notice the results on the oscilloscope between the phase of the two signals—adjusting the scope time base as necessary. Also, be sure to check the 4 V p-p and adjust it if necessary. Notice what happens and keep this in mind for answering the questions—there is no need to record the results.

11. Increase the signal generator to 1 kHz with 4 V p-p applied—notice how you adjust the signal generator (up or down in amplitude) and adjust the scope to display about 2 periods of each signal. Draw the two signals on the graph and indicate which signal is the input and which is point C to ground. Also include the amplitude of V_C and the difference in time between the two signals (x axis time) in Table 21-1.6.

12. Increase the signal generator to 10 kHz, adjusting the applied voltage. Again, in Table 21-1.6, record the traces on the graph with amplitude, time and phase differences.

13. Return the signal generator to 1 kHz with 4 V p-p applied. Now, connect the scope Channel 2 probe to point B (to ground). Notice how the signal at point B differs from that at point C due (at 1 kHz). Move the scope probe back and forth between points B and C to verify the difference between inductor values (33 mH versus 100 mH). Use this observation in your report.

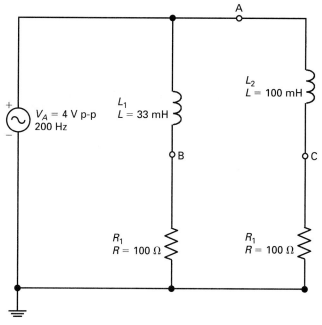

Fig. 21-1.4 Step 8.

QUESTIONS FOR EXPERIMENT 21-1

1. Why is the dc voltage across the load resistor always zero for this experiment?

2. Why is the ac voltage across the load resistor the same as the generator?

3. What would happen if the inductor value (100 mH) increased greatly, for example to 1H, in the first two circuits measured?

4. In the last circuit, what was the effect on the phase (time axis) of the two traces as the applied frequency was decreased?

5. In the last circuit, was the applied voltage (4 V p-p) decreased or increased as you increased the frequency for the last steps? Explain why you think this was necessary.

6. In the last step, was the signal across the resistor (point B) greater or lesser for the branch with 33 mH, instead of 100 mH and why?

TABLES FOR EXPERIMENT 21-1

TABLE 21–1.1 $X_L = 400\ \Omega$

Resistance and Q of 33 mH inductor	Resistance and Q of 100 mH inductor
$R =$ _____	$R =$ _____
$Q =$ _____	$Q =$ _____

NOTE: Calculate Q as X_K/R.

TABLE 21–1.2

Applied dc voltage	DC voltage point B	DC voltage point C
5.0	_____	_____

TABLE 21–1.3

Frequency	A: Peak Volts	B: Peak Volts	C: Peak Volts
0 Hz	_____	_____	_____
20 Hz	_____	_____	_____
40 Hz	_____	_____	_____
60 Hz	_____	_____	_____
80 Hz	_____	_____	_____
100 Hz	_____	_____	_____
300 Hz	_____	_____	_____
500 Hz	_____	_____	_____
700 Hz	_____	_____	_____
1 kHz	_____	_____	_____
3 kHz	_____	_____	_____
5 kHz	_____	_____	_____
7 kHz	_____	_____	_____
9 kHz	_____	_____	_____
10 kHz	_____	_____	_____

TABLE 21–1.4

Applied dc voltage	DC voltage point B	DC voltage point C
5.0	_____	_____

TABLE 21–1.5

Frequency	A: Peak Volts	B: Peak Volts	C: Peak Volts
0 Hz			
20 Hz			
40 Hz			
60 Hz			
80 Hz			
100 Hz			
300 Hz			
500 Hz			
700 Hz			
1 kHz			
3 kHz			
5 kHz			
7 kHz			
9 kHz			
10 kHz			

TABLE 21-1.6

STEP 9: 200 Hz

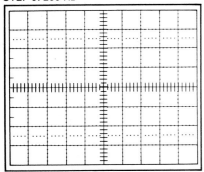

V_A = 4 V p-p applied

V_C = _____ p-p

Time difference = _____

Phase shift = _____

STEP 11: 1 kHz

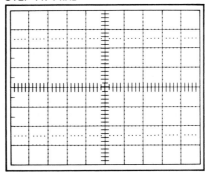

V_A = 4 V p-p applied

V_C = _____ p-p

Time difference = _____

Phase shift = _____

STEP 12: 10 kHz

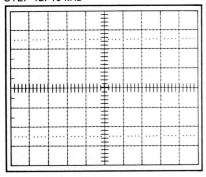

V_A = 4 V p-p applied

V_C = _____ p-p

Time difference = _____

Phase shift = _____

EXAMPLE: 1 kHz

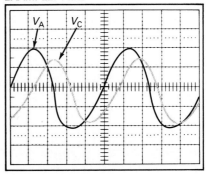

V_A = 4 V p-p applied

V_C = 400 mV p-p

Time difference = 0.1 ms

at 1 ms per division

Phase shift = $\frac{0.1 \text{ ms}}{1 \text{ ms}} \times 360 = 36°$

Use this formula:

$\frac{\text{time difference}}{\text{time for 1 cycle}} \times 360$

EXPERIMENT RESULTS REPORT FORM

Experiment No: _____   Name: _____
                                 Date: _____
Experiment Title: _____ Class: _____
_____   Instr: _____

Explain the purpose of the experiment:

List the first Learning Objective:
OBJECTIVE 1:

After reviewing the results, describe how the objective was validated by this experiment?

List the second Learning Objective:
OBJECTIVE 2:

After reviewing the results, describe how the objective was validated by this experiment?

List the third Learning Objective:
OBJECTIVE 3:

After reviewing the results, describe how the objective was validated by this experiment?

Conclusion:

If required, attach to this form: ☐ Answers to Questions, ☐ Tables, and ☐ Graphs.

EXPERIMENT 21-1

CHAPTER 22

EXPERIMENT 22-1

RC TIME CONSTANT

OBJECTIVES

At the completion of this experiment, you will be able to:
- Understand the concept of $T = RC$.
- Validate the decaying charge current in an RC series circuit.
- Plot a graph of V_C and I_C versus time for equal values of single, series, and parallel capacitance.

SUGGESTED READING

Chapters 18 and 22, *Basic Electronics,* Grob/Schultz, Tenth Edition

INTRODUCTION

A capacitor in series with a resistance will charge to 63.2 percent of the applied voltage in one time constant. This time constant T (in seconds) is applied to nonsinusoidal waveforms as a transient response, where R is the series resistance in ohms and C is the value of the capacitance in farads, or

$$T = RC$$

For example, the circuit of Fig. 22-1.1 has the following time constant:

$$\begin{aligned} T = RC &= (1000)(4 \times 10^{-6}) \\ &= 4000 \times 10^{-6} \\ &= 4 \times 10^{-3} \\ &= 4 \text{ ms} \end{aligned}$$

Notice that the applied voltage does not affect the value of T. The time constant T is the time for the voltage across C to change by 63.2 percent. If V_A in Fig. 22-1.1 were 100 V, then C would change to 63.2 V in 4 ms. In another 4 ms, the capacitor would change to 63.2 percent of the remaining voltage, or 63.2 percent of 36.8 V. This would repeat until the voltage across C equaled the applied voltage. Also, during this time, the circuit current would steadily decay until V_C reached its maximum voltage, or V_A.

Similarly, if the capacitor were discharging, RC would specify the time it takes C to discharge 63.2 percent of the way down, to the value equal to 36.8 percent of the initial voltage across C at the start of discharge. A capacitor is usually considered charged after approximately 5 time constants.

EQUIPMENT

High-voltage power supply
Microammeter (VOM)
Grease pencil (erasable)
Tissue, paper towels, or cloth
Leads; protoboard
Graph paper

COMPONENTS

Either A: ($V_A = 90$ V)
    (2) 4 μF capacitors and (1) 1.2 MΩ resistor
or B: ($V_A = 100$ V)
    (2) 1 μF capacitors and (1) 3 MΩ resistor
    (2) SPST switches

PROCEDURE

1. Connect the circuit of Fig. 22-1.2. Use either component values (A) or (B). (S_1 can be a switch, a wire lead, or the power supply on/off switch. S_2 is a wire lead.)

Caution: Do not turn power on.

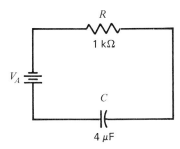

Fig. 22-1.1 *RC* time constant circuit.

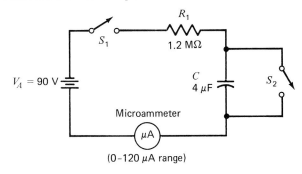

Fig. 22-1.2 *RC* series circuit for time constant measurement shown with (A) component values. The (B) values are $V_A = 100$ V, $R = 3$ MΩ, and $C = 1$ μF.

EXPERIMENT 22-1 **341**

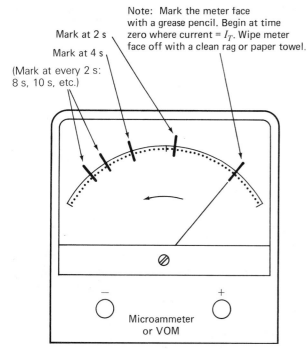

Fig. 22-1.3 Microammeter in *RC* time constant circuit.

the word *mark*, make your grease pencil mark. Try a practice run first.

Keep time for about 30 s. As the current decays (capacitor charges), the marks will get closer and closer together. Use approximate values when you record the results.

Record the results in a table. Record the value of circuit current at each 2-s interval, beginning at the zero until the 30-s limit. After about 16 s, the grease marks will be difficult to read. Approximate.

5. Repeat the procedure three times. Afterward, add the values for each 2-s interval and divide by 3 to get an average value.

6. Plot a graph of time versus circuit current and V_C (voltage across the capacitor). Refer to Fig. 22-1.4. The traces should have a similar appearance.

7. Calculate the time constant $T = RC$ and indicate those points on the graph for both V_C and I_T. Show T points on the x axis.

8. Repeat steps 1 to 7, two capacitors in parallel.

9. Repeat steps 1 to 7 with two capacitors in series.

2. Calculate and record in Table 22-1.1 the amount of circuit current I_T by using Ohm's law. Do not forget to indicate which component values, (A) or (B), you use.

3. Close S_2 in order to discharge C_1 (keep it closed). Close S_1 and measure and record the circuit current I_T in Table 22-1.1.

4. Refer to Fig. 22-1.3. Prepare to mark the meter face with a grease pencil by making the first mark at the point of I_T now showing on the meter.

When you start, mark the meter face every 2 s with a fine, straight mark. The 2-s interval can be called out by using the term *mark*. Thus, either your lab partner or a group timekeeper should start by announcing, "ready, go." Every 2 s, when you hear

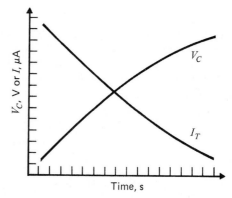

Fig. 22-1.4 Typical plot of time versus amplitude for *RC* time constant. Label both axes as necessary and indicate *RC* values in the title.

QUESTIONS FOR EXPERIMENT 22-1

Answer TRUE (T) or FALSE (F) to the following:

____ 1. The RC time constant is expressed in units of farads.

____ 2. The formula for the RC time constant is $T = 1/RC$.

____ 3. The time constant T is the time for the voltage across R to change by 36.8 percent.

____ 4. The change of 63.2 percent is constant for all values of R and C as long as a sine wave is the applied voltage.

____ 5. For a 1000 Ω R in series with an 18 μF C, the time constant would be 180 ms.

____ 6. The time constant formula is the same for a discharging capacitor.

____ 7. After approximately five time constants, a capacitor is considered charged to the applied dc voltage.

____ 8. After two time constants in an RC circuit where $R = 2$ kΩ and $C = 15$ μF, the capacitor would be charged to about 86 V with an applied voltage of 100 V.

____ 9. For question 8, if C were increased to 30 μF and R were increased to 4 kΩ, the capacitor would charge twice as fast.

____ 10. An RC charge or discharge curve usually has the same shape, regardless of the values of R and C.

TABLES FOR EXPERIMENT 22-1

TABLE 22–1.1: $C =$ ____, $R =$ ____, $V_A =$ ____

Procedure Step	Measurement	Value	Calculations
2	I_T		____ A
3	I_T	____ A	
7	T		____ s

Note: Remember that you will have three data tables when you are finished:
    Table 1. For (1)4 μF.
    Table 2. For (2)4 μF in series (2 μF total).
    Table 3. For (2)4 μF in parallel (8 μF total).

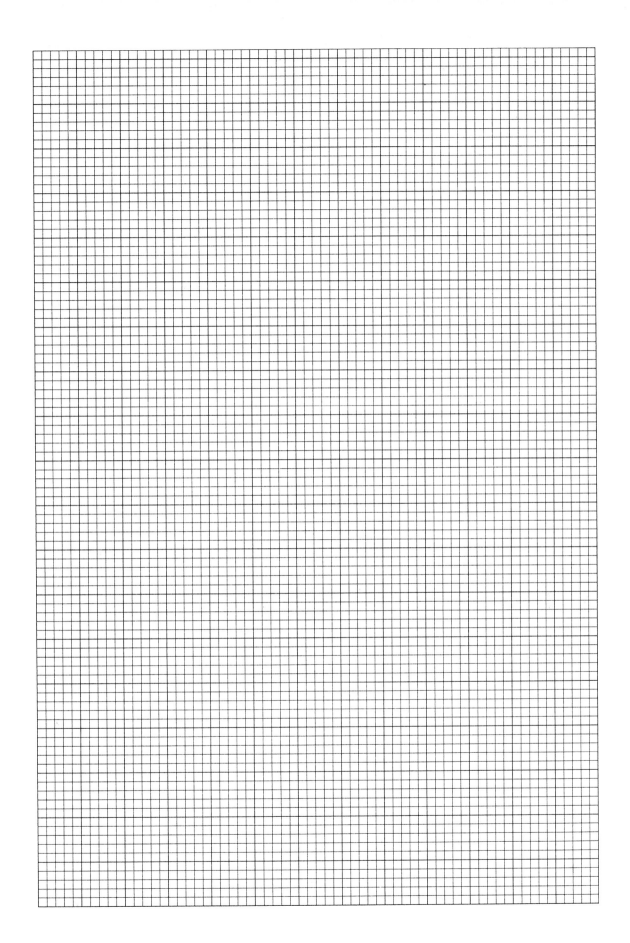

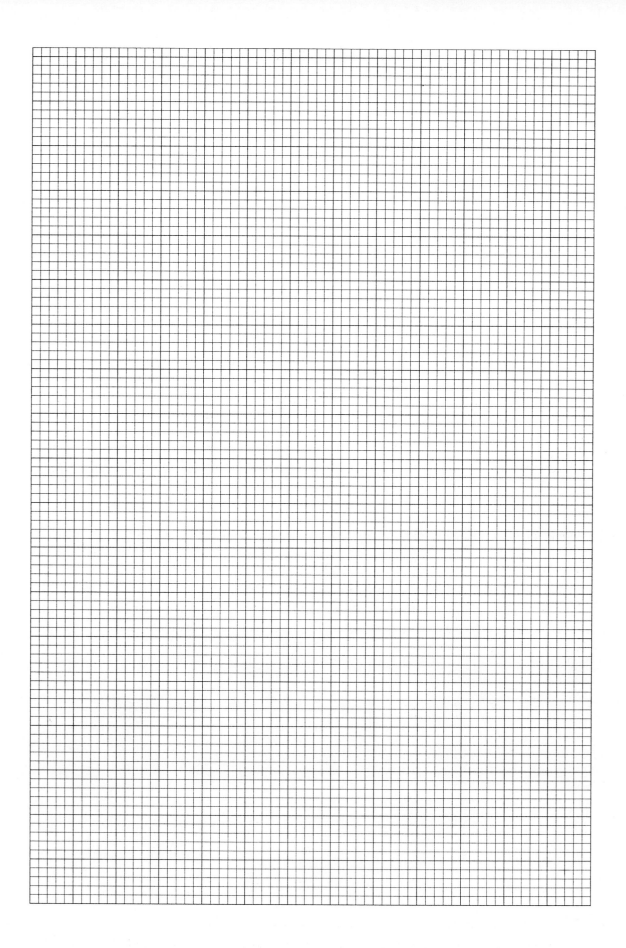

EXPERIMENT RESULTS REPORT FORM

Experiment No: _____ Name: _____
                                        Date: _____
Experiment Title: _____ Class: _____
_____   Instr: _____

Explain the purpose of the experiment:

List the first Learning Objective:
OBJECTIVE 1:

After reviewing the results, describe how the objective was validated by this experiment?

List the second Learning Objective:
OBJECTIVE 2:

After reviewing the results, describe how the objective was validated by this experiment?

List the third Learning Objective:
OBJECTIVE 3:

After reviewing the results, describe how the objective was validated by this experiment?

Conclusion:

If required, attach to this form: ☐ Answers to Questions, ☐ Tables, and ☐ Graphs.

CHAPTER 23

EXPERIMENT 23-1

AC CIRCUITS: *RLC* SERIES

OBJECTIVES

At the completion of this experiment, you will be able to:
- Understand series reactance and resistance.
- Determine the net reactance, phase angle, and impedance of an *RLC* circuit.
- Determine the real power of an ac circuit as opposed to apparent power.

SUGGESTED READING

Chapter 23, *Basic Electronics,* Grob/Schultz, Tenth Edition

INTRODUCTION

Previous experiments covered *RL* series circuits and *RC* series circuits. In both of these ac circuits, the concept of reactance was investigated. The results showed that the reactive component had its own voltage drop that was out of phase with the resistive component, although the circuit current was the same in all parts of the circuit.

In an *RLC* series circuit, the same concepts apply. However, when X_L and X_C are both in the circuit, the opposite phase angles enable one to cancel the effect of the other. For X_L and X_C in series, the net reactance is the difference between the two series reactances, resulting in less reactance than either one alone. In parallel circuits, the branch currents cancel, resulting in less total current.

Consider the circuit of Fig. 23-1.1. Notice that the circuit current is found by dividing the applied voltage by the total net reactance of the circuit. Here,

$$\frac{120 \text{ V}}{20 \text{ }\Omega} = 6 \text{ A}$$

The net reactance is the difference between $X_L = 60 \text{ }\Omega$ and $X_C = 40 \text{ }\Omega$. In the same manner, the difference between the two voltages is equal to the applied voltage, because the IX_L and IX_C voltages are opposite. If the values were reversed, the net reactance would be 20 Ω X_C. The current would still be 6 A, but it would have a lagging ($-90°$) instead of a leading ($+90°$) phase angle.

When resistance is added to the circuit, the total effect is determined by phasors. The phasor for the circuit of Fig. 23-1.1, if *R* were added, would be as shown in Fig. 23-1.2. Or, if X_L were greater, it would be as shown in Fig. 23-1.3.

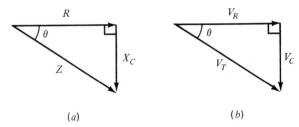

Fig. 23-1.2 Phasor diagrams for net capacitive reactance. (*a*) Impedance. (*b*) Voltage.

When X_L and X_C are equal, the net reactance is 0 Ω, and the result is called *resonance.* Because resonance has a specific application, it will be studied separately in later experiments.

Fig. 23-1.1 *LC* circuit reactances. Net reactance = 60 Ω − 40 Ω = 20 Ω.

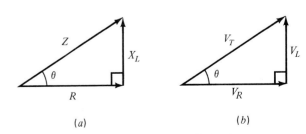

Fig. 23-1.3 Phasor diagrams for net inductive reactance. (*a*) Impedance. (*b*) Voltage.

EXPERIMENT 23-1 **349**

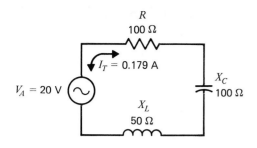

Net reactance = $X_C - X_L$ = 100 Ω − 50 Ω = 50 Ω
Apparent power = VI = 20 V × 0.179 A = 3.58 VA
Real power = I^2R = 3.20 W

Fig. 23-1.4 *RLC* series circuit.

The procedure that follows will investigate an *RLC* circuit where there will be a net reactance. In that case, there will be a difference between real power and apparent power. Consider the example of the circuit shown in Fig. 23-1.4. With *V* and *I* out of phase because of reactance, the product of the applied voltage and the circuit current is *apparent power,* and the unit is the voltampere (VA). However, the *real power* is the resistive power, dissipated as heat, and can always be found as I^2R. Finally, the ratio of real power over apparent power is called the *power factor* and is equal to the cosine of the phase angle of the circuit. The power factor is used commercially to bring about efficient distribution of energy in power lines.

EQUIPMENT

Signal generator
Oscilloscope
VTVM
Leads
Protoboard

COMPONENTS

(1) 1 kΩ 0.25 W resistor
(1) 0.01 μF capacitor
(1) 33 mH inductor

PROCEDURE

1. Connect the circuit of Fig. 23-1.5.
2. Use an oscilloscope. Measure and record in Table 23-1.1 the peak-to-peak voltage across the capacitor (V_C), the inductor (V_L), and the resistor (V_R).

Note: Avoid ground loops when measuring voltage with the oscilloscope by moving the resistor and inductor to keep the same ground point.

3. Calculate and record X_L and X_C in Table 23-1.1.
4. Calculate and record the circuit current I_T for V_R/R, V_C/X_C, and V_L/X_L.
5. Calculate and record in Table 23-1.1 the net reactance and the impedance *Z*. On a separate sheet, draw a phasor diagram for these values and determine the phase angle.
6. Determine the apparent power, real power, and power factor and record in Table 23-1.1.
7. Decrease the frequency to 100 Hz and repeat steps 1 to 6. Record in Table 23-1.1.

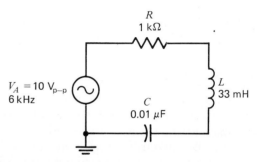

Fig. 23-1.5 *RLC* series circuit for measuring net reactance.

OPTIONAL PROCEDURE

Use different values of *R*, *L*, and *C*. Record the values and repeat steps 1 to 7. Also, try different frequencies: 10 kHz, 100 kHz, and 1 MHz. Use a separate sheet to record and discuss the results.

QUESTIONS FOR EXPERIMENT 23-1

1. In an *RLC* series circuit, why can an inductor have more measured voltage than the applied voltage?

2. If X_L and X_C were zero, what would the phase angle of Fig. 23-1.5 be?

3. Explain the difference between real power and apparent power.

4. If an *RLC* series circuit had four inductors, five resistors, and seven capacitors, all in series, would the value of current be the same in all parts of the circuit?

5. At what frequency would the circuit of Fig. 23-1.5 (as shown) have the opposite net reactance?

TABLES FOR EXPERIMENT 23-1

TABLE 23–1.1

Procedure Step	Circuit Component	Steps 2–6: Value at $f = 6$ kHz	Step 7: Value at $f = 100$ Hz
2	V_C measured		
	V_L measured		
	V_R measured		
3	X_L calculated		
	X_C calculated		
4	$I_T = V_R/R$		
	$I_T = V_C/X_C$		
	$I_T = V_L/X_L$		
5	X_O calculated (net reactance)		
	Z calculated		
6	Apparent power		
	Real power		
	Power factor		

EXPERIMENT RESULTS REPORT FORM

Experiment No: _____   Name: _____
                                        Date: _____
Experiment Title: _____   Class: _____
_____   Instr: _____

Explain the purpose of the experiment:

List the first Learning Objective:
OBJECTIVE 1:

After reviewing the results, describe how the objective was validated by this experiment?

List the second Learning Objective:
OBJECTIVE 2:

After reviewing the results, describe how the objective was validated by this experiment?

List the third Learning Objective:
OBJECTIVE 3:

After reviewing the results, describe how the objective was validated by this experiment?

Conclusion:

If required, attach to this form: ☐ Answers to Questions, ☐ Tables, and ☐ Graphs.

CHAPTER 24

EXPERIMENT 24-1

COMPLEX NUMBERS FOR AC CIRCUITS

OBJECTIVES

- Use complex numbers to describe circuit impedances
- Measure and record complex data
- Calculate the complex impedance of measured circuits

SUGGESTED READING

Chapter 24, *Basic Electronics,* Grob/Schultz, Tenth Edition

INTRODUCTION

Complex numbers are most often used to describe the impedance (Z) of a circuit. Consider an amplifier for example as shown in Fig. 24-1.1. It may have both resistance and reactance to the input signal that is to be amplified over a specific frequency range. The ac signal is also coming from the output of another stage or circuit, such as a preamp. Therefore, to achieve maximum power transfer, the output impedance of one stage must be matched to the input of the next stage. To do this, you need to determine the amplifier's input impedance (resistance and reactance) so that it can be modified to match the output impedance of the previous stage. Therefore, the use of complex numbers becomes an important way to describe Z and obtain a match.

In general, many *RF* circuits are designed for off-the-shelf use to have a standard value of impedance: $50 + j0$. Of course, there is some tolerance. For this reason, you will see test equipment often made with $50\,\Omega$ of impedance.

The complex number describes the amount of resistance (real part) and the amount of reactance (imaginary part) due to X_L or X_C in the circuit. The reactance or imaginary part is referred to as the j operator and is often plotted on a polar plot. By describing the impedance Z as a complex number, the test technician can work with the circuit designer to achieve a better match from one output to the input of another stage.

In *RF* and microwave circuits, the measurement of circuit impedance is usually done with a **network analyzer.** The network analyzer sends a signal into a circuit and determines the amount of signal that is reflected from the input. Thus, by knowing the input signal and the reflected signal, the impedance can be calculated automatically and plotted on the display. The network analyzer can also measure transmission to determine the amount of signal passing through (gain or loss) a circuit compared to the injected signal. However, this type of expensive test equipment is usually reserved for high frequency (microwave) study and not often used in the first year of electronics for technicians. But it is possible to measure circuits that represent impedances and then describe them using complex numbers.

As shown in Fig. 24-1.2, the impedance of a circuit can be plotted as a phasor or put on a polar plot.

Using complex numbers to describe impedance has this advantage: the real part from the resistance stays the same regardless of the frequency applied. However, the reactive part of $+j$ or $-j\,\Omega$ is specific to a given frequency. Also, remember that an impedance of $25 - j100\,\Omega$ for example is not meaningful unless the frequency is specified. This is because the reactive component (j) is calculated from X_L or X_C.

The circuit of Fig. 24-1.3 shows a mismatch between two circuits. With two different impedances, the ac signal will not transfer with maximum power from one circuit stage to the next, unless the impedances are matched exactly equal (real and imaginary). In addition, there will always be a phase shift between the real and imaginary parts unless there is no reactance at all.

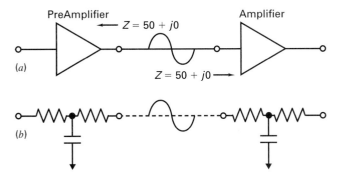

Fig. 24-1.1 Amplifier circuit; (*a*) equivalent circuit.

EXPERIMENT 24-1 **355**

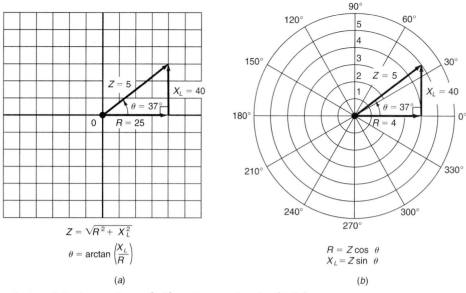

Fig. 24-1.2 Impedance plotted as a phasor (left) and as a polar plot (right).

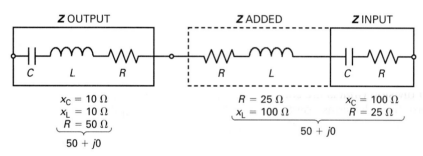

Fig. 24-1.3 A mismatch between two circuits. Solution in text below.

For the circuit with the mismatch (Z input), a resistance of 25 Ω is added to the input; along with an inductor that had 100 Ω of inductive reactance at the frequency of interest to cancel out the 100 Ω of capacitive reactance. Also, keep in mind that when L and C are in series or parallel, they can also resonate at some frequency so this is another factor that must be considered.

For this experiment, without using a network analyzer, the use of complex numbers to describe circuit impedance will be obtained from measuring voltages and then calculating reactance and writing the complex number to describe the circuits. Afterward, the circuit will be modified to achieve the desired impedance.

EQUIPMENT

Ohmmeter
Dual trace oscilloscope
Breadboard
Signal generator
Leads as needed

COMPONENTS

(1) 50 Ω resistor or (2) 100 Ω resistors connected in parallel
(1) 1 μF capacitor
(1) 33 mH capacitor

PROCEDURE

1. With an ohmmeter, measure the 50 Ω resistor or the two 100 Ω resistors in parallel to be sure you have 50 Ω of series resistance within 5 percent tolerance.

2. Connect the circuit of Fig. 24-1.4, opposite.

Note: As you study electronics, you will learn that 50 Ω is a standard value of impedance for electronic circuits. With air being about 300 Ω and a perfect conductor being 0 Ω, 50 Ω is a reasonable standard for all high frequency applications.

3. Connect the scope across the signal generator and adjust input signal V_A to 2 volt p-p with a frequency of 3 kHz the input.

4. Measure the p-p signal voltage across the capacitor at V_C and record the value in Table 24-1.1.

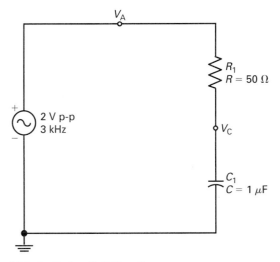

Fig. 24-1.4 Series *R-C*. Step 2.

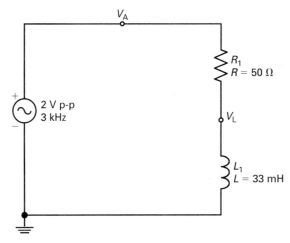

Fig. 24-1.5 Series *R-L*. Step 13.

5. Calculate the reactance of the 1 μF capacitor at 3 kHz and record the value in Table 24-1.1. Use the formula: $X_C = 1/(6.28 \times 3\text{ kHz} \times 1\text{ μF})$.

6. Knowing the R and X_C values, record the complex impedance Z in Table 24-1.1.

7. Look at the two signals V_C and V_A on the scope and approximate the difference in time between the V_A peak and the V_C peak. For example, if the scope is set to 0.1 ms per division (that is the same as 100 μs), and the peaks are three divisions apart, then the difference in time would be about 300 microseconds. Record the value between peaks in microseconds in Table 24-1.1 as delta time.

8. Notice which peak, V_A or V_C, is leading or lagging and note this on a separate sheet of paper—even drawing it if you want. You will use this information to answer the questions for this lab.

9. Reduce the frequency of the signal generator to 300 Hz, adjusting the applied voltage from the signal generator to 2 V p-p input.

10. Measure and record the p-p signal of V_C in Table 24-1.1.

11. Calculate X_C for the signal at 300 Hz and record the complex impedance.

12. Approximate the distance in time between peaks V_A and V_C and record that value also—you may see very little change but try to record it anyway. You will use this information in the report. Also, you may want to move the traces up and down or take the scope out of CAL to line up the peaks for a better view. You may be adjusting the time-base also. Always remember to check your scope, signal generator, and circuit connections carefully at each step and reset them as needed. Do not make the mistake of assuming the scope is correctly set—always check your settings before recording a measured value.

13. Connect the *R-L* circuit of Fig. 24-1.5 by replacing the capacitor with a 33 mH inductor.

14. Repeat the steps for this circuit by measuring and recording V_L, calculating X_L and writing the complex impedance, and also approximating and recording the distance between peaks V_L and V_A at both 3 kHz and 300 Hz according to Table 24-1.2.

Note: At this point in the experiment, you should see that the complex numbers for the impedance of each of the two circuits have some similarities and some differences. Consider these concepts and the information in the introduction for your report and also for the next steps.

15. Connect the circuit of Fig. 24-1.6 where both L and C are now in series with the 50 Ω of resistance. Notice that point V_X is across the combined series *L-C* combination. In this case, both L and C components are combined to achieve the purpose of

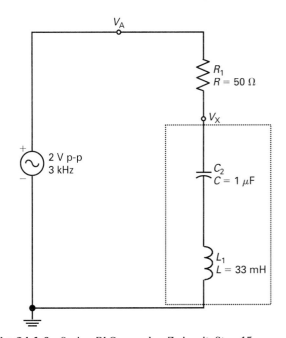

Fig. 24-1.6 Series *RLC* complex Z circuit. Step 15.

EXPERIMENT 24-1 **357**

making the impedance $50 + j0$ at a specific frequency. Also, this is really the resonant frequency where Z is at a minimum (zero) and this results in maximum current through L and C.

16. With 2 V p-p applied at each frequency, measure and record the values of V_X at both 300 Hz and 3 kHz. Also, calculate the net reactance X_O (difference between X_L and L_C) for each frequency and write the complex number that describes the impedance at both frequencies. Remember that the complex number for the impedance uses the net reactance as the imaginary part. Record all the values in Table 24-1.3.

17. Look at the data you have collected, especially the complex impedance values. You may have concluded that somewhere between 300 Hz and 3 kHz there may be a frequency where X_C and X_L cancel for a net reactance of zero. Carefully adjust the signal generator to a frequency where the V_X signal is at a minimum value. Record this frequency value in Table 24-1.4. This is the frequency where L and C are series resonant and X_L and X_C cancel. At this point, the circuit impedance is effectively: $50 + j0 \ \Omega$.

18. Record the approximate p-p value of point V_X in Table 24-1.4 at its minimum.

QUESTIONS FOR EXPERIMENT 24-1

1. In the circuit R-C of Figure 24-1.4, which signal was leading and which was lagging in time? Explain your answer and remember that the oscilloscope x axis is in the time domain. Therefore, a signal peak that appears first (starting from the left or earlier time point) is really the leading signal—the one that follows later in time is lagging.

2. In the circuit of Fig. 24-1.5, which signal was leading and which was lagging in time?

3. Explain and phase difference between the measure signals at V_A and either V_C or V_L in the RC and RL circuits.

4. In the last circuit (Fig. 24-1.6), why do you think the p-p values of V_X were equal to V_A? Use your data to explain this. HINT: Compare the reactance to the 50 Ω of resistance.

5. Would it be possible to use parallel L-C, instead of series L-C, to achieve the same result? Explain your answer.

TABLES FOR EXPERIMENT 24-1

TABLE 24–1.1 *R-C* Complex Impedance Circuit (Fig. 24-1.4)

Frequency	V_A p-p	V_C p-p	X_C	Delta Time	Complex Z
3 kHz	2 V p-p	_____	_____	_____	_____
300 Hz	2 V p-p	_____	_____	_____	_____

TABLE 24–1.2 *R-L* Complex Impedance Circuit (Fig. 24-1.5)

Frequency	V_A p-p	V_L p-p	X_L	Delta Time	Complex Z
3 kHz	2 V p-p	_____	_____	_____	_____
300 Hz	2 V p-p	_____	_____	_____	_____

TABLE 24–1.3 *RLC* Complex Impedance Circuit (Fig. 24-1.6)

Frequency	V_A p-p	V_X p-p	X_O	Delta Time	Complex Z
3 kHz	2 V p-p	_____	_____	_____	_____
300 Hz	2 V p-p	_____	_____	_____	_____

NOTE: X_O is the net reactance.

TABLE 24–1.4 Minimal *Z* Values (Fig. 24-1.6)

Frequency (within 10 Hz) for $50 + j0$	Approx p-p value of V_X

360 EXPERIMENT 24-1

EXPERIMENT RESULTS REPORT FORM

Experiment No: _____ Name: _____
                                       Date: _____
Experiment Title: _____   Class: _____
_____   Instr: _____

Explain the purpose of the experiment:

List the first Learning Objective:
OBJECTIVE 1:

After reviewing the results, describe how the objective was validated by this experiment?

List the second Learning Objective:
OBJECTIVE 2:

After reviewing the results, describe how the objective was validated by this experiment?

List the third Learning Objective:
OBJECTIVE 3:

After reviewing the results, describe how the objective was validated by this experiment?

Conclusion:

If required, attach to this form: ☐ Answers to Questions, ☐ Tables, and ☐ Graphs.

CHAPTER 25

EXPERIMENT 25-1

SERIES RESONANCE

OBJECTIVES

At the completion of this experiment, you will be able to:
- Understand the concepts of series resonance.
- Validate the formula for the resonant frequency $f_r = 1/[2\pi(LC)^{1/2}]$.
- Plot a graph of frequency versus circuit current.

SUGGESTED READING

Chapter 25, *Basic Electronics*, Grob/Schultz, Tenth Edition

INTRODUCTION

The resonant effect takes place when $X_L = X_C$ for a series inductor and capacitor. The frequency at which the opposite reactances are equal is called the *resonant frequency*, and it can be calculated as

$$f_r = \frac{1}{2\pi(LC)^{1/2}} \text{ or } f_r = \frac{1}{2\pi\sqrt{LC}}$$

Because of the canceling effect of X_L and X_C (opposite in phase), the resonant effect can only occur at one specific frequency for a given combination of L and C. In general, large values of L and C provide a relatively low resonant frequency. Smaller values of L and C provide a higher value for f_r. Figure 25-1.1 shows (*a*) a series resonant circuit and (*b*) its response curve.

Notice that X_L and X_C are equal at 1000 kHz. This can be proven by using the formulas for $X_C = 1/(2\pi fC)$ and $X_L = 2\pi fL$. If the frequency of the signal generator (V_T) were to change, the circuit would no longer be resonant. Although the opposite reactances (180° out of phase) would still give a canceling effect, there would still be some net reactance remaining: X_C for a decrease in frequency and X_L for an increase in frequency. In Fig. 25-1.1, the only opposition to current flow is the resistance r_S, which is the resistance of the coil.

The main characteristic of a series resonant circuit is that the circuit current is maximum at the resonant frequency, as shown in Fig. 25-1.1*b*. This is called a *resonant rise* in current. Also, because of the canceling of X_C and X_L, the circuit impedance is minimum at the resonant frequency. And the circuit current is in phase with the generator voltage, meaning that the circuit phase angle is 0° at resonance.

Finally, because the circuit current is maximum, the voltage across either L or C is maximum at resonance. Thus, if the output of Fig. 25-1.1 were taken across either L or C, the result would be maximum voltage.

In addition, the quality of a resonant circuit is determined by the sharpness of the resonant rise in voltage across L or C, called Q, where Q is calculated as X_L/r_S. The greater the ratio of the reactance to the series resistance, the higher the Q and the sharper the resonant effect. Note that the X_L reactance, rather than X_C, is used to determine Q, due to the dc resistance of the coil. Also, the greater the L/C ratio, the greater the circuit Q. Thus, increasing L and decreasing C can produce a higher Q for any given resonant frequency. Q can also be measured as $Q = V_{in}/V_{out}$, where V_{out} is equal to the voltage across either L or C and V_{in} equals the generator voltage.

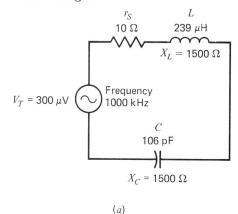

Fig. 25-1.1 Series resonant circuit. (*a*) Schematic diagram. (*b*) Response curve.

Remember, the main purpose of resonant circuits is for tuning, that is, to tune to a desired frequency, as in radio, television, and other forms of communication instrumentation.

EQUIPMENT

Audio signal generator
Oscilloscope
Protoboard
Leads

COMPONENTS

(1) 56 Ω 0.25 W resistor
(1) 0.1 µF capacitor
(1) 33 mH inductor

PROCEDURE

1. Connect the circuit of Fig. 25-1.2.

Note: Avoid ground loops by moving R and L as required.

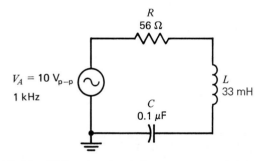

Fig. 25-1.2 *RLC* series circuit for measurement data.

2. Calculate and record the resonant frequency in Table 25-1.1.
3. Measure the voltages across R, L, and C, using an oscilloscope, and record the results in Table 25-1.1.

Note: For steps 4 through 8, create your own data table and graph the results in a response curve.

4. Increase the frequency in 250 Hz steps from 1 to 5 kHz and repeat step 3 at each frequency. Be sure to keep the applied voltage constant at each step. Thus, measure V_A at each step to be sure there is no change.

Note: Knowing the calculated resonant frequency, measure and record V_R, V_L, and V_C at several (3 or 4) extra frequency points on either side of that frequency.

5. Calculate the Q of the circuit. Now determine Q by measurement, and compare both values.
6. Calculate and record the circuit current for each frequency step as $I = V_R/R$.
7. Calculate and record the values of X_L and X_C at the two edges of the bandwidth. This is the same as the two half-power points on either side of the resonant frequency (f_r), where the voltage decreases to 70.7 percent.
8. Calculate and record the circuit impedance Z and the phase angle at f_r and at both half-power points.
9. Plot a graph of frequency versus the voltages V_R, V_L, and V_C. Indicate the bandwidth (half-power points on either side of f_r), where each half-power point is 70.7 percent of I_{max}.

QUESTIONS FOR EXPERIMENT 25-1

Answer TRUE (T) or FALSE (F) to the following:

_____ 1. The resonant effect occurs when X_L is greater than X_C.

_____ 2. The resonant frequency of a series L_C circuit must always be above 60 Hz.

_____ 3. The circuit impedance Z of a series circuit is maximum at f_r.

_____ 4. X_L and X_C are 90° out of phase at resonance.

_____ 5. The highest value of series circuit current is at resonance.

_____ 6. The greater the circuit Q, the higher the resonant frequency.

_____ 7. $f_r = 1/[2\pi(LC)^{1/2}]$ and $Q = X_L/X_C$.

_____ 8. The resonant frequency of a series RLC circuit with $R = 100\ \Omega$, $L = 8$ H, and $C = 4\ \mu F$ is 28 Hz.

_____ 9. The Q of the circuit values in question 8 is 14.

_____ 10. The resonant frequency of the circuit values of question 8 would not change if $L = 4$ H and $C = 8\ \mu F$.

TABLE FOR EXPERIMENT 25-1

TABLE 25–1.1

Procedure Step	Measurement	Measured Value	Calculation
2	f_r	_____ Hz	_____ Hz
3	V_R	_____ V at 1 kHz	
	V_L	_____ V at 1 kHz	
	V_C	_____ V at 1 kHz	

GRAPH FOR EXPERIMENT 25-1

EXPERIMENT RESULTS REPORT FORM

Experiment No: _____   Name: _____
                                       Date: _____
Experiment Title: _____   Class: _____
_____   Instr: _____

Explain the purpose of the experiment:

List the first Learning Objective:
OBJECTIVE 1:

After reviewing the results, describe how the objective was validated by this experiment?

List the second Learning Objective:
OBJECTIVE 2:

After reviewing the results, describe how the objective was validated by this experiment?

List the third Learning Objective:
OBJECTIVE 3:

After reviewing the results, describe how the objective was validated by this experiment?

Conclusion:

If required, attach to this form: ☐ Answers to Questions, ☐ Tables, and ☐ Graphs.

CHAPTER 25

EXPERIMENT 25-2

PARALLEL RESONANCE

OBJECTIVES

At the completion of this experiment, you will be able to:
- Understand the concept of parallel resonance.
- Note the differences between a series and a parallel LC resonant circuit.
- Plot a graph of frequency versus amplitude.

SUGGESTED READING

Chapter 25, *Basic Electronics*, Grob/Schultz, Tenth Edition

INTRODUCTION

Similar to a series resonant circuit, the resonant effect takes place in a parallel LC circuit, when $X_L = X_C$. The formulas for f_r and $Q = X_L/R$ are also the same. The cancellation of X_L and X_C, due to the 180° phase difference at the resonant frequency, is similar for both series and parallel circuits. However, the parallel LC circuit has differences due to the source being outside the parallel branches of L and C.

At resonance, the reactive branch currents cancel in the main line and produce minimum current only in the main line. Because the main-line current is minimum, its impedance is maximum at f_r. The parallel branches of L and C (also called a *tank circuit*), however, have maximum current at resonance, because they are in separate branches. In a circulating manner, the capacitor discharges into the inductor, which in turn has its field current collapse into the other side of the capacitor. This is called the *flywheel effect* and is only possible because of the ability of L and C to store energy.

Therefore, a parallel resonant circuit, or tank circuit, has both main-line and tank currents. Because the tank current is maximum at resonance, the greatest voltage drop will occur across either L or C at f_r. In the circuit of Fig. 25-2.1a, note that the resistances r_{S_1} and r_{S_2} are used to measure either line current or tank current for comparison ($V/R = I$).

Remember that in a parallel resonant circuit, low frequencies will take the path of least resistance, or the L branch. Likewise, high frequencies will take the path of least reactance, or the C branch. Therefore, when X_L and X_C are equal, the tank impedance will be maximum. Thus, as a tuning circuit, the maximum voltage can be taken across the tank at the resonant frequency.

A higher circuit Q will result in a sharper response curve or narrower bandwidth. Because the bandwidth is determined by the half-power points (70.7 percent of I_{max}) on either side of the resonant frequency, the bandwidth is equal to f_r/Q. Thus, for f_1 and f_2 in Fig. 25-2.1b, the difference between the two frequencies is the bandwidth.

EQUIPMENT

Audio signal generator Protoboard
Oscilloscope Leads

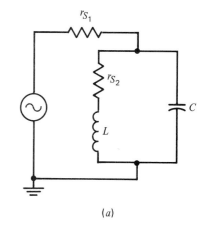

(a)

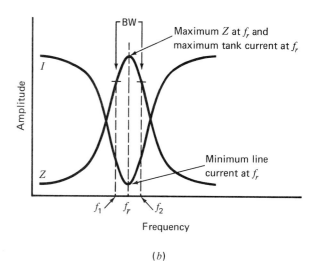

(b)

Fig. 25-2.1 Parallel resonant circuit. (a) Schematic. (b) Graph.

COMPONENTS

(1) 1 MΩ potentiometer (connected as rheostat)
(1) 100 Ω 0.25 W resistor
(1) 47 kΩ 0.25 W resistor
(1) 0.01 μF capacitor
(1) 100 mH inductor

PROCEDURE

1. Calculate and record in Table 25-2.1 the resonant frequency of the circuit of Fig. 25-2.2. Show your work.
2. Calculate and record in Table 25-2.1 the Q. Show your work.
3. Calculate and record in Table 25-2.1 the bandwidth (equals f_r/Q).
4. Connect the circuit of Fig. 25-2.2.

Note: Keep the input voltage constant for all frequencies, and avoid ground loops by moving the components as necessary.

5. Measure and record in Table 25-2.1 the voltage across r_{S_1}, r_{S_2}, L, and C. Calculate and record main-line and tank currents ($V/R = I$).
6. Increase the frequency in 250 Hz steps from 3500 to 6500 Hz, and repeat step 5 for each frequency step. Create your own data table to record the values.
7. Determine Z_T as follows and record the results in Table 25-2.1. Replace r_{S_1} (main-line resistor) with

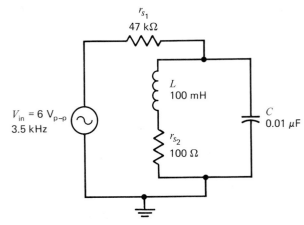

Fig. 25-2.2 Parallel resonant circuit for measurement.

a 1 MΩ rheostat. Adjust the circuit to the resonant frequency and adjust r_{S_1} so that its voltage equals the voltage across the tank. Record the value.

Note: When the voltage across the rheostat equals the voltage across the tank, its ohmic value is equal to the tank impedance. This is due to the laws of series circuits, where equal resistances or impedances divide the applied voltage equally.

8. Plot a graph of frequency versus amplitude for both tank current and main-line current. Indicate the bandwidth and any other significant points on the graph.

QUESTIONS FOR EXPERIMENT 25-2

Answer TRUE (T) or FALSE (F) to the following:

____ 1. The formula for resonance is $1/(2\pi fC)$.

____ 2. The higher the Q, the sharper the bandwidth response.

____ 3. One difference between a series resonant circuit and a parallel resonant circuit is that the parallel resonant circuit has minimum impedance at f_r, while the series has maximum impedance at f_r.

____ 4. Greater values of L and C will result in increased bandwidth.

____ 5. If a tank circuit had a resonant frequency of 8 kHz with a 2 kHz bandwidth, the bandwidth could not be decreased without changing the resonant frequency.

____ 6. At the resonant frequency, a tank circuit requires minimum input power from the source.

____ 7. In a tank circuit, line current is maximum and tank current is minimum at resonance.

____ 8. If a voltage were applied to a tank circuit, like the circuit in Fig. 25-2.1, no current would flow.

____ 9. For a tank circuit with $L = 2$ H and $C = 10$ μF, the resonant frequency is approximately 36 Hz.

____ 10. For question 9, a resonant frequency of 3600 Hz could be obtained by increasing L to equal 200 H.

TABLE FOR EXPERIMENT 25-2

TABLE 25–2.1

Procedure Step	Measurement	Measured Value	Calculation
1	f_r		_____ Hz
2	Q		_____ (X_L = _____ Ω)
3	BW		_____ Hz
5	$V_{r_{S1}}$	_____ Vp-p at 3.5 kHz	
	$V_{r_{S2}}$	_____ Vp-p at 3.5 kHz	
	V_L	_____ Vp-p at 3.5 kHz	
	V_C	_____ Vp-p at 3.5 kHz	
	I_{tank}		_____ A
	I_{line}		_____ A
7	Z_T	_____ Ω	

GRAPH FOR EXPERIMENT 25-2

EXPERIMENT RESULTS REPORT FORM

Experiment No: _____   Name: _____
                                  Date: _____
Experiment Title: _____ Class: _____
_____   Instr: _____

Explain the purpose of the experiment:

List the first Learning Objective:
OBJECTIVE 1:

After reviewing the results, describe how the objective was validated by this experiment?

List the second Learning Objective:
OBJECTIVE 2:

After reviewing the results, describe how the objective was validated by this experiment?

List the third Learning Objective:
OBJECTIVE 3:

After reviewing the results, describe how the objective was validated by this experiment?

Conclusion:

If required, attach to this form: ☐ Answers to Questions, ☐ Tables, and ☐ Graphs.

CHAPTER 26

EXPERIMENT 26-1

FILTERS

OBJECTIVES

At the completion of this experiment, you will be able to:
- Understand how filters separate different frequency components.
- Learn the difference between high-pass, low-pass, bandstop, and bandpass filters.
- Plot a graph of frequency versus amplitude for different filter types.

SUGGESTED READING

Chapter 26, *Basic Electronics*, Grob/Schultz, Tenth Edition

INTRODUCTION

Filters are used to separate wanted from unwanted signals. For example, a radio can receive many different station broadcasts. But somehow, it must isolate the desired broadcast signal and filter out the frequencies that are not part of the broadcast. Thus, filters are used to allow only those desired frequencies to pass through certain parts of a circuit.

Inductors and capacitors are used, in various configurations, to build filters. Generally, capacitors allow high frequencies to pass, while inductors allow low frequencies to pass through them with very little reactance or opposition to current flow. Also, the manner in which a filter's components are placed topology determines the type of filter it is.

Low-pass filters allow low frequencies to pass through them, while higher frequencies are sent to ground. High-pass filters allow only high frequencies to pass, while blocking low frequencies by sending them to ground.

The following procedures examine four simple filter types.

EQUIPMENT

Audio signal generator
Oscilloscope
Protoboard
Leads; graph paper

COMPONENTS

(1) 1 kΩ resistor
(1) 56 Ω resistor
(1) 0.01 μF capacitor
(1) 0.1 μF capacitor
(1) 10 μF capacitor
(2) 33 mH inductors

PROCEDURE

1. Connect the circuit of Fig. 26-1.1.

Note: Remember to keep the input voltage constant at each frequency for all steps in this procedure.

2. Increase the frequency from 1 kHz to 10 kHz in 1 kHz steps. Measure and record in Table 26-1.1 the peak-to-peak output voltage at each step.
3. At 10 kHz, place another 33 mH inductor across points C and D. Measure and record the output voltage.
4. At 10 kHz, remove one of the inductors and replace the 0.01 μF capacitor with a 0.1 μF capacitor. Measure and record the output voltage.
5. Connect the circuit of Fig. 26-1.2.
6. Decrease the frequency from 10 kHz to 100 Hz in the steps shown in Table 26-1.1. Measure and record the output voltage at each step.

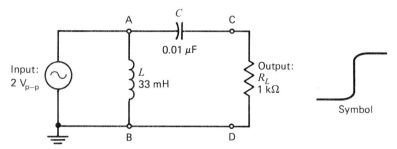

Fig. 26-1.1 High-pass filter.

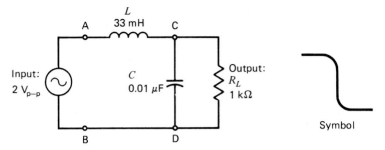

Fig. 26-1.2 Low-pass filter.

7. Place a second capacitor of 10µF across points A and B. Measure and record the output voltage for each step shown in Table 26-1.1.
8. Connect the circuit of Fig. 26-1.3.
9. Increase the frequency from 2 to 18 kHz in the steps shown in Table 26-1.1. Measure and record the output voltage at each step.
10. Connect the circuit of Fig. 26-1.4.
11. Increase the frequency from 2 to 18 kHz in the steps shown in Table 26-1.1. Measure and record the output voltage at each frequency step.
12. Plot a graph of frequency versus load voltage for each of the previous four filter circuits.

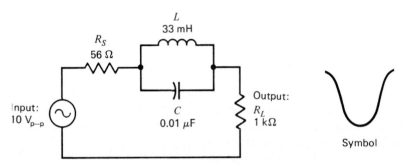

Fig. 26-1.3 Bandstop filter.

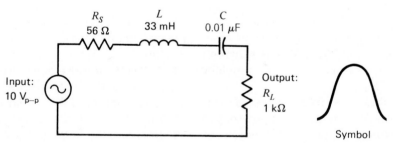

Fig. 26-1.4 Bandpass filter.

376 EXPERIMENT 26-1

QUESTIONS FOR EXPERIMENT 26-1

1. Explain what happens to low frequencies in the circuit of Fig. 26-1.1. Why don't they reach the load R_L

2. Explain what happens to high frequencies in the circuit of Fig. 26-1.2. Why don't they reach the load R_L?

3. Explain how the resonant effect works as a filter in the circuit of Fig. 26-1.3.

4. Explain how the resonant effect works as a filter in the circuit of Fig. 26-1.4.

5. Design a filter circuit that (a) passes frequencies from approximately 8 kHz to 12 kHz and (b) stops or rejects frequencies from 8 kHz to 12 kHz. Show all values and calculations, where $V_A = 1$ Vp-p and $R_L = 8\ \Omega$ (similar to a radio speaker).

TABLE FOR EXPERIMENT 26-1

TABLE 26–1.1

Procedure Step	Measurement Frequency	Load Voltage, V_{p-p}
2	1 kHz	_____
	2 kHz	_____
	3 kHz	_____
	4 kHz	_____
	5 kHz	_____
	6 kHz	_____
	7 kHz	_____
	8 kHz	_____
	9 kHz	_____
	10 kHz	_____
3	10 kHz	_____
4	10 kHz	_____
6	10 kHz	_____
	9 kHz	_____
	8 kHz	_____
	7 kHz	_____
	6 kHz	_____
	5 kHz	_____
	4 kHz	_____
	3 kHz	_____
	2 kHz	_____
	1 kHz	_____
	500 Hz	_____
	200 Hz	_____
	100 Hz	_____
	50 Hz	_____
7	10 kHz	_____
	5 kHz	_____
	2 kHz	_____
	1 kHz	_____
	500 Hz	_____
	100 Hz	_____
	50 Hz	_____

TABLE 26–1.1 (continued)

Procedure Step	Measurement Frequency	Load Voltage, V_{p-p}
9	2.0 kHz	_____
($C = 0.001\ \mu F$)	3.0 kHz	_____
$f_c \approx 9$ kHz	4.0 kHz	_____
	5.0 kHz	_____
	6.0 kHz	_____
	6.5 kHz	_____
	7.0 kHz	_____
	7.5 kHz	_____
	8.0 kHz	_____
	8.5 kHz	_____
	9.0 kHz	_____
	9.5 kHz	_____
	10.0 kHz	_____
	12.0 kHz	_____
	14.0 kHz	_____
	16.0 kHz	_____
	18.0 kHz	_____
11	2.0 kHz	_____
($C = 0.01\ \mu F$	3.0 kHz	_____
$f_o \approx 9$ kHz)	4.0 kHz	_____
	5.0 kHz	_____
	6.0 kHz	_____
	7.0 kHz	_____
	8.0 kHz	_____
	8.5 kHz	_____
	9.0 kHz	_____
	9.5 kHz	_____
	10.0 kHz	_____
	11.0 kHz	_____
	12.0 kHz	_____
	13.0 kHz	_____
	14.0 kHz	_____
	15.0 kHz	_____
	18.0 kHz	_____

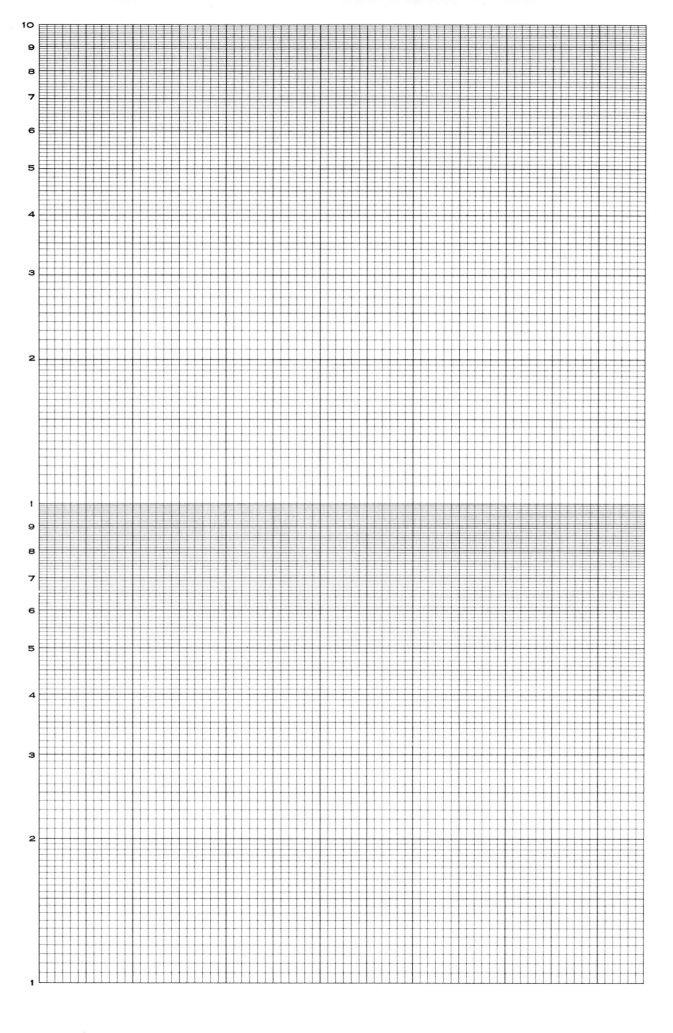

EXPERIMENT RESULTS REPORT FORM

Experiment No: _____ Name: _____
                                 Date: _____
Experiment Title: _____ Class: _____
_____ Instr: _____

Explain the purpose of the experiment:

List the first Learning Objective:
OBJECTIVE 1:

After reviewing the results, describe how the objective was validated by this experiment?

List the second Learning Objective:
OBJECTIVE 2:

After reviewing the results, describe how the objective was validated by this experiment?

List the third Learning Objective:
OBJECTIVE 3:

After reviewing the results, describe how the objective was validated by this experiment?

Conclusion:

If required, attach to this form: ☐ Answers to Questions, ☐ Tables, and ☐ Graphs.

CHAPTER 26

EXPERIMENT 26-2

FILTER APPLICATIONS

OBJECTIVES

At the completion of this experiment, you will be able to:

- Design a resonant filter circuit.
- Measure BW and determine Q for resonant filters.
- Troubleshoot bandstop and bandpass filters.

SUGGESTED READING

Chapters 25 and 26, *Basic Electronics*, Grob/Schultz, Tenth Edition

INTRODUCTION

A resonant circuit can be used to tune to a specific resonant frequency. Therefore, it has the effect of filtering out all signals except the resonant frequency. This applies to both series and parallel resonant circuits. Therefore, depending on the circuit configuration and where the resistive load is connected, resonant filters can be bandpass or bandstop in their effects.

Series resonant filters have a maximum current and a minimum impedance at resonance. When connected in series with a load, the series-tuned LC circuit allows the maximum current to flow to the resistive load at resonance. This is a bandpass filter (see Fig. 26-2.1a). However, when connected across the load, the tuned LC-R circuit becomes a low-impedance shunt path for frequencies at and near resonance, and little or no current flows to the load. This is a bandstop filter (see Fig. 26-2.1b).

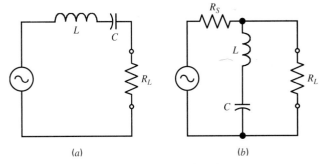

Fig. 26-2.1 Resonant filters (series). (*a*) Bandpass filter. (*b*) Bandstop filter.

Parallel resonant filters (tanks) have a maximum impedance and a minimum current at resonance. When connected in series with a load, the parallel-tuned LC circuit provides maximum impedance in series with the load, stopping the current flow to the resistive load at resonance. This is a bandstop filter (see Fig. 26-2.2a). When the load is connected across the parallel resonance, however, the LC circuit becomes an impedance for frequencies at and near resonance and allows maximum current flow to the load. This is a bandpass filter (see Fig. 26-2.2b).

Designing a Filter

To design a bandpass or bandstop filter, you need to calculate the resonant frequency that you want to stop or pass. This is done using the resonant formula:

$$\text{Resonant frequency } (f_r) = \frac{1}{2\pi\sqrt{LC}}$$

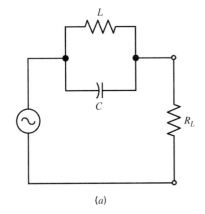
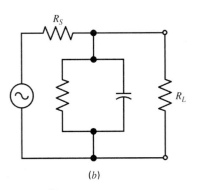

Fig. 26-2.2 Resonant filters (parallel). (*a*) Bandstop filter. (*b*) Bandpass filter.

In addition, the bandwidth (BW) of the resonance can become critical if it is too wide or too narrow. Bandwidth is measured practically as 70.7 percent of the peak of the response. Bandwidth is related to the circuit Q, where Q is a ratio: $Q = X_L/r$ (coil). The idea usually is to have a sharp (narrow) response with reasonable bandwidth: high Q. In a high-Q circuit, the ratio of L to C is also greater, and this means that you can adjust different values of L and C to achieve the same resonance, but with different bandwidths, and different Q rises in voltage or current. Figure 26-2.3 shows the response of a filter with different values of Q and varying BW.

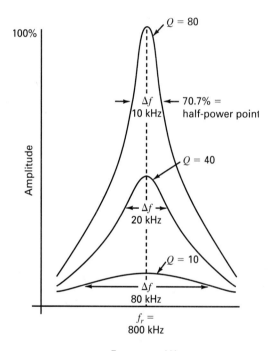

Fig. 26-2.3 Resonant filter response: BW and Q.

EQUIPMENT

Signal generator
Ohmmeter
Voltmeter
Oscilloscope

COMPONENTS

Resistors: 100 Ω and 1 kΩ
Capacitors: 0.01 μF and one other to be determined by design
Inductors: 0.01 H and one other to be determined by design or any available value

PROCEDURE

1. Refer to the circuit in Fig. 26-2.4. Calculate X_L and record the value in Table 26-2.1.

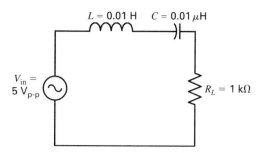

Fig. 26-2.4 Series-tuned LC filter.

2. Measure the resistance of the coil L and record the value in Table 26-2.1.

3. Calculate the circuit Q, the resonant frequency, and the bandwidth. Record the values in Table 26-2.1.

4. Connect the circuit of Fig. 26-2.4, but do not turn on power.

5. Apply power and adjust the signal generator frequency so that the maximum voltage appears across the load resistor. Record both the measured resonant frequency and the peak-to-peak voltage across the load resistor in Table 26-2.1.

6. Measure the BW as 70.7 percent of the value in step 5 and record the measurement in Table 26-2.1.

7. Insert a 10 Ω resistor between L and C. Do not make any adjustments to the signal generator voltage. This resistor will represent additional series resistance of the coil, which should change the circuit Q. Add 10 Ω to the measured value and record this number as R coil measured in Table 26-2.1.

8. Calculate the new values of Q and BW with the added resistance and record the values in Table 26-2.1.

9. Measure the peak-to-peak voltage across the load resistor and record the value in Table 26-2.1.

10. Measure the BW as 70.7 percent of the value in step 9 and record it in Table 26-2.1.

11. Refer to the circuit in Fig. 26-2.5. Calculate X_L and record the value in Table 26-2.2.

12. Measure the resistance of the coil L and record the value in Table 26-2.2.

13. Calculate the circuit Q, the resonant frequency, and the bandwidth and record the values in Table 26-2.2.

14. Connect the circuit of Fig. 26-2.5 but do not turn on the power.

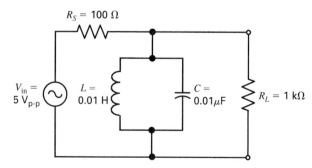

Fig. 26-2.5 Parallel-tuned LC filter.

15. Apply power and adjust the signal generator to so that the maximum voltage appears across the load resistor. Record both the measured resonant frequency and the peak-to-peak voltage across the load resistor in Table 26-2.2.

16. Measure the BW as 70.7 percent of the value in step 15 and record the measurement in Table 26-2.2.

17. Insert a 10 Ω resistor in series with the coil *L*. Do not make any adjustments to the signal generator voltage. This resistor will represent additional series resistance of the coil, which should change the circuit *Q*. Add 10 Ω to the measured value and record this number as the measured value in Table 26-2.2.

18. Calculate the new values of *Q* and BW with the added resistance and record the values in Table 26-2.2.

19. Measure the peak-to-peak voltage across the load resistor and record the value in Table 26-2.2.

20. Measure the BW as 70.7 percent of the value in step 19 and record the measurement in Table 26-2.2.

21. *Filter design*. Design a bandstop filter with a resonant frequency of either 5 kHz or 50 kHz, or do both if you have time. The goal is to have a high-*Q* circuit if possible. On a separate sheet of paper, show all your calculations and draw the circuit. Be sure your lab can supply the values of *L* and *C* you need—if not, use different values of *L* and *C*. Show all your calculations and then build the circuit. Test your design by measuring 10 frequency points on either side of the resonant frequency, with enough spacing so that you have the data to plot the response on a graph. Graph the results and indicate BW, resonant frequency, and any other values you want.

Note: V_{in} = 5 Vp-p or less.

QUESTIONS FOR EXPERIMENT 26-2

1. What elements control the BW of a resonant filter?

2. What is meant by the Q of a resonant circuit?

3. Why is a high-Q circuit desirable?

4. When would you expect one inductor to have a greater resistance than another?

5. What would happen if the series resistor R_S were not in the circuit of Fig. 26-2.5?

TABLES FOR EXPERIMENT 26-2

TABLE 26–2.1 (Steps 1 through 10.)

	Freq. Calc.	Freq. Meas.	V_{RL} Meas.	BW Meas.	R_{coil} Meas.	X_L Calc.	Q Calc.	BW Calc.
With 10-Ω R_{coil} added								

TABLE 26–2.2 (Steps 11 through 20.)

	Freq. Calc.	Freq. Meas.	V_{RL} Meas.	BW Meas.	R_{coil} Meas.	X_L Calc.	Q Calc.	BW Calc.
With 10-Ω R_{coil} added								

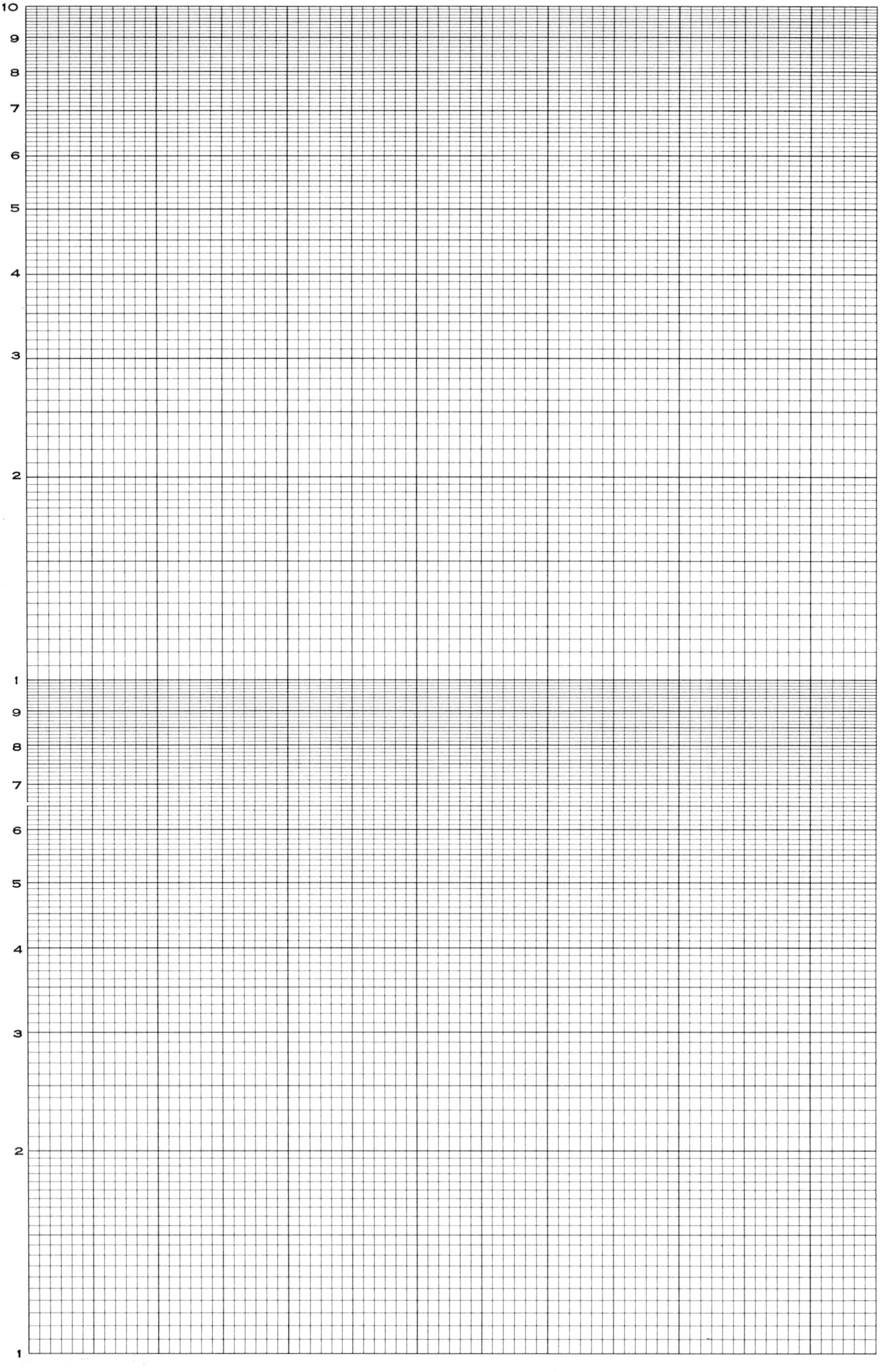

EXPERIMENT RESULTS REPORT FORM

Experiment No: _____   Name: _____
                                   Date: _____
Experiment Title: _____   Class: _____
_____   Instr: _____

Explain the purpose of the experiment:

List the first Learning Objective:
OBJECTIVE 1:

After reviewing the results, describe how the objective was validated by this experiment?

List the second Learning Objective:
OBJECTIVE 2:

After reviewing the results, describe how the objective was validated by this experiment?

List the third Learning Objective:
OBJECTIVE 3:

After reviewing the results, describe how the objective was validated by this experiment?

Conclusion:

If required, attach to this form: ☐ Answers to Questions, ☐ Tables, and ☐ Graphs.

Appendix A Applicable Color Codes

TABLE A-1 Color Code for Carbon Composition Resistors

COLOR	DIGIT 1st	DIGIT 2nd	MULTIPLIER	TOLERANCE
Black	0	0	1	—
Brown	1	1	10	—
Red	2	2	100	—
Orange	3	3	1,000	—
Yellow	4	4	10,000	—
Green	5	5	100,000	—
Blue	6	6	1,000,000	—
Violet	7	7	10,000,000	—
Gray	8	8	100,000,000	—
White	9	9	1,000,000,000	—
Gold	—	—	0.1	± 5%
Silver	—	—	0.01	±10%
No band	—	—	—	±20%

TABLE A-2 Color Code for Carbon Film Resistors

COLOR	DIGITS 1st	DIGITS 2nd	DIGITS 3rd	MULTIPLIER	TOLERANCE
Black	0	0	0	1	
Brown	1	1	1	10	±1%
Red	2	2	2	100	±2%
Orange	3	3	3	1,000	
Yellow	4	4	4	10,000	
Green	5	5	5	100,000	±5%
Blue	6	6	6	1,000,000	±0.25%
Violet	7	7	7	10,000,000	±0.10%
Gray	8	8	8		±0.05%
White	9	9	9		±5%
Gold	—	—	—	0.1	±10%
Silver	—	—	—	0.01	

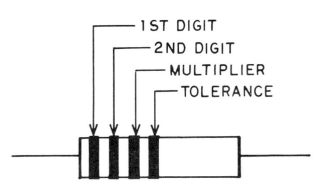

FIG. A-1 Color coding on a carbon composition resistor.

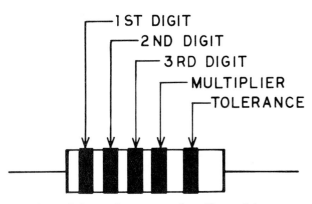

FIG. A-2 Color coding on a carbon film resistor.

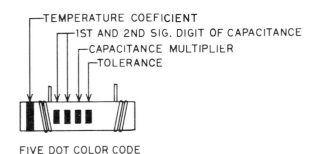

FIVE DOT COLOR CODE
RADIAL LEAD

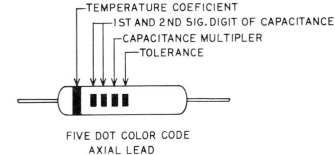

FIVE DOT COLOR CODE
AXIAL LEAD

SIX DOT COLOR CODE
RADIAL LEAD

FIG. A-3 Color coding on tubular ceramic capacitors.

Reproduced from Baer/Ottaway, *Electrical and Electronic Drawing*, Fifth Edition, McGraw-Hill, 1986.

TABLE A-3 Color Codes for Tubular and Disc Ceramic Capacitors

BAND OR DOT COLOR	CAPACITANCE IN PICOFARADS			TOLERANCE		TEMPERATURE COEFFICIENT pp °C (5-DOT SYSTEM)	TEMPERATURE COEFFICIENT 6-DOT SYSTEM SIG. FIG.	TEMPERATURE COEFFICIENT MULTIPLIER
	SIGNIFICANT DIGITS		CAPACITANCE MULTIPLIER					
	1st	2nd		≤10 pF	>10 pF			
Black	0	0	1	±2.0 pF	±20%	0	0.0	−1
Brown	1	1	10	±0.1 pF	±1%	−33		−10
Red	2	2	100		±2%	−75	1.0	−100
Orange	3	3	1000		±3%	−150	1.5	−1000
Yellow	4	4				−230	2.0	−10000
Green	5	5		±0.5 pF	±5%	−330	3.3	+1
Blue	6	6				−470	4.7	+10
Violet	7	7				−750	7.5	+100
Gray	8	8	0.01	±0.25 pF		+150 to −1500		+1000
White	9	9	0.1	±1.0 pF	±10%	+100 to −75		+10000
Silver	—	—						
Gold	—	—						

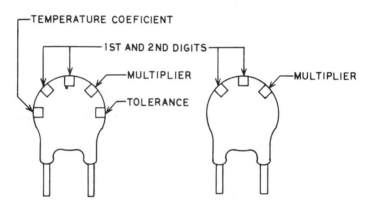

FIVE-DOT COLOR COLOR   THREE-DOT COLOR CODE

FIG. A-4 Color coding on a ceramic disk capacitor.

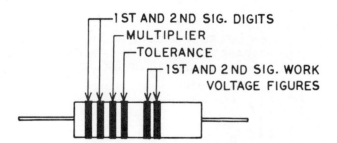

FIG. A-5 Color coding on a paper tubular capacitor.

TABLE A-4 Color Code for Molded Paper Tubular Capacitors

COLOR	(CAPACITANCE IN PICOFARADS)			TOLERANCE	VOLTAGE
	SIGNIFICANT DIGITS				
	1st	2nd	MULTIPLIER		
Black	0	0	1	±20%	—
Brown	1	1	10		100
Red	2	2	100		200
Orange	3	3	1000	±30%	300
Yellow	4	4	10000		400
Green	5	5			500
Blue	6	6			600
Violet	7	7			700
Gray	8	8			800
White	9	9			900
Gold	—	—			1000
Silver	—	—		±10%	—

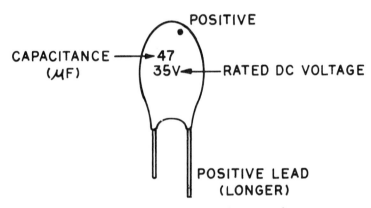

FIG. A-6 Marking of a tantalum capacitor.

Appendix B   Lab Report Preparation

Appendix B-1: How to Write Lab Reports

1. Write short sentences whenever possible.
2. Use a dictionary and spell all words correctly.
3. Write neatly in blue or black ink.
4. Avoid personal pronouns: I, you, we, etc.
5. Never discuss results unless they are part of your data.
6. Do not copy or paraphrase textbook theory.
7. Label all data tables and graphs with titles, numbers, proper units, and column headings.
8. Never write in the margins of your paper.
9. Leave enough space between sentences for the instructor to make corrections.
10. Technical accuracy and completeness are the most important part of a lab report. Unless the report is well organized and easy to read, it is of little value.

Every instructor will have different standards and different ideas about report writing. However, most lab reports reflect the scientific method as follows:

- A hypothesis is formulated. This is like a statement of purpose.
- Data is collected and analyzed. This is like the procedure and results.
- The hypothesis is proven or disproven based upon the results. This is like the discussion and conclusion.

Refer to the two sample reports that follow in Appendixes B-3 and B-4. One is a poor report that does not follow these suggestions. The other is a good report that does use the suggestions.

EXPERIMENT RESULTS REPORT FORM

Experiment No: _____   Name: _____
                                  Date: _____
Experiment Title: _____   Class: _____
_____   Instr: _____

Explain the purpose of the experiment:

List the first Learning Objective:
OBJECTIVE 1:

After reviewing the results, describe how the objective was validated by this experiment?

List the second Learning Objective:
OBJECTIVE 2:

After reviewing the results, describe how the objective was validated by this experiment?

List the third Learning Objective:
OBJECTIVE 3:

After reviewing the results, describe how the objective was validated by this experiment?

Conclusion:

If required, attach to this form: ☐ Answers to Questions, ☐ Tables, and ☐ Graphs.

Appendix B-3: An Example of a Poorly Written Report.

EXPERIMENT RESULTS REPORT FORM

Experiment No: __3__
Experiment Title: __Ohm's Law__

Name: __J. Doe__
Date: _____
Class: __Elec.__
Instr: __Jones__

Explain the purpose of the experiment:

To do the Ohm's Law experiment

List the first Learning Objective:
OBJECTIVE 1:

After reviewing the results, describe how the objective was validated by this experiment?

The more voltage I had the more current I had when I measured.

List the second Learning Objective:
OBJECTIVE 2:

After reviewing the results, describe how the objective was validated by this experiment?

When we changed the resistors I saw more current but it was hard to measure because the needle was bent a little.

List the third Learning Objective:
OBJECTIVE 3:

After reviewing the results, describe how the objective was validated by this experiment?

$I = \dfrac{V}{R}$ is ohm's law.

Conclusion:

Ohm's law works good.

If required, attach to this form: ☐ Answers to Questions, ☐ Tables, and ☐ Graphs.

Appendix B-4: An Example of a Better Written Student Report.

EXPERIMENT RESULTS REPORT FORM

Experiment No: __3__   Name: __John Doe__
Date: _____
Experiment Title: __Ohm's Law__   Class: __Elec. 101__
Instr: __Mrs. R. Jones__

Explain the purpose of the experiment:
To Validate the Ohm's Law Expressions of $V = I \cdot R$, $I = \dfrac{V}{R}$ and $R = \dfrac{V}{I}$

List the first Learning Objective:

OBJECTIVE 1: Validate $I = \dfrac{V}{R}$

After reviewing the results, describe how the objective was validated by this experiment?

With Resistance "R" held constant, current varied in direct proportion to any changes in applied voltage.

List the second Learning Objective:

OBJECTIVE 2: Validate $V = I \cdot R$

After reviewing the results, describe how the objective was validated by this experiment?

With the voltage held constant, the current was inversely proportional to any changes in circuit resistance.

List the third Learning Objective:

OBJECTIVE 3: $R = \dfrac{V}{I}$

After reviewing the results, describe how the objective was validated by this experiment?

Resistance increased as current decrease, when the voltage was held constant.

Conclusion:

Ohm's Law is valid based upon the results of this experiment. Current is directly proportional to voltage and inversely proportional to resistance.

If required, attach to this form: ☐ Answers to Questions, ☐ Tables, and ☐ Graphs.

Appendix C-1   Blank Graph Paper

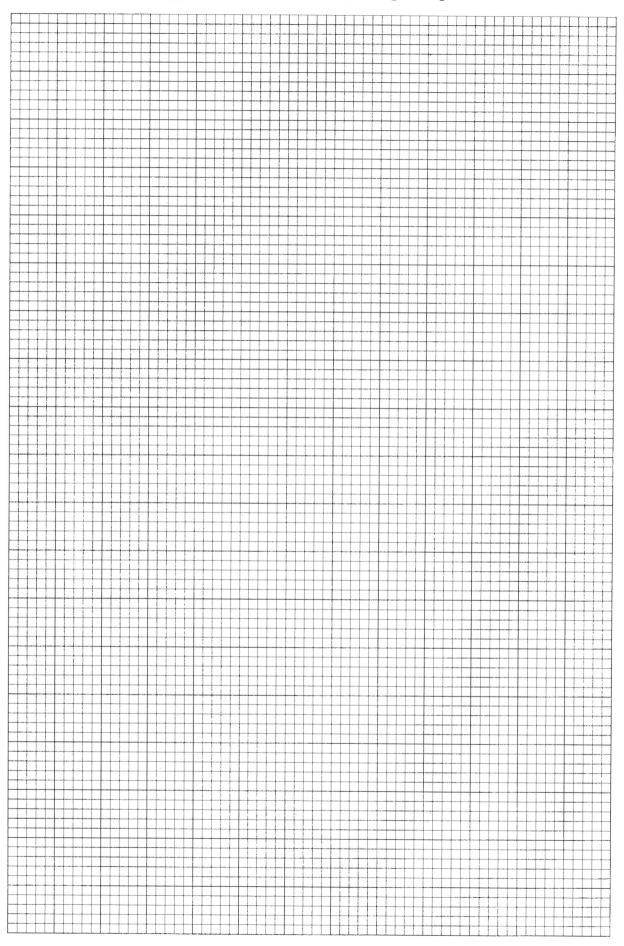

Appendix C-1   Blank Graph Paper

Appendix C-1   Blank Graph Paper

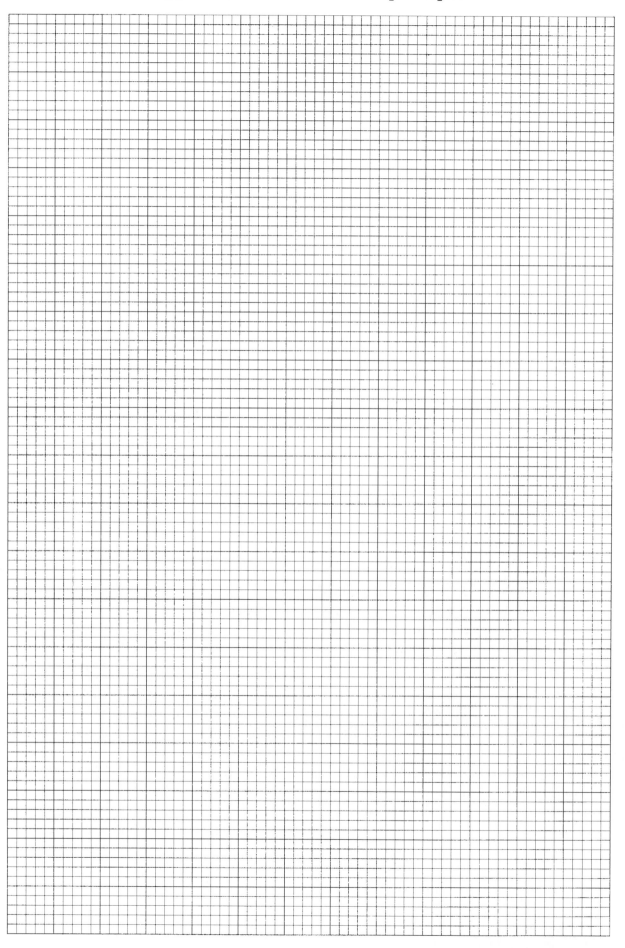

Appendix C-1   Blank Graph Paper

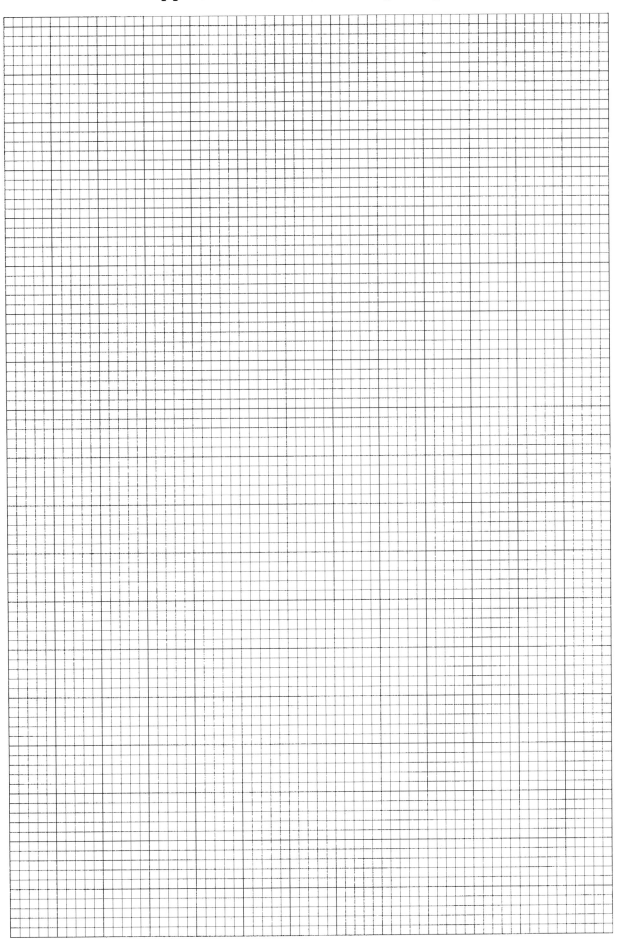

Appendix C-1   Blank Graph Paper

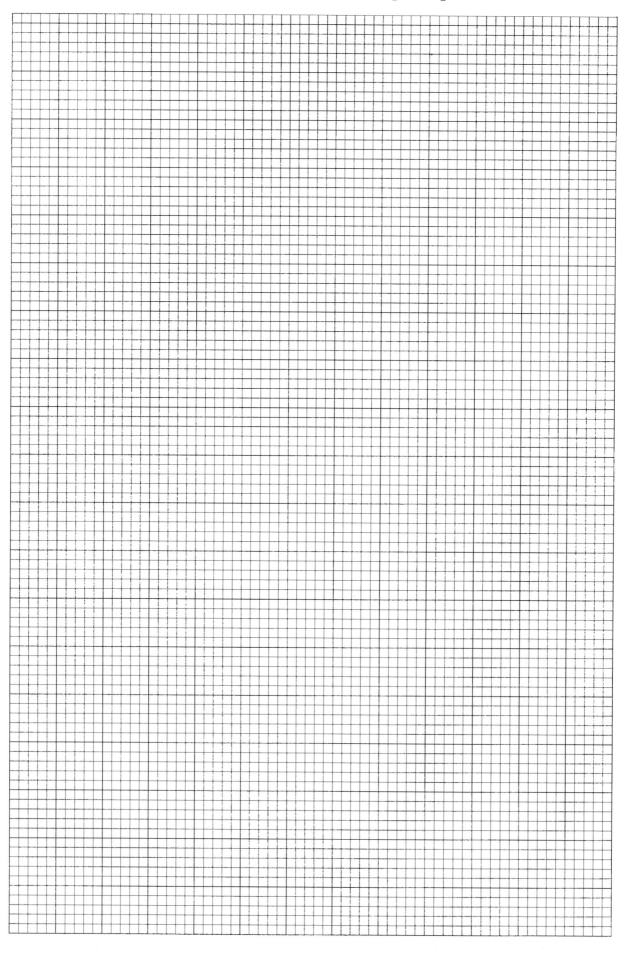

Appendix C-2   Blank Log Graph Paper

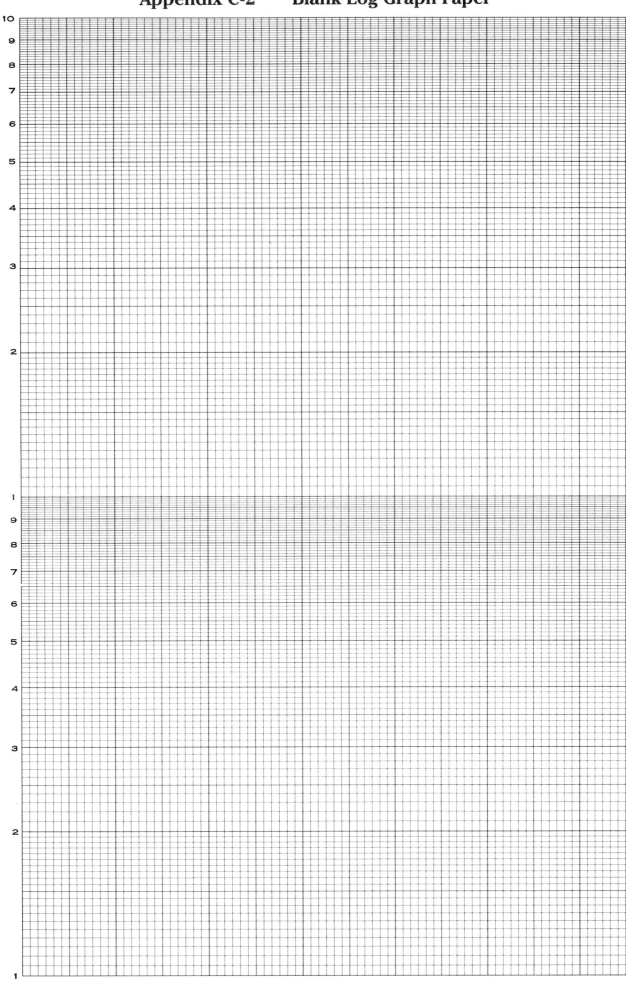

Appendix C-2   Blank Log Graph Paper

Appendix C-2 Blank Log Graph Paper

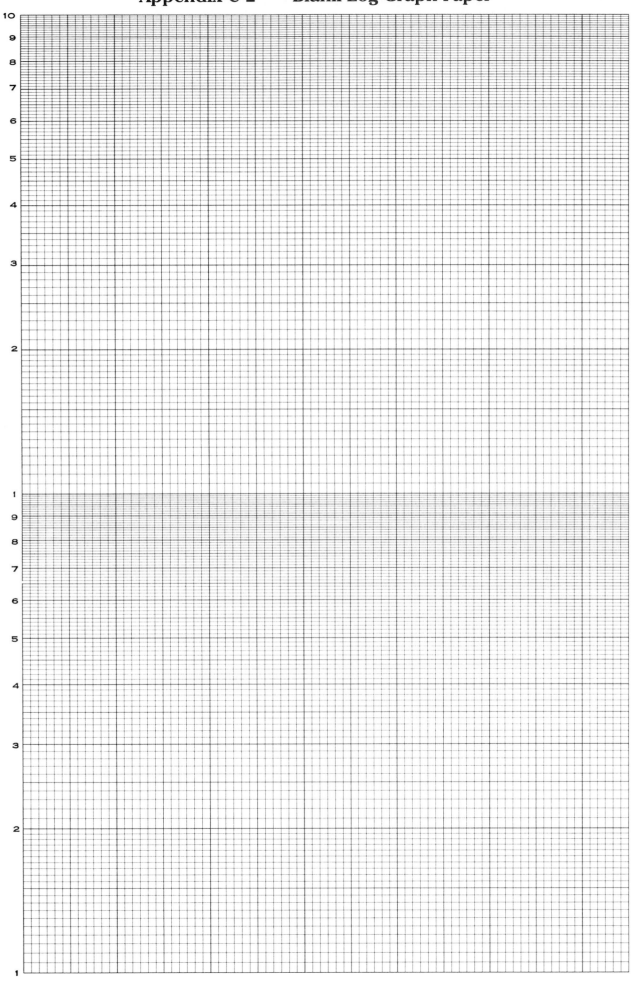

Appendix C-2   Blank Log Graph Paper

Appendix C-2   Blank Log Graph Paper

Appendix D   How To Make Graphs

Graphs usually show the relationships of two or more variables. The relationship is actually the curve or line that results. Here are some things to remember when making graphs.

1. Be neat and complete.
2. Never connect points. Always show the characteristic of the curve.
3. There should be room in the margins to title the graph.
4. Use the fullest scales possible. Do not confine a graph to one corner of the paper.
5. Use semilog graph paper for exponential quantities. Label the X axis as 1×10^2 for 100s, 1×10^3 for 1000s, etc.

Examples

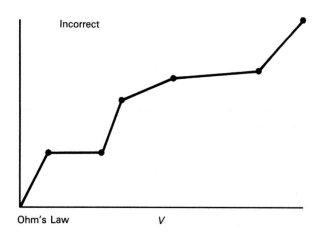

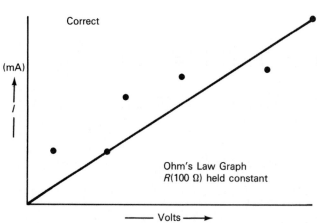

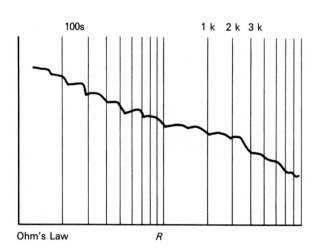

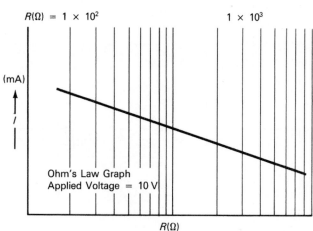

Appendix E Oscilloscope Graticules

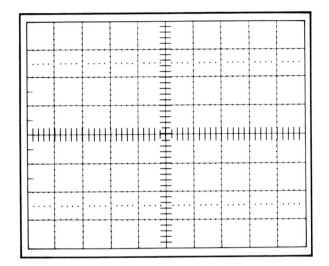
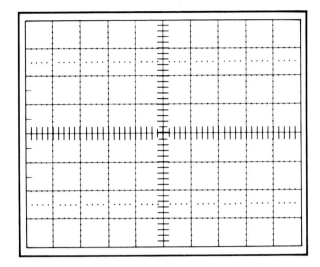
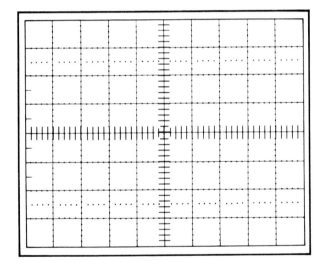
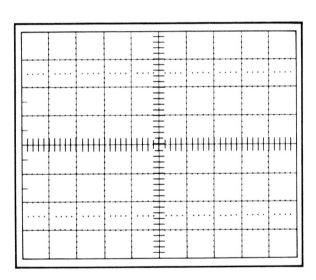

Appendix E   Oscilloscope Graticules

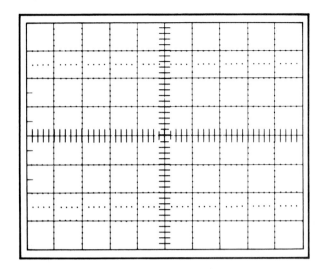

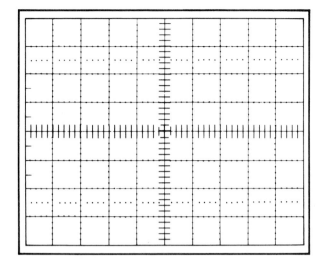

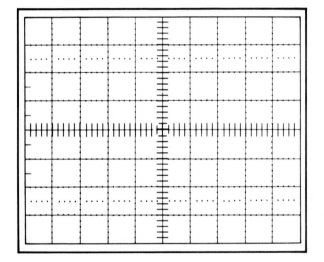

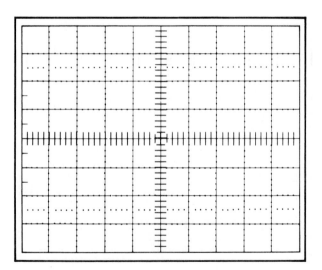

Appendix E Oscilloscope Graticules

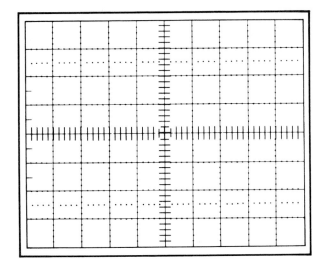

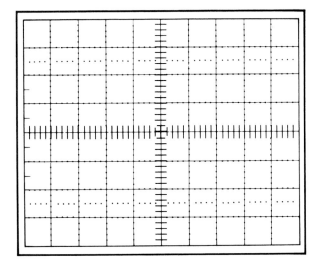

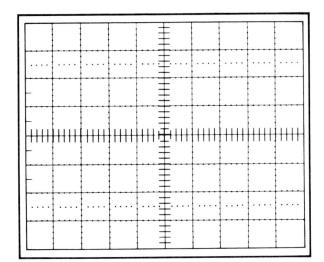

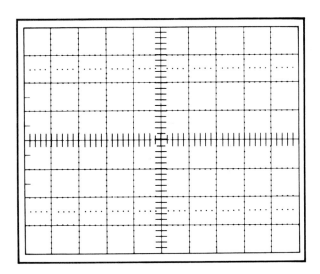

Appendix E   Oscilloscope Graticules

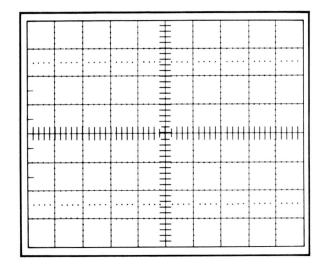
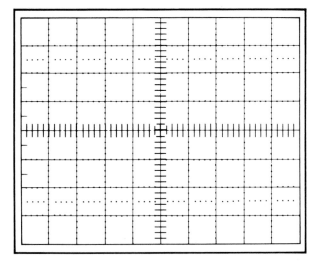
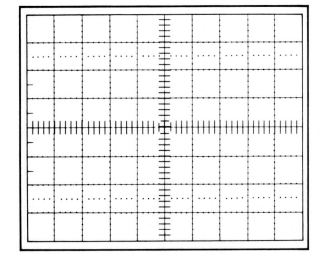
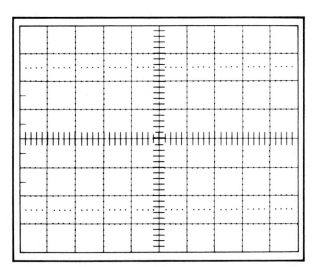

Appendix E   Oscilloscope Graticules

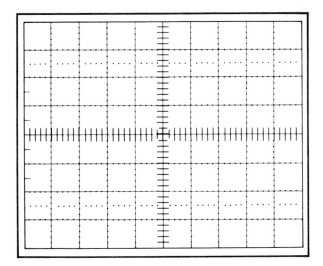
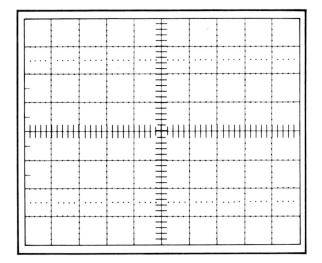
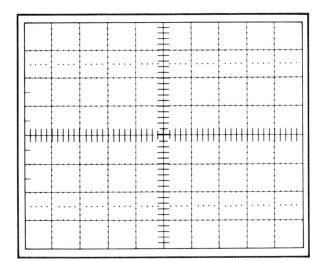
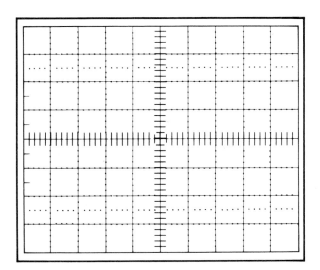

Appendix F   The Oscilloscope

The purpose of this appendix on oscilloscopes is to help you learn to use this measurement tool accurately and efficiently. This introduction is divided into two sections: the first section describes the functional parts of the basic oscilloscope, and the second section describes probes.

The oscilloscope is the most important tool to an experienced electronics technician (see Fig. F-1). While working through this appendix, it is best to have your laboratory oscilloscope (or the one you will be using in your studies of electronics) in front of you so that you can learn, practice, and apply your newly acquired knowledge. This appendix will discuss fundamentals which apply to any oscilloscope; your instructor or laboratory aid will help you with the fine details and special provisions that apply to your particular oscilloscope.

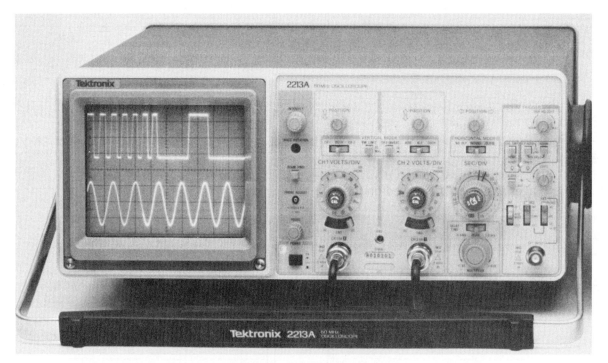

FIG. F-1 The oscilloscope.

Part 1: Functional Oscilloscope Sections

There are four functional parts to the basic oscilloscope: the display system, the vertical system, the horizontal system, and the trigger system.

Display System. Ths display system is a coordinated system of controls. It includes a cathode-ray tube (CRT), intensity control, and focus control. The CRT has a phosphor coating inside it. As an electron beam is moved across the phosphor coating, a glow is created which follows the beam and persists for a short time. A grid, also known as a graticule, is etched or painted on the screen of the CRT. This grid serves as a reference for taking measurements. Figure F-2 shows a typical oscilloscope graticule. Note the major and minor divisions.

The intensity control adjusts the amount of glow emitted by the electron beam and phosphor coating. The beam-trace should be adjusted so that it is easy to see and produces no halo. The focus control adjusts the beam for an optimum trace.

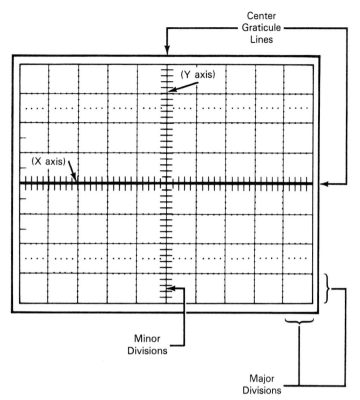

FIG. F-2 Oscilloscope CRT.

Vertical System. The vertical system controls and develops the deflection voltages which are displayed on the CRT. In this system, you typically find controls for vertical position, vertical sensitivity, and input coupling. The vertical position controls the placement of the trace. Adjusting this control will move the entire trace either up or down along the Y axis of the CRT. This control is adjusted as needed to help the operator make accurate measurements. The vertical-sensitivity control, also known as the volts-per-division switch, controls the Y axis sensitivity. The range is usually controlled from 1 mV to 50 V per division. For example, if you were to observe a trace occupying 4 divisions on the CRT, and the volts-per-division switch was turned to 1 mV/div., then the measured voltage would equal 1 mV × 4 divisions, or 4 mV. If the switch were turned to the 50 V/div. position and a 4 division trace was observed, then the oscilloscope would be measuring 200 V. The input-coupling switch lets the operator determine how the circuit under examination is connected to the oscilloscope. The three positions of this switch are ground, dc coupling, and ac coupling. When the switch is in the ground position, the operator can adjust the position of the trace with no input signal applied to the oscilloscope. This function is used primarily to align the oscilloscope to a reference point prior to taking a measurement. When in the dc-coupling position, the oscilloscope allows the operator to see the entire signal. However, when the input-coupling switch is in the ac-coupling position, only the ac signal components are displayed on the CRT. All dc components are blocked when the switch is in the ac position.

Horizontal System. As the oscilloscope trace is moved across the CRT (from left to right), it moves at a rate of speed which is related to frequency. The horizontal system is dominated by two main controls: the horizontal-position control and the time-base control. The horizontal-position control performs the same task as the vertical-position, but utilizes the X axis. The time-base, or seconds-per-division control, is used to select the appropriate sweep necessary to see the input signal. Ranges typically found on the time-base control extend from $0.1\,\mu s$ to $0.5\,\mu s$ per division on the CRT.

Trigger System. The trigger system allows the operator to select a part of the input signal and synchronize it with the trace displayed on the CRT. Normally, a trigger-level control is available. The position of the trigger-level control determines where on the selected trace the oscilloscope triggering will occur.

Each oscilloscope has different features. Your instructor is the best source for varying operational procedures.

Part 2: Oscilloscope Probes

Probes should accurately reproduce the signal for your oscilloscope. Probes can be divided by function into two main areas: current sensing and voltage sensing. Voltage-sensing probes can be further divided into passive and active types. For most applications, the probes that were supplied with your oscilloscope are the ones you should use. An operator picks the type of probe based on the voltage intended to be measured. For example, if you are measuring a 50 V signal, and the largest vertical sensitivity available is 5 V, then that particular signal will occupy 10 divisions on the CRT. This is a situation where attenuation is needed, and a $\times 10$ probe would reduce the amplitude of your signal to a reasonable proportion. The best way to ensure that your oscilloscope and probe measurement system have the least effect on the accuracy of your measurements is to use the probe recommended for your oscilloscope.

Appendix G   Component List

Resistors

All resistors 0.25 W, 5% unless indicated otherwise.
- (1) 10 Ω
- (1) 15 Ω
- (1) 47 Ω, 2 W
- (1) 56 Ω
- (1) 56 Ω, 2 W
- (1) 68 Ω
- (3) 100 Ω
- (1) 100 Ω, 1 W
- (1) 120 Ω
- (2) 150 Ω
- (2) 150 Ω, 1 W
- (5) 220 Ω
- (1) 270 Ω
- (2) 330 Ω
- (1) 330 Ω, 1 W
- (7) 390 Ω
- (3) 470 Ω
- (5) 560 Ω
- (3) 680 Ω
- (3) 820 Ω
- (4) 1 kΩ
- (3) 1.2 kΩ
- (2) 1.5 kΩ
- (3) 2.2 kΩ
- (1) 2.7 kΩ
- (1) 3.3 kΩ
- (1) 3.9 kΩ
- (3) 4.7 kΩ
- (1) 5.6 kΩ
- (1) 8.2 kΩ
- (4) 10 kΩ
- (4) 22 kΩ
- (1) 22 kΩ, 2 W
- (1) 33 kΩ
- (2) 47 kΩ
- (1) 68 kΩ
- (1) 86 kΩ
- (2) 100 kΩ
- (1) 150 kΩ
- (1) 220 kΩ
- (1) 470 kΩ
- (1) 1 MΩ
- (1) 1.2 MΩ
- (1) 3 MΩ
- (1) 3.3 MΩ

Capacitors

All capacitors 25 V or greater.
- (1) 0.0068 μF
- (1) 0.068 μF
- (4) 0.01 μF
- (2) 0.1 μF
- (2) 1 μF
- (4) 10 μF
- (2) 25 μF
- (2) 47 μF electrolytic
- (2) 47 μF
- (2) 4 μF
- (1) 100 μF

Inductors

- (2) 33 mH
- (1) 100 mH
- (1) 1 H (or optional value)
- (4) Varying values from 10 mH to 0.5 H
- (1) 0.5 H
- (1) 0.01 H

Potentiometers

- (1) 1 kΩ, 1 W Linear Taper
- (1) 5 kΩ, 1 W Linear Taper
- (1) 100 kΩ, 1 W Linear Taper
- (1) 1 MΩ, 1 W Linear Taper

Batteries

- (4) D cells
- (4) D-cell holders

Diodes

- (1) 1N4004 or equivalent

Bench Equipment

Ammeter with 30-mA capacity; VOM/DMM
Voltmeter: DVM, VTVM, VOM, DMM
DC power supply, 0–30 V
High-voltage power supply with 120: 12.6 V center-tap @ 6.3 V filament transformer
Galvanometer or microammeter movement
Signal generator (sine wave, to 1 MHz preferred)
Oscilloscope (solid-state, auto-trigger, dual-trace, with operator's manual preferred)
Frequency counter
Breadboard
AC signal generator

Miscellaneous Parts

- (1) Circuit board; proto springboard or breadboard
- (2) SPST switch

- (1) SPDT switch
- (1) 0–1 mA meter movement
- (1) 50 μA meter movement
- (6) Test leads 3 Red, 3 Black
- (1) Decade box
- (1) Magnetic compass
- (1) Heavy-duty horseshoe magnet, 20 lb plus pull
- (1) Sheet-metal shield, 6 × 6 in
- (1) Grease pencil
- 2–3 ft, thin insulated wire
- Iron filings
- No. 18 steel nail
- (4) Light bulbs: 25, 60, 100, 150 W
- (1) Clear glass functioning fuse (any current rating)
- (1) Clear glass nonfunctioning fuse (any current rating)
- (1) Neon bulb, approximately 60 V

Notes

Notes